B. sauteri

Ptychomnion

B. dichotomum

Sphagnum

Lord Howe

Buxbaumia

Catagonium

Thamnobryum

QLD

NSW

VIC

TAS

The Mosses of Southern Australia

FRONTISPIECE. *Hypnodendron comosum* VIC × 2

The Mosses of Southern Australia

by

GEORGE A. M. SCOTT

Monash University

and

ILMA G. STONE

University of Melbourne

With illustrations by
CELIA ROSSER

Monash University

1976

ACADEMIC PRESS

London New York San Francisco

A Subsidiary of Harcourt Brace Jovanovich, Publishers

ACADEMIC PRESS INC. (LONDON) LTD.
24/28 Oval Road,
London NW1

United States Edition published by
ACADEMIC PRESS INC.
111 Fifth Avenue
New York, New York 10003

Library of Congress Catalog Card Number: 75 19671
ISBN: 0 12 633850 7

PRINTED IN GREAT BRITAIN BY
WILLIAM CLOWES & SONS, LIMITED
LONDON, BECCLES AND COLCHESTER

Preface

The resolute botanist, faced with identifying Australian mosses, has hitherto had little encouragement. For texts and illustrations he has had to rely on handbooks of New Zealand, South African and British mosses, supplemented by a wide range of papers, mostly out of date and hard to obtain; workable keys he has had to do without. It is not surprising that so few have felt it worthwhile to persist. For tropical Australia this is still the case, but we hope that this book will make the richness of the temperate moss flora much more easily accessible to both amateur and professional botanists, and bring to others the joy and excitement of working with these fascinating and challenging plants. Their role in soil stabilization, their significance as ecological indicators and their usefulness in experimental work are not yet sufficiently widely appreciated in Australia; but the diversity of the flora is a continual lure to overseas bryologists, tempted by the richness of these unworked resources. The continuing destruction of Australia's moist coastal forests has removed many choice bryological habitats, but knowledge of the moss flora is still too meagre to assess the extent of the losses. More workers, both in the field and in the herbarium, are urgently needed. It was for that reason that we undertook the production of this book, somewhat prematurely, when our knowledge of the plants was (and is) still far from definitive. Any imperfections in the book are to be laid at our door and not that of the many friends and organizations who helped it on its way.

We offer our gratitude to Professor Kevin Westfold, Dean of the Faculty of Science at Monash University, without whose generous support, in the first year, the endeavour would have died still-born; to the Australian Research Grants Committee for subsequent support, and to Monash University for a Senior Research Fellowship to whose tenure one of us (G.A.M.S.) owes his participation. To Professor M. J. Canny and Professor T. C. Chambers, and colleagues in the Botany Departments of Monash and Melbourne Universities, we offer our thanks for encouragement and assistance liberally given. Miss Lois Davey of Melbourne University Botany School Library, and the library staff of both universities and the

National Herbarium, gave much assistance with the bibliography; Dr W. Weber, Colorado, and Dr A. Touw, Leiden, corrected and supplemented some esoteric references. Mrs Anne de Corona, Mr J. Scrimgeour, Mr A. Stone and Mr B. Fuhrer helped with photography, and Miss Anna Lucarelli and Mrs Margaret Robertson earned our warmest gratitude for typing the manuscript. We are grateful to the Directors and staff of herbaria in the British Museum, Sydney and Melbourne, in particular Mrs Val Jones, for readily preparing and lending specimens; to the National Parks authorities in New South Wales, Victoria and Tasmania, the Western Australian Department of Agriculture, and the Forests Commission of Victoria, for permission to collect. Our indebtedness to Dr Helen Ramsay for data on chromosome numbers will be evident to all. Like all bryologists nowadays we have come to depend heavily on the Index Muscorum as a source book, and we record our thanks to its authors, Van der Wijk, Margadant and Florschütz. We are also greatly indebted to all those who tried out the keys and commented on them, and to many friends who have given unstintingly of encouragement and help, especially Mr K. W. Allison, Professor D. G. Catcheside, Mr A. Eddy, Mr J. T. Linzey, Dr D. N. McVean and Dr J. H. Willis. Above all we are grateful to Mr A. C. Crundwell, Dr W. B. Schofield and Dr E. V. Watson who read and criticized the first version of the text, suggested many improvements, and continually helped with advice and specimens.

Finally it is a pleasure to record our thanks to Mr O. R. Johnson and Miss Jenny Pollard for technical assistance, to Miss Carol Davies for continual help in many ways but especially with the Index, and to our families for endless tolerance and encouragement.

December, 1975   G.A.M.S.
I.G.S.
C.R.

Conspectus

Classification:

Sphagnidae
  Sphagnales
    Sphagnaceae   1   *Sphagnum*
Andreaeidae
  Andreaeales
    Andreaeaceae   2   *Andreaea*
Bryidae
  Buxaumiales
    Buxbaumiaceae   3   *Buxbaumia*
  Polytrichales
    Polytrichaceae   4   *Polytrichum*
                                    5   *Atrichum*
                                    6   *Pogonatum*
                                    7   *Polytrichadelphus*
                                    8   *Psilopilum*

    Dawsoniaceae   9   *Dawsonia*
  Fissidentales
    Fissidentaceae   10 *Fissidens*
                                        Moenkemeyera
                                        (*Octodiceras*—see 10
                                            Fissidens fontanus)

  Grimmiales
    Grimmiaceae   11 *Grimmia*
                                    12 *Rhacomitrium*

  Archidiales
    Archidiaceae   13 *Archidium*

vii

Code for Collectors

MOSSES ARE VULNERABLE

It is our experience that recovery of a moss population after sampling is often a matter of years, and rare species could easily be lost by over-collecting. Already, in other countries, this is happening.

Small specimens are ample for most purposes and collectors in Australia, as elsewhere, must never forget the need to preserve their heritage unimpaired.

Whether for serious research, for exchange, or merely for horticulture, indiscriminate or wasteful collecting is unethical, immoral, and altogether to be deplored.

Introduction

The discoverers and explorers of Australia included a few mosses among their botanical collections and these found their way into the early bryological literature (e.g. Hedwig, 1801; Labillardière, 1806; Smith, 1808). Further investigation seems to have been concentrated on Western Australia (e.g. Hooker, W. J., 1840; Taylor, 1846; Hampe, 1844) and Tasmania, culminating in the production of Wilson's account in J. D. Hooker's "Botany of the Antarctic Voyage. Flora Tasmaniae", and the associated papers. Until about 1850 most of the bryophytes collected in "New Holland" (Australia) came from the general region of Swan River in Western Australia, mostly collected by James Drummond. Thereafter bryological activity switched to eastern Australia where resident botanists were beginning to investigate the local floras. Of these, Victoria's first Government Botanist, Ferdinand Müller, later Baron Ferdinand von Mueller ("the Baron") was by far the most assiduous. He dominated botany in Victoria (indeed even Australia) for most of the rest of the nineteenth century although his contributions to bryology were mainly as a collector, apart from some published species lists and the first fascicle, never continued, of a set of moss illustrations (1864). He was followed by a number of local collectors such as Sullivan, Reader and Bastow, as described by Willis (1949, 1955e) and later by Willis himself.

In Tasmania, Bastow, Weymouth and Rodway were the best-known bryologists about the turn of the century. Their activity culminated in the "Tasmanian Bryophyta" of 1913–17 by Rodway, the Tasmanian Government Botanist, which is best read in conjunction with Sainsbury's invaluable later commentaries on Rodway's herbarium (1953b–1956b).

In New South Wales during the same period, Watts and Whitelegge were the principal local bryologists (later followed by Burges who compiled, for New South Wales only, a complement to their incomplete but invaluable "Census of Australian Mosses") and in Queensland, Wild and Bailey. In South Australia, the driest of the States and with the least rich moss flora, there was little bryological activity until the middle of this century when Catcheside began a revision of the State moss flora.

With few exceptions the mosses collected in the early days were sent to

Europe, for identification mainly by Hampe, Carl Müller, and later Brotherus. Of these Müller in particular described so many ill-founded species that Australian bryology has been crippled by the burden, especially as most of his voucher specimens are believed destroyed in the bombing of Berlin. Dixon, and later Sainsbury, Willis and others, have done much to clarify the resulting confusion but there still remain very many redundant species names which should be relegated to synonymy. We have made little attempt to clear up this muddle; it will take the work of many years and many bryologists to do so and, in any case, we feel that world-wide generic revisions usually present the most appropriate opportunities for such a task. We have, instead, concentrated on describing and discussing the species which we know to exist, leaving all others as hypothetical in the meantime.

Plant Geographical Relationships
In Tasmania the rich and exciting moss flora is extraordinarily similar to that of New Zealand in both the total floristic list and the appearance of the moss communities. It is no accident that the only two published bryophyte floras in Australia, up to the present day, have both been for Tasmania. The wetter regions of Victoria share much the same flora also, although less rich. In both these regions more and more New Zealand plants are being discovered. That same part of the flora has its affinities mainly with South America.

In the drier regions of Victoria, South Australia and Western Australia the floristic affinities are perhaps rather closer with South Africa and fresh links between the two floras are continually being discovered. Several of these species, such as *Gigaspermum repens*, also extend across to New Zealand where they are rare or very rare.

The affinities of the Queensland flora, on the other hand, are rather with New Guinea and through that with the great riches of the Indo-Malaysian flora. This tropical flora we do not yet know at all and have not dealt with beyond listing the species given by the Index. The two kinds of flora, northern tropical and southern temperate, meet in the Sydney region and we have taken this as being a reasonable limit to the area we deal with. Tropical species penetrate south of this point, and temperate species north of it, but a break has to be made somewhere and that seems the most useful point at which to make it. The southern, temperate, region of Australia, for which this book is primarily intended, is thus arbitrarily held to end somewhere in the region of Sydney in the east, and Geraldton, north of Perth, in the west, although some of the species we deal with penetrate far north of that line, and a few of the tropical species, which we ignore, extend just to the south of it.

Life History of Mosses
The basic life cycle of mosses—spore, protonema, leafy gametophyte, fertilization, sporophyte, meiosis, spores—is too well known to require amplification here. Much more information can be found in texts such as Parihar's "Bryophyta" and E. V. Watson's "Structure and Life of Bryophytes". But there are extraordinarily few mosses in the world for which the actual life history is known in any detail: time of fertilization, spore discharge and vegetative growth; rate of growth; longevity; kind of protonema; method of perennation; relative importance of reproduction by spores or gemmae or plant fragments; morphology of different stages in the development of the mature growth form, etc. In Australia it would be true to say that there is no species of moss for which we have this simple biological information in any but the broadest terms. There are not many fields of biology in which the amateur naturalist of ordinary attainments can participate with the certainty of being able to make such a significant contribution as this one. Thus, although the basic life cycle is presumed to be the same for all our mosses, nothing is known about the relative importance and timing of the parts of the cycle for any of our mosses.

Guide to Use of the Keys, Text and Illustrations

Morphological Terminology

The great majority of mosses have the leaves, or at least their points of insertion, distributed evenly round the stem. In such cases the leaf base is, perhaps invariably, wider than it is thick so that one surface faces the stem (*adaxial*) and one away from the stem (*abaxial*). We have preferred these unambiguous terms to the more usual, but sometimes confusing, *ventral* and *dorsal*, respectively. A number of mosses, on the other hand, especially in the Australasian flora, have the typical liverwort leaf arrangement with the leaves restricted to two or three longitudinal rows (orthostichies) along the stem; we have then used the customary liverwort terminology referring to the leaves of the median row as *underleaves* and the others as *lateral leaves*. The structure, size and shape of the leaves are very diverse indeed, and there are many technical terms used to describe this diversity. We have tried to keep them to a minimum and those used are explained in the glossary. There is no hard and fast line between terms and some of them will cause difficulty and are scarcely capable of precise definition. *Subula*, in particular, is hard to circumscribe, and has been used broadly to include leaves which are only apparently subulate, by contraction of the leaves when dry. Similarly the cells at the bottom outer corners of the leaf, at the point of insertion on the stem, are in many cases specialized in form or size or colour, or any combination of all three, and are then said to be distinct *alar cells*, but all intermediates can occur. When the alar group, as well as being distinct, bulges outwards from the leaf outline to form rounded lobes, they become *auricles*. We have used the term *channelled* to mean U-shaped in profile across the leaf and *keeled* to mean V-shaped.

For the leaves surrounding the female reproductive organs we have used the terms *perichaetial bracts* and *perichaetial leaves* completely interchangeably, for variety. The seta, on which the capsule is carried, is often twisted like a corkscrew when dry; there is always difficulty in describing the direction of this twist in such a way that it can be easily and unambiguously visualized. The easiest, though not customary, way is to imagine the seta as a screw being driven, capsule-end leading, into a block of wood. A normal screw requires a clockwise rotation of the hand to drive it in and is said to have a right hand thread; a seta with a similar twist we have designated right hand. A left hand thread would require an anti-clockwise rotation to drive it in (e.g. Plate 31).

Where possible we have tried not to use bryological jargon if there was an alternative, and have sometimes created our own descriptive terms instead. The word *cucullate*, for example, we have used only to describe the hooded, boat-shaped apex found on the leaves of some species. The same term is also widely used for an analogous but quite different shape of calyptra; there we have used *side-split* as a graphic, unambiguous replacement. The terms *apophysis* and *neck* refer to adjacent regions of the sporophyte; they are scarcely separable and we have tended to use them interchangeably.

The terms *monoicous, dioicous* etc. have been used in preference to *monoecious, dioecious* etc, following the long-established bryological custom to distinguish the condition in gametophytes (haploid) from the analogous state in the sporophyte generation.

Where we know of specialized propagules by which the plant can spread vegetatively, i.e. without spore formation, we have recorded the fact but have not tried to distinguish the different kinds of *gemmae*, preferring to use the word in its broadest sense, except for the rhizoid gemmae for which the term *tuber* is gaining wide acceptance.

Text

The terms "Handbook", "Index" and "Studies" refer to Sainsbury (1955a), Wijk, Margadant and Florschütz (1959-69) and Dixon (1913-29) respectively, throughout. The overall classification into families, genera etc. which we have used is an individual one, based on that of Sainsbury's "Handbook" with modifications to accommodate the results of recent work and to suit our own convictions. All the genera known to occur in Australia (mostly on the authority of the Index) are listed numerically in the Conspectus, but those which we do not recognize as both valid and useful are bracketed, and those which we have considered to be tropical and beyond the scope of the book are not numbered. Their species are listed in the text at the beginning of each family for the sake of completeness, to make the book more useful as a platform for further work, but neither discussed nor described. Illustrations of some of them are to be found in Bartram (1933) Fleischer (1902-8) and Brotherus (1924-5).

The species described in the text comprise nearly all the species in the southern Australian region, as far as our experience goes. Those that are sufficiently common, and/or that we know well, are fully described and are distinguished by bold type; the others, in italics, are more briefly described. Throughout we have concentrated on providing the means of identification, rather than describing every part of the plant. In general we have not dealt with taxa below the rank of species. Generic names are cross-referenced by the Conspectus number, where we have felt it useful.

Cytology

The chromosome numbers are all quoted from Ramsay (1967b, 1974) with few exceptions noted in the text. No attempt has been made to give counts based on overseas material.

Measurements

Measurements have, where practicable, been based on our own collections and observations, and are typical for these. We cannot pretend that they are therefore average for the species. Cell measurements are taken over the whole cell (not just the cell cavity, or lumen), at about ½–⅔ of the way up the leaf from the base and half way between the nerve and the margin in the case of single-nerved leaves, or in a corresponding position in other leaves. Cells in the plates come from the same region, except where other-otherwise stated in the captions. All spore measurements have been made on spores mounted in water.

Keys

For economy and clarity we have chosen the artificial, non-indented lay-out of dichotomous keys. The reader is presumed to be familiar with the method of using them. We have made a deliberate effort to base these keys on vegetative features where possible, since capsules are often not readily found, especially by those most likely to need the keys. Since vegetative characters are usually less definitive and unambiguous than fruiting characters we have consequently had to allow for keying out many of the genera repeatedly, in different positions, taking account also of likely mistakes in using the keys. We have tried to make the alternatives in each couplet mutually exclusive and exactly paired, character for character. Despite this there will be many occasions when the correct choice at a couplet is not clear and the user then will have to try both alternatives to find which leads to the right species. It is important to read the description of any species keyed out, as confirmation. Where there are not more than one or two main species in a genus, in our area, or where the species discrimination is based on a whole suite of characters, keys within a genus have sometimes been omitted.

Plates

These were drawn in pencil. After many trials the most effective method discovered was to photograph the specimen, with an appropriate scale, trace the leaf outlines from the photographic enlargement on to drawing

paper, and then to draw in the detail under a binocular dissecting microscope. This ensures accuracy of scale and morphology. The great range in
total size of moss plants made it impracticable to have a single scale of magnification, but the number of different scales chosen has been as low as
possible, mostly × 2, 5, 10, 15, 20, 30, 40, 50. All drawings are reduced by
¼ in printing, giving final magnifications of approximately 1½, 4, 7, 11, 15,
22, 30, 37. In all cases the scale subdivisions on the plate itself are millimetres.
Isolated leaves have been drawn where greater detail was required. These
have usually been drawn to give a final magnification of × 50 on the printed
page. We have deliberately chosen to draw all cells at × 1333, giving a
final magnification in print of × 1000. This has imposed several constraints
on the lay-out but the comparison of cells of different species has been so
illuminating that it seemed to us important to treat all cells alike. All whole
specimens have been drawn dry except where otherwise noted in the captions, and all cells and leaves drawn mounted in water. This is partly because the majority of our mosses are at their most distinctive when dry,
showing the kind of twisting the leaves undergo, and partly because the
optical problems associated with photographing and drawing moist
specimens under strong illumination are often insuperable. In a few cases
restrictions on time and human eyesight have made accurate drawings of
the complete plant impracticable. In these cases the photographic outline
has been blocked in to give an impression, and only a portion finished in
detail. The outlines of cell contents and walls have been left unshaded where
there is no doubt which is which. In very long narrow cells shading has
been needed to show the walls, and in very thick-walled short cells, as in
Andreaea, there is a difficult technical problem of ambiguous or reversible
images which can only be overcome by rather elaborate shading.

Techniques

The photography of intricately surfaced objects in the size range 1–100 mm
presents formidable technical difficulties of lighting and depth of field.
Most of our working photographs were taken on 120 Ilford PanF film in a
Leitz "Aristophot" photo-macrographic apparatus, against a black background, with a single oblique spot light focused on the specimen from above
at about 45°. For very difficult objects we have had unstinted help from
various friends who have mostly used 35 mm cameras with close-up lenses,
illuminating the specimens by diffuse light in a translucent "tent"; this

produces superior photographs in the hands of an expert but we have found it less easy to use for routine work.

In a few cases, such as *Andreaea*, it has been profitable to highlight the leaf outlines of dark leaves with a cloud of ammonium chloride in a closed chamber, using the mixed vapours of concentrated hydrochloric acid and strong ammonia.

Staining can also be useful to show up the edges of translucent leaves in greater contrast, and this alone made it possible to draw *Cratoneuropsis relaxa*. The best stain that we know is extremely dilute Toluidine Blue used just long enough for only the leaf outlines to stain up. Such a dye is also useful in highlighting the narrow decurrent strips from the leaf bases of some species, and is almost essential in examining *Sphagnum* leaves and stems since it shows up the pores in the cells with great clarity. To make water mounts permanent, the easiest way is to place a large drop of gum chloral solution (e.g. Watson, 1955, p. 14) at the side of the coverslip and set the slide aside; as the water dries out the gum is drawn in gradually to replace it. Staining beforehand will help to keep the specimens visible even when the chloral makes them transparent. With big thin-walled cells, such as in Funariaceae, this technique leads to excessive shrinkage which can be avoided by treating the leaves with lactic acid, washing in water, and then mounting.

Transverse sections of leaves and stems are very easily made provided the section does not have to be cut from a specified part of a specified leaf. With thick, stiff plants like *Polytrichum* the quickest and most satisfactory way is usually to shave off fine dust from the cut surface across a dry leafy stem, on to a glass slide, using a new razor blade and cutting down on to the slide. The resulting dust, when soaked out, will usually contain superbly thin sections somewhere among the debris. With more delicate fleshy plants it may be preferable to cut the stems wet in a similar fashion. For precise sections of fresh material, cutting with a razor blade or very sharp fine scalpel under the dissecting microscope is sometimes best; at other times traditional sectioning in pith or carrot will be needed. Polystyrene foam and the pith of the common weed, Fennel (*Foeniculum vulgare*), are ideal.

Precise dissection of leaves is fortunately not often necessary and most commonly the leaves are stripped off soaked-out plants by running a pair of lightly clasped forceps down the stem against the line of the leaves. When the stems are too soft to stand this treatment the leaves have to be pulled off, downwards, individually. Very fine jewellers' forceps are by far the easiest tool for dissections, at least for beginners.

For taxonomic work, mosses are usually adequately preserved by being allowed to dry out at room temperature and most of them, except the species with large thin-walled cells, such as *Tayloria*, recover very well when soaked out in water; many of them show no undue deterioration after a century or more in the herbarium. Cellophane packets or paper packets folded out of a single sheet are alternatives for storing the dried specimens. Ordinary envelopes are not satisfactory, nor are plastic bags. The best technique for storage is a matter for hot dispute in bryological circles. Our own preference is for cellophane packets within folded paper packets stored like filing cards when the herbarium is small, and mounted on herbarium sheets when the collection is big enough to justify it. The inner cellophane packet, for small species especially, does a lot to keep the specimens intact and undamaged. Pressing, as for flowering plants, can make specimens easier to package but tends to alter their appearance.

Symbols and Abbreviations

ACT — Australian Capital Territory
c — circa (approximately)
cm — centimetre
mm — millimetre
NSW — New South Wales
NT — Northern Territory
QLD — Queensland
q.v. — quod vide (which see)
SA — South Australia
TAS — Tasmania
TS — transverse section
VIC — Victoria
WA — Western Australia
μm — micron (formerly μ)
± — more or less
† — further information to be found in Addenda.

Keys

General Key
Numbers in brackets refer to position of genus in text.

1. Leaves whitish, consisting of large empty hyaline (colourless) and small, narrow green cells, evenly distributed throughout 2
   Hyaline cells, if present, restricted in distribution e.g. to nerve or border or apex 3

2. Leaf 1 cell thick; hyaline cells with conspicuous helical thickenings (fibrils) internally. Branches in fascicles   *Sphagnum* [1]
   Leaf consisting of a network of green cells sandwiched between 2 layers of larger hyaline cells which lack helical fibres. Branches single . . . . . .   *Leucobryum* [28]

3. Plants dendroid; at least some parts consisting of an unbranched erect stem surmounted by a cluster of branches . . . .   Group H
   Plants not dendroid 4

*4. Leaves clearly in ranks (orthostichies), or *flattened* into one plane (complanate) i.e. as if pressed strongly . 5
   Leaves spirally or irregularly arranged, not in orthostichies, not complanate 7

5. Leaves inserted on the stem in 2–6 ranks 6
   Leaves spirally inserted on the stem but complanately flattened . . . .   Group D

*6. Leaves clearly in 3–6 ranks . . . .   Group B
   Leaves clearly in 2 ranks . . . .   Group C

---

* In doubtful cases they should key out either way.

7. Gemmae obviously present on the adaxial surface of the leaf apex, or at the tip of the stem, or as small bulbils in the axils of the leaves (not including either deciduous shoots or rhizoid gemmae) . .   Group E

    Gemmae absent or not obvious   8

8. Lamellae present, running longitudinally down the adaxial surface of the nerve which is usually very wide, occupying virtually the whole of the width of the lamina; leaves thus with a very opaque, solid texture (except *Atrichum*) . . . . . .   Group A

    Lamellae absent; leaves usually thinner in texture .   9

9. Blackish, or dark red-brown, fragile mosses of sub-alpine and alpine siliceous rocks. Stems in cushions usually not more than 1–2 cm tall. Cell walls very thick. Capsule opening by 4 longitudinal slits .   *Andreaea* [2]

    Plants without such a combination of characters. Capsule never opening by 4 slits   10

10. Minute plants, stems commonly less than 1 mm tall, growing on soil. Capsule sessile or nearly so, the seta no longer than the capsule . . . . . .   Group J

    Plants mostly more than 1 mm tall; habitats various. Seta usually distinct or else fruit not present . .   11

11. Nerve single, usually extending more than ½ way up the leaf (excluding hair point); rarely forked at the end   12

    Nerve mostly none or double but not usually reaching ½ way up the leaf . . . .   Group M

12. Upper part of nerve, adaxially, covered or obscured by filaments or granules. Small mosses of dry areas .   Group N

    Upper part of nerve may be papillose but never obscured by granules or filaments   13

13. Leaves subulate, at least 8 times as long as the width in mid-leaf and with the nerve reaching well into the subula . . . .   Group F

    Leaves not subulate, less than 10 times as long as wide, or with a nerveless subula   14

14. Nerve excurrent as or into a hair-point or long mucro   15
    Nerve ceasing shortly beyond or in the apex, or below
      it   16

15. Leaf clearly bordered, at least in part, by narrow cells
     different in size and shape from the adjacent cells of
     the lamina .   .   .   .   Group I
    Leaf not bordered .   .   .   .   Group G

16. At least some cells papillose or with high mamillae .   Group K
    All cells smooth or with a finely wrinkled surface, but
     no definite papillae .   .   .   .   Group L

Group A
Polytrichum and related genera

1. Peristome absent or too imperfect or immature . .   2
   Mature peristome and calyptra present   16

2. Lamellae on leaves few (less than 10)   3
   Lamellae numerous (more than 20)   4

3. Leaves undulate, toothed (both on margins and on the
   backs of the leaves on the crests of the undulations),
   mostly over 5 mm long. Stems long, over 1 cm.
   Moist regions . . . .   *Atrichum androgynum*
   [5]

   Leaves not undulate. Stems mostly less than 3 mm
   long. Dry regions . . . .   Group N

4. Leaves crisped or curled when dry   5
   Leaves ± straight and appressed to the stem when
   dry, or with the tips incurved or recurved . . .   7

5. Margins of leaves narrow, 1–2 cells wide; lamellae
   usually 6 cells high . . . .   *Polytrichadelphus*
   magellanicus [7]

   Margins of leaves wider, 3–5 or more cells wide;
   lamellae usually 3 cells high   6

6. Usually lowland or upland habitat (under 4000 ft★).
   Terminal cells of lamellae smooth or scarcely
   papillose . . . .   *Pogonatum subulatum*
   [6]

   Usually montane to alpine habitat (over 5000 ft★).
   Terminal cells of lamellae distinctly papillose . . .   (*Psilopilum*) 9

★ These altitudes do not apply in Tasmania

7. Leaf margins entire or at most slightly denticulate;
   leaf apices usually prolonged as short, reddish points   8
   Leaf margins clearly toothed   10

8. Margins of leaves wide, inflexed so as almost to cover
   the lamellae; terminal cells of lamellae smooth . . *Polytrichum juniperinum*
   [4]

   Margins of leaves narrow, not at all covering the
   lamellae; terminal cells of lamellae papillose . .   9

9. Lamellae loosely arranged, 2–4 cells high; leaves
   usually toothed . . . . *Psilopilum crispulum*
   [8]

   Lamellae closely arranged, 5–10 cells high; leaves
   entire or toothed . . . . *Psilopilum australe*
   [8]

10. Border of empty, thin-walled, hyaline cells on at least
    part of margins of sheathing leaf bases. NB: Border
    usually not conspicuous   11
    Leaf base unbordered   12

11. *Terminal cells of lamellae smooth*, grooved in TS. Low-
    land or alpine moss . . . . *Polytrichum commune*
    [4]

    Terminal cells of lamellae papillose, not grooved. Almost
    exclusively alpine moss, usually at over 5000 ft. . . *Polytrichum alpinum*
    [4]

12. Stem less than 3 cm tall (excluding seta), often less than
    1 cm   13
    Stem usually more than 3 cm tall   14

13. Stem over 400 µm across, measured just below the
    leafy part. Densely leafy: 30–45 leaves visible in TS
    across mid-stem. Teeth sharp, spiny. . . . . *Dawsonia longiseta* [9]
    Stem less than 250 µm across. Less densely leafy with
    with less than 25 leaves in TS. Teeth short, blunt . *Pogonatum subulatum*
    [6]

14. Leaves more than 1·5 cm long, usually much more.
     Stem in section with abundant stereids in central
     core. Terminal cells of lamellae in side view mostly
     wider than high and 2× the width of the cells
     below . . . .   *Dawsonia superba* [9]
     Leaves less than 1·5 cm long. Stem core with few or no
     stereid cells. Terminal cells of lamellae in side view
     mostly either isodiametric or taller than wide, not
     wider than cells below   15

15. Dry stems, including leaves, usually *c* 0·5 cm wide,
     the axis itself up to 500 μm across at the base of the
     leafy part. *Most leaves*, except on male plants,
     slightly but *consistently falcate and secund. Apical cells
     of lamellae* mostly higher than wide *in side view*, with
     a *domed*, often almost hemispherical top, the lamel-
     lae thus appearing strongly crenulate in side view.
     Plant usually sparsely branched. Lamellae usually
     6 cells high. Margins toothed right to the top of the
     leaf sheath . . . .   *Polytrichadelphus*
                                                  magellanicus [7]

     Dry stems + leaves nearly 1 cm wide, the axis itself
     over 600 μm across. *Leaves evenly arranged* all round
     the stem. *In side view, apical cells of lamellae mostly* at
     least as wide as tall, *flat* or slightly convex but never
     hemispherical; at least some *terminal cells* of lamellae
     asymmetrical, with off-centre thickenings. Lamellae
     usually 4 cells high. Margins toothed except just
     above the leaf sheath . . . .   *Dawsonia poly-*
                                                        trichoides [9]

16. Calyptra glabrous or slightly bristly   17
     Calyptra densely hairy   19

17. Old capsules persisting, with 2–3 generations of sporo-
     phytes present on one plant; capsules crescent-
     shaped in TS, with one very convex and one slightly
     concave side . . . .   *Polytrichadelphus*
                                                  magellanicus [7]

     Old capsules not persisting; capsules circular or bi-
     convex in TS   18

18. Capsule long, erect, narrow, terete, slightly curved, more than 4× as long as wide; calyptra smooth at apex. Leaves undulate . . . . *Atrichum androgynum* [5]

    Capsule short, usually inclined, biconvex, inflated; not more than twice as long as wide; calyptra rough at apex. Leaves may be crisped, occasionally slightly undulate . . . . (*Psilopilum.* Go to 9)

19. Peristome with teeth and epiphragm 20
    Peristome of a whitish brush of hairs . . . . . (*Dawsonia*) 23

20. Capsule cylindrical, slightly inclined, terete, or with narrow longitudinal ribs; no obvious apophysis . 21
    Capsule 4-angled; apophysis prominent, separated from rest of capsule by a constriction 22

21. Capsule surface papillose, dry or wet. Terminal cells of lamellae usually smooth and green . . . . *Pogonatum subulatum* [6]

    Capsule surface smooth. Terminal cells of lamellae papillose and hyaline . . . . *Polytrichum alpinum* [4]

22. Margins of leaves wide, entire, inflexed over the lamellae . . . . *Polytrichum juniperinum* [4]

    Margins of leaves toothed, not inflexed 23

23. Leaves usually over 1·5 cm long, usually much more. Leafy part of stem more than 6 cm long. Capsule large, 5–8 mm long × 4–7 mm wide . . . . *Dawsonia superba* [9]
    Leaves usually up to 1·5 cm long. Leafy part of stem up to 5 cm long. Capsule 4–5 mm long × 3 mm wide 24

24. Leafy stem 1–5 cm long; leaves long and gradually narrowed. *Seta no longer than leafy part of stem* . . *Dawsonia polytrichoides* [9]

    Leafy stem less than 1 cm long. Leaves short and wide, tapering abruptly at apex. *Seta longer than leafy stem*, often twice as long . . . . *D. longiseta* [9]

Group B
Leaves in 3, 4, 5, or 6 ranks

1. Leave all alike, in 3 or 5 equal ranks, sometimes rather irregular but not complanate 2

   Leaves of two sorts, with 2 (or 4) rows of lateral leaves and one or more rows of smaller dorsal and/or ventral leaves; sometimes complanate 9

2. Leaves in 5 ranks. 3

   Leaves in 3 ranks. 5

3. Shoots richly branched. Usually epiphytic . . . *Papillaria (crocea)* [97]

   Shoots unbranched or almost so. Usually terrestrial . 4

4. Leaves mostly consisting of empty hyaline cells (cf. couplet 2, General Key) . . . . *Leucobryum* [28]

   Leaves densely chlorophyllose throughout, lacking or almost lacking hyaline cells . . . . *Conostomum* [84]

5. Cells smooth 6

   Cells papillose 7

6. Stems under 5 mm tall . . . . *Eccremidium arcuatum* [16]

   Stems over 1 cm (usually 5 cm) . . . . . . *Meesia triquetra* [74]

7. Leaves erect when dry, closely appressed to the stem, usually clearly in 3 ranks. Papillae tall, usually bifid . . . . *Triquetrella papillata* [49]

   Leaves spirally twisted round the stem when dry, usually not very clearly in 3 ranks. Papillae low, usually unbranched 8

8. Leaves scarcely more than 1 mm long. Nerve failing well below apex . . . . *Anomodon tasmanicus* [49]

   Leaves usually more than 2 mm long. Nerve reaching apex or slightly excurrent . . . . *Leptodontium* [48]

9. Leaves in 4 or 6 ranks 10

   Leaves in 3 ranks. 12

10. Leaves acute with nerve excurrent in a hair-point.
Shoots not complanate . . . . *Rhacopilum convo-lutaceum* [87]

Leaves very obtuse, with nerve not excurrent. Leaf
arrangement complanate 11

11. Nerve strong, single . . . . *Mesochaete undulata* [80]

Nerve absent or very short . . . . *Eriopus* [108]

12. Stem black, usually unbranched; robust and stiff . . *Cyathophorum bulbosum* [112]

Stem usually greenish, not stiff 13

13. Stem branched 14
Stem unbranched . . . . *Mittenia plumula* [76]

14. Growth form dendroid; row of smaller leaves on
underside of stem 15
Growth form creeping or tufted; smaller leaves on
upper side of stem . . . . *Rhacopilum convo-lutaceum* [87]

15. Cells rounded, isodiametric. Frond flattened in plane
of stem . . . . *Lopidium concinnum* [113]

Cells rhomboidal, somewhat elongated. Frond flattened in a plane at right angles to stem, giving
umbrella-like growth form . . . . *Hypopterygium* [111]

Group C
Distichous mosses

1. Leaf base sheathing, at least part of it clasping the
stem on both sides 2
Leaf base not sheathing, not clasping the stem . . 5

2. Leaf lamina slender, almost hair-like, from a wider, sheathing base. Usually forming dense tufts on limestone . . . . *Distichium capillaceum* [14]

   Leaf lamina wide, not hair-like 3

3. Leaf lamina apparently mostly single, only the uppermost half or quarter next to the stem being double and clasping the stem . . . . *Fissidens* [10]
   Lamina folded tightly along the midline and thus appearing double throughout 4

4. Cells very long and narrow, over 50 μm long. Plants usually very glossy . . . . *Catagonium politum* [129]

   Cells isodiametric, about 10 μm across. Plants usually matt . . . . *Fissidens* (especially *taylorii* and *vittatus*) [10]

5. Nerve lacking or very short and faint . . . . . *Hampeella pallens* [93]

   Nerve extending to mid-leaf or beyond, sometimes faint 6

6. Cells each with a single dome-shaped papilla; leaf ending in a long hair-point which may or may not contain the nerve . . . . *Hymenodon pilifer* [79] (mis-keyed)

   Cells flat, not dome-shaped; mostly without hair-points 7

7. Leaves toothed or denticulate at least at the apex or, if entire, having the nerve excurrent in a point. Protonema not persistent 8
   Leaves entire, nerve failing ½–¾ way up leaf. Protonema usually persistent and appearing as a glistening, apparently luminous, weft on the ground *Mittenia plumula* [76]

8. Cells isodiametric or somewhat elongated, less than 30 μm long; cell walls thick 9

Cells elongated, over 50 μm long; cell walls thin. Leaves shrivelling when dry . . . . *Goniobryum sub-basilare* [78]

9. Border strong; teeth strong, multi-cellular; margins undulate when dry . . . . *Mesochaete undulata* [80]

Borders various; teeth unicellular; margins at most slightly flexuose 10

10. Cells in upper ¼ of leaf squarish or irregular in shape, or elongated across the leaf; not pointed . . . *Rhizogonium* [77]

Cells in upper ¼ of leaf rhomboidal, pointed at both ends, slightly elongated parallel to nerve . . . *Thamnobryum* [101]

Group D
Complanate mosses

1. Leaves margined with long hairs . . . . . . *Fabronia hampeana* [126]

Leaf margins not ciliate 2

2. Leaves with a well defined large group of differentiated alar cells at basal margin 3

Leaves without such cells 6

3. Cells very elongated, more than 8× as long as wide; leaves sometimes decurrent at base and lateral leaves usually asymmetric 4

Cells less than 8× as long as wide [usually much less, 2–5×] leaves neither decurrent nor asymmetric . . 5

4. Leaf margins plane . . . . *Plagiothecium* [128]

Leaf margins inrolled or incurved at the apex to form a channel or cone, or boat-shaped point . . . *Hampeella pallens* [93]

5. Alar cells quadrate, thin-walled; leaves less than 1 mm
   long, not plicate; cells not porose .   .   .   .   *Fabronia* [126]
   Alar cells very incrassate; leaves 2–3 mm long, plicate;
   cells porose .   .   .   .   *Glyphothecium*
   sciuroides [92]

6. Nerve single, extending to mid-leaf or beyond, sometimes forked above   7
   Nerve double or absent, or else single but not reaching
   mid-leaf   14

7. Leaves distinctly bordered with narrow cells .   .   .   8
   Leaves unbordered   9

8. Cells more than 4× as long as wide; leaves very
   narrow .   .   .   .   *Daltonia splachnoides*
   [106]

   Cells isodiametric or slightly elongated; leaves very
   wide .   .   .   .   *Distichophyllum*
   (exc. *microcarpum*)
   [107]

9. Branches closely rolled-up (circinate) when dry .   .   *Leptodon smithii*
   [100]

   Branches not rolled-up when dry   10

10. Leaves toothed   11
    Leaves entire   13

11. Cells in distal ¼ of leaf isodiametric   12
    Cells in distal ¼ of leaf elongated, 3 × 1 or longer .   .   *Goniobryum* [78]

12. Stems unbranched or sparsely branched; nerve forked;
    upper cells collenchymatously thickened .   .   .   *Pterygophyllum*
    [109]

    Stems pinnately branched; nerve and cells not as
    above .   .   .   .   *Thamnobryum*
    (= *Thamnium*) [101]

13. Leaves less than 1 mm long, basal margin often de-
    current down stem. Sex organs terminal .   .   .   *Mittenia plumula*
    [76]

    Leaves 2–4 mm long, not decurrent, insertion narrow.
    Sex organs on short lateral branches .   .   .   .   *Distichophyllum*
    microcarpum [107]

14. Cells in mid-leaf isodiametric or slightly elongated, not more than 4× as long as wide 15

    Cells in mid-leaf narrow and elongated, over 5× as long as wide, mostly much more 16

15. Leaves bordered with narrow cells . . . . . *Eriopus* [108]

    Leaves unbordered . . . . *Pterygophyllum* [109]

16. Fronds flat, pinnately branched. Leaves large, nearly 1 mm wide 17

    Stems simple or irregularly branched, not forming flat fronds. Leaves small, usually less than 0·5 mm wide 18

17. Leaves strongly transversely corrugated, golden or olive . . . . *Neckera* [99]

    Leaves flat, some usually with a dead, whitish-silver sheen . . . . *Trachyloma planifolium* [96]

18. Leaves, at least at the stem apex, falcate 19

    Leaves all straight . . . . (return to 4)

19. Leaf axils with tufts of usually white branched rhizoids . . . . *Sauloma tenella* [110]

    Leaf axils lacking rhizoids . . . . *Isopterygium* [130]

Group E
Mosses with gemmae (not including deciduous shoots or rhizoid gemmae)

1. Gemmae on surface of leaf or at leaf apex 2

    Gemmae on stem or in leaf axils 5

2. Shoots complanate; nerve forked . . . . . (mis-keyed *Pterygophyllum dentatum* [109])

    Shoots not complanate; nerve single 3

3. Tall plants, 2–3 cm high. Leaves highly contorted
with very undulate margins when dry, conspicuously bordered . . . . *Calyptopogon mnioides*
[35]

  Short plants, less than 1 cm high. Leaves twisted
spirally but not undulate when dry, not conspicuously bordered 4

4. Plants, excluding fruit, scarcely more than 1 (at most
2) mm high. On soil . . . . (mis-keyed Group N)
Plants rarely less than 2 mm high. On trees, more
rarely on rock . . . . *Tortula papillosa*
[39]

5. Gemmae brown 6
Gemmae green or yellow 8

6. Stems branched, more or less dendroid. Nerve faint
or absent . . . . *Trachyloma planifolium*
[96]

  Stems usually simple, never dendroid. Nerve strong,
excurrent 7

7. Gemmae dark brown filaments in axils of upper
leaves . . . . *Leptotheca gaudi-*
chaudii [75]

  Gemmae red-brown, globular, in axils of lower
leaves . . . . *Leptobryum pyriforme*
[68]

8. Gemmae more than 1 cell wide; cell balls, bulbils or
leaf-like 9
Gemmae consisting of uniseriate filaments of cells . 11

9. Gemmae in dense clusters like tiny narrow leaves at
the apex of the stem 10
Gemmae minute bulbils in axils of leaves . . . *Bryum dichotomum,*
pachytheca etc. [66]

10. Gemmae on tip of stem well above leaves . . . *Aulacomnium palustre*
[75]

  Gemmae nested in terminal leaves . . . . . *Tortula pagorum* [39]

11. Stems bi-pinnately branched, bearing paraphyllia . (mis-keyed *Thuidium* [114])

    Stems scarcely branched 12

12. Gemmae in terminal head at apex of stem, not covered by leaves . . . . *Tetraphidopsis pusilla* [94]

    Gemmae covered by leaves 13

13. Gemmae of 5–6 cells in a row 14
    Gemmae of *c* 20 cells in a row . . . . *Hampeella pallens* [93]

14. Leaf nerve single, strong, extending almost to apex . *Zygodon* [58]
    Leaf nerve very faint, short, double . . . . . *Glyphothecium* [92]

Group F
Subulate, nerved leaves

1. Apical leaves curved in an arc (falcate) and mostly pointing towards one side of the plant (secund) . 2
    Apical leaves straight or twisted but not secund . . 12

2. Leaves less than 4 mm long; cells smooth 3
    Leaves more than 4 mm long, usually much more; or, if less, cells papillose 5

3. Leaves sheathing at base; alar cells absent. Stems erect, seldom branched. Plants not aquatic . . . *Ditrichum* [14]
    Leaves not sheathing; alar cells usually distinct. Stems usually prostrate and branched. Plants more or less aquatic 4

4. Cells predominantly pointed, not in longitudinal rows . . . . *Drepanocladus* [120]
    Cells mostly blunt or square-ended, in longitudinal rows . . . . *Blindia* [18]

5. Alar cells conspicuous, enlarged and usually coloured .   6
    Alar cells not distinct   8

6. Nerve very wide, taking up more than ½ of the leaf
    base .   .   .   .   *Campylopus* (esp.
        pallidus) [22]

    Nerve narrower, less than ½ the width of the leaf   .   7

7. Plants on rock in water, in montane or sub-alpine
    regions .   .   .   .   *Blindia* [18]
    Plants not aquatic   .   .   .   .   *Dicranoloma* [25]
        (including *Dicranum*
        [20])

8. Leaves strongly toothed with twinned teeth   .   .   .   *Rhizogonium mnioides*
        or *parramattense* [77]

    Leaves toothed or not, teeth never twinned   .   .   9

9. Cells papillose   10
    Cells smooth   29

10. Leaves squarrose when dry, like a grass leaf, with an
    evident sheathing base of long smooth cells clasping
    the stem, and an abruptly diverging limb of papil-
    lose cells   .   .   .   .   *Bartramia* [81] (see also
        Dicranella cardotii
        [24])

    Leaves erect or diverging but not squarrose; without
    a distinct leaf base; cells papillose throughout   .   .   11

11. Stems very tomentose. Rhizoids usually papillose.
    Leaves plicate .   .   .   .   *Breutelia* [83]
    Stems scarcely tomentose. Rhizoids usually smooth.
    Leaves not plicate .   .   .   .   *Philonotis* [85]

12. Leaves tightly twisted or strongly contorted when
    dry, or wound part way round the stem   .   .   .   13
    Leaves straight or flexuose when dry, but neither
    strongly contorted nor twisted round the stem   .   21

13. Leaves strongly toothed with double teeth . . . *Rhizogonium mnioides* or *parramattense* [77]

Leaves entire or slightly denticulate but teeth never double 14

14. Cells papillose 15
Cells smooth 18

15. Leaves when dry twisted, rope-like, part way round the stem . . . . *Barbula* [46]
Leaves individually contorted, not wound round the stem 16

16. Nerve widening in the upper part and usually granulose on the adaxial surface there . . . . . . (mis-keyed *Desmatodon convolutus* [37])

Nerve narrowing in the upper part; not granulose . 17

17. Leaf margins usually inrolled; nerve cells covered adaxially by lamina cells . . . . *Weissia controversa* [45]

Leaf margins usually plane or recurved; nerve cells exposed adaxially . . . . *Amphidium cyathicarpum* [54]

18. Stems tomentose below . . . . *Holomitrium perichaetiale* [27]

Stems not tomentose 19

19. Alar cells enlarged and sometimes coloured . . . *Dicranoweisia* [26]
Alar cells not distinct 20

20. Hair-point absent. Cell cavities not constricted . . *Ptychomitrium* [52]
Hyaline hair-point usually present. Some cell cavities in mid-leaf constricted, dumbell-shaped . . . *Grimmia trichophylla* [11]

21. Nerve very wide, usually taking up more than ½ of the width of the leaf base (occasionally also in *Pleuridium*) . . . . *Campylopus* [22]
Nerve less than ½ the width of the leaf, at the base . 22

22. At least some cells papillose   23
     All cells smooth   26

23. Leaves bristle-like, stiffly erect and pressed against
        the stem. Upper part of leaf margins and back of
        nerve toothed with strong, spiny teeth . . . .   *Bartramia stricta* [81]
     Leaves not stiffly erect; entire or serrulate but usually
        not strongly toothed   24

24. Stems tightly packed together into a hard turf, bright
        green in upper 5 mm or so. On calcareous rock
        or mortar   25
     Stems often loosely matted together but never tightly.
        Upper part of stems yellowish or silvery, usually
        more than 1 cm . . . .   return to 10

25. Leaf quickly narrowed from a wide, sheathing base
        to a long subula. Leaves arranged in 2 ranks in some
        parts . . . .   *Distichium
                                                        capillaceum* [14]

     Leaf gradually narrowed; no distinct sheathing leaf
        base. Leaves never in 2 ranks . . . . . .   *Gymnostomum
                                                        calcareum* [41]

26. Fruit with a very long neck, equalling or exceeding
        the capsule . . . .   *Trematodon* [29]
     Neck much shorter than capsule or fruit not present .   27

27. Cells of limb very long and narrow, reaching more
        than 10× as long as wide   28
     Cells isodiametric or oblong, usually not more than
        5× as long as wide   29

28. Most leaves slightly denticulate at the apex; leaves
        similar in length to the stem. Inner peristome with
        a conspicuous basal membrane *c* ½ length of the
        processes. A common plant of flower pots, garden
        beds, waste places . . . .   *Leptobryum pyriforme*
                                                                [68]

Leaves almost completely entire, several times the length of the stem. Inner peristome consisting of fine filaments with almost no basal membrane. Usually on tree trunks, and especially charred wood, in montane or sub–alpine regions . . . . . *Orthodontium lineare* [71]

29. Plants large, over 4 cm long, stiff and bristly. Cells mostly isodiametric . . . . *Echinodium hispidum* [102]

Plants usually less than 2 cm tall. Cells 2–6× as long wide (except *Brachydontium*) 30

30. Leaves strongly bordered with elongated narrow cells, especially in the upper half. Leaf base not sheathing.. *Daltonia splachnoides* [106]

Leaves bordered, if at all, with hyaline cells little different from adjacent cells of lamina; border confined to sheathing base of leaf or absent 31

31. Capsule on an elongated seta, well above the leaves . 32
Capsule nearly sessile, overtopped by leaves . . . *Pleuridium* [17] (also mis-keyed Group J)

32. Capsule almost spherical, with a small mouth . . *Conostomum* [84]
Capsule elongated, or short; mouth almost as wide as the capsule 33

33. Seta cygneous when moist. Stem usually less than 1 mm . . . . *Brachydontium* [19]
Seta straight. Stem usually more than 5 mm . . . 34

34. Capsule smooth or lightly grooved. Operculum not much more than ½ length of the capsule . . . *Ditrichum* [14]
Capsule strongly ribbed (not in *jamesonii*). Operculum long, about as long as the capsule . . . . . *Dicranella* [24]

Group G
Unbordered leaves. Nerve excurrent in a hair or long point. Leaves not subulate

1. Nerve expanded above and densely granular on adaxial surface . . . . (mis-keyed Group N)
   Nerve not expanded above, nor granular 2

2. Cell walls, at least below, with corrugated thickenings. Dry leaves, when soaking out, flexing strongly backwards and then slightly forwards 3
   Cell walls straight or sinuose. Leaves not returning forwards at all when soaking out 4

3. Basal corrugated cells long, to 4×1 or longer. Plants usually forming wefts. Seta erect when moist . . *Rhacomitrium* [12]
   Basal corrugated cells usually shortly rectangular, *c* 2×1. Plants forming cushions. Seta cygneous when moist . . . . *Grimmia* [11]

4. Minute mosses (up to 2 mm) with almost sessile, often cleistocarpous capsules, overtopped by the leaves . [Group J]
   Mostly more than 2 mm high. Capsules neither sessile nor cleistocarpous 5

5. Leaf widest above the middle 6
   Leaf widest in the middle or below 11

6. Cells papillose 7
   Cells smooth . . . . 8 (sometimes *Leptostomum* [69])

7. Leaf margins plane . . . . *Pottia* [32] (sometimes *Tortula* [39])
   Leaf margins incurved or recurved 15

8. Moist leaves concave (U-shaped in TS) at least in the upper half; concave or flat at base. Capsule usually asymmetric . . . . *Funaria* [60]
   Moist leaves keeled (folded), the wings often somewhat convex, (V-shaped in TS) at least below. Capsule always symmetrical 9

9. Cells rhomboidal to hexagonal, pointed at both ends — *Bryum* [66]
   Cells isodiametric or oblong, squarish at both ends . — 10

10. Stem tomentose. Cells in mid-leaf oblong, 20–40 μm
    wide . . . . — *Tayloria* [65]
    Stem not tomentose. Cells in mid-leaf roughly isodia-
    metric, 15–20 μm wide . . . . — *Pottia truncata* [32]

11. Cells papillose or at least roughened by projecting cell
    tips . . . . — 12
    Cells smooth . . . . — 20

12. Leaves straight or slightly flexuose when dry. Cells
    usually only slightly papillose . . . . . . — 13
    Leaves twisted part way round the stem or strongly
    curled when dry. Cells usually densely papillose . — 14

13. Nerve excurrent in hyaline hair . . . . . — *Grimmia* [11]
    Excurrent nerve not hyaline . . . . . . — 18

14. Leaf margins recurved . . . . — 15
    Leaf margins plane or incurved . . . . . — 17

15. Peristome absent or very short . . . . . . — *Pottia* [32] (also
    Desmatodon [37])
    Peristome teeth long and twisted . . . . . — 16

16. Leaves widest below the middle; usually less than 1
    mm wide. Hair-point, if any, yellowish . . . — *Barbula* [46]
    Leaves usually widest in the middle or above; leaves
    often more than 1·5 mm wide. Hair-point hyaline — *Tortula* [39]

17. Leaves flat when moist. Adaxial surface of nerve in
    upper half of leaf completely obscured by papillae.
    Capsule cleistocarpous, scarcely longer than the seta — *Tetrapterum
    cylindricum* [42]

    Leaves undulate when moist. Adaxial surface of nerve
    papillose but the cells not obscured. Seta very long,
    capsule with peristome . . . . — *Tortella* [43]

18. Rhizoids usually smooth. Leaves not plicate; alar cells
     not distinct.   19
     Rhizoids papillose, in part. Leaves plicate; alar cells
     distinct, subquadrate .   .   .   .   *Breutelia* [83]

19. Operculum with a curved beak. Peristome teeth
     joined   at   apex.   Stems   usually   unbranched;
     epidermal cells usually intact .   .   .   .   .   .   *Conostomum* [84]
     Operculum almost flat. Peristome teeth separate.
     Stems usually with whorls of branches above;
     rough with collapsed large epidermal cells .   .   .   *Philonotis* [85]

20. Nerve very wide occupying at least ⅓ (and often much
     more) of the width of the leaf base .   .   .   .   .   *Campylopus* [22]
     Nerve less than ⅓ the width of the leaf base .   .   .   21

21. Cells isodiametric or nearly so, except sometimes
     in the leaf base.   22
     Cells at least twice as long as wide   27

22. Plants creeping. Leaves in 4 ranks, those in the upper
     two ranks smaller; crisped when dry   .   .   .   .   (mis-keyed
                                                            *Rhacopilum
                                                            convolutaceum* [87])

     Plants tufted. Leaves not in 4 ranks; straight or
     slightly twisted when dry   23

23. Nerve-point toothed   24
     Nerve-point smooth   25

24. Teeth on leaf margins twinned .   .   .   .   .   .   *Rhizogonium* [77]
     Teeth single .   .   .   .   *Grimmia* [11]

25. Cell cavities square. Plants not epiphytic .   .   .   .   *Ceratodon purpureus*
                                                                [15]
     Cell cavities rounded or angular. Plants epiphytic   .   26

26. Leaf large *c* 2·5 mm long; hair-point long and flexuose
     c 0·5 mm long .   .   .   .   *Leptostomum inclinans*
                                                            [69]

     Leaf usually small 1·0–1·5 mm long; point short and
     stiff, *c* 0·1 mm long   .   .   .   .   *Leptotheca
                                                            gaudichaudii* [75]

27. Cells in mid-leaf mostly elongated (4× as long as
    wide). Rhizoids papillose . . . . *Pohlia* [72]
    Cells in mid-leaf mostly less than 4× as long as wide.
    Rhizoids smooth or papillose 28

28. Stems pinnately branched; bearing paraphyllia . . *Cratoneuropsis*
    relaxa [119]
    Stems scarcely branched; without paraphyllia . . 29

29. Inner peristome with narrow processes and no cilia . *Brachymenium*
    preissianum [67]
    Inner peristome with cilia and wide processes . . *Bryum* [66]

Group H
Dendroid mosses

1. Frond flattened into the same plane as the main stem
   (espalier-like) 2
   Frond not flattened, or flattened in a plane at right
   angles to the main stem so that the growth form is
   umbrella-like 6

2. Leaves arranged in 3 ranks, flattened in one plane; 2
   rows of lateral leaves and 1 of underleaves. Leaves
   strongly bordered . . . . *Lopidium concinnum*
   [113]
   Leaves spirally or irregularly arranged, spreading
   evenly round the branches or complanate or, if in 3
   ranks, leaves unbordered 3

3. Nerve double, faint or absent . . . . *Trachyloma planifolium*
   [96]
   Nerve single, strong 4

4. Cells short, *c* 2 × 1. Plants delicate, feathery . . . *Thamnobryum* [101]
   Cells long, 5 × 1 or more. Plants robust 5

5. Leaves densely imbricate on the branches; obtuse   .   *Braithwaitea sulcata*
                                                              [86]

    Leaves rather distant on branches; acute   .   .   .   .   *Hypnodendron vitiense*
                                                              and *spininervium*
                                                              [86]

6. Leaves of ultimate branches arranged in 2 or 3 ranks
       and flattened into one plane (complanate) .   .   .   7
    Leaves of branches spirally inserted, often not com-
       planate   9

7. Leaves in 2 ranks .   .   .   .   *Rhizogonium bifarium*
                                                              [77]

    Leaves in 3 ranks; 2 rows of lateral leaves and 1 row of
       underleaves or upper leaves   8

8. Leaves usually clearly bordered   .   .   .   .   .   .   *Hypopterygium*
                                                              [111]
    Leaves unbordered   .   .   .   .   *Hypnodendron* [86]

9. Stem densely tomentose   10
    Stem not tomentose   11

10. Leaves glaucous, less than 2 mm long .   .   .   .   .   *Philonotis*
                                                              scabrifolia [85]

    Leaves green or yellowish, not glaucous, more than
       3 mm long .   .   .   .   *Hypnodendron*
                                                              (*Mniodendron*
                                                              comosum [86])

11. Stem black or dark. Leaves toothed. Cells usually
       papillose abaxially (single papillae at ends of cells)   *Hypnodendron*
                                                              vitiense and
                                                              spininervium [86]
    Stem greenish, pale. Leaves entire. Cells smooth .   .   *Camptochaete*
                                                              arbuscula [104]

Group I
Bordered leaves with excurrent nerves

1. Leaf cells papillose 2
    Leaf cells smooth 6

2. Nerve excurrent in a long hyaline hair-point, usually
    visible with the naked eye 3
    Hair-point usually too short to see without a lens,
    coloured or hyaline only at the tip 4

3. Cells with single high papillae . . . . . . *Dicnemoloma*
    pallidum [23]
    Cells with low, multiple papillae . . . . . *Tortula* [39]

4. Leaves strongly undulate at the margins and therefore
    contorted when dry 5
    Leaves not undulate, somewhat twisted when dry . *Encalypta vulgaris*
    [51]

5. Border formed from outermost cells of leaf . . . *Dicnemoloma*
    pallidum [23]
    Border intramarginal, in part, with small quadrate
    cells outside it . . . . *Calyptopogon*
    mnioides [35]

6. Cells isodiametric . . . . *Mnium rostratum* [73]
    Cells in mid-leaf at least 2 × as long as wide . . . 7

7. Leaves ovate or obovate, widest above or just below
    the middle. Border of constant width throughout 8
    Leaves narrowly triangular. Border widening at base *Daltonia*
    splachnoides [106]

8. Inner peristome rudimentary, consisting of linear,
    thread-like teeth. Border weak . . . . . . *Brachymenium*
    preissianum [67]
    Inner peristome well-developed with wide teeth and
    cilia . . . . *Bryum* [66]

Group J
Minute mosses with sessile or almost sessile capsules

1. Seta short, arcuate 2
   Seta not arcuate 3

2. Capsule cleistocarpous with no line of dehiscence.
   Spores papillose, *c* 30 μm. Outermost cells of capsule
   wall not thickened . . . . *Pleuridium arnoldii*
   [17]

   Capsule with dehiscent lid, outermost cells collenchymatously thickened. Spores larger than 30 μm
   (much larger) . . . . *Eccremidium* [16]

3. Apophysis very large 4
   Apophysis small or absent 6

4. Capsule without dehiscence line; apophysis truncate
   at the base; seta usually shorter than capsule . . *Bruchia brevipes* [21]
   Capsule with dehiscent lid; apophysis tapered to the
   seta; seta 2–6 mm long 5

5. Calyptra large, inflated at the base, split on one side . *Funaria apophysata*
   [60] (+perhaps
   Trematodon amoenum
   [29])
   Calyptra not inflated, symmetrical, 2–3-lobed at base *Physcomitrium*
   conicum [63]

6. Capsule cleistocarpous, elongated, more or less
   cylindrical, often weakly 4-angled . . . . . *Tetrapterum*
   cylindricum [42]
   Capsule globose to oval, cleistocarpous or dehiscent . 7

7. Stems more or less julaceous with tiny appressed
   leaves; perichaetials much larger and subulate. Capsule cleistocarpous. 8
   Plants without this combination of characters . . 9

*8. Capsule globose without any apiculus; columella and seta absent; spores few, yellow, very large (over 100 µm) . . . .   *Archidium* [13]

Capsule globose to oval, with columella and apiculus; seta short; spores numerous, relatively small, (20–30 µm) . . . .   *Pleuridium nervosum* [17]

9. Calyptra very large, covering capsule   10

Calyptra very small, usually covering less than half the capsule   11

10. Calyptra plicate; capsule orange, dehiscent. Leaves smooth . . . .   *Goniomitrium* [61]

Calyptra not plicate, completely enveloping the ripe capsule and not tearing away at the base; capsule cleistocarpous. Leaves papillose . . . .   *Bryobartramia novae-valesiae* [50]

11. Calyptra campanulate, delicate and sometimes lacerated at base. Leaves narrow, long in relation to width   12

Calyptra not campanulate. Leaves usually ovate or obovate   13

12. Leaves strongly spiny-toothed. Capsule cleistocarpous, red-brown when ripe. Green protonema abundant, persistent . . . .   *Ephemerum cristatum* [64]

Leaves entire or rough with projecting cell ends. Capsule with line of dehiscence. No obvious persistent protonema . . . .   *Eccremidium* [16]

13. Capsule dehiscent, wide-mouthed when empty; spores more than 100 µm, brown, angled. Leaves nerveless; perichaetial leaves silvery, much larger than vegetative leaves . . . .   *Gigaspermum repens* [59]

Capsule cleistocarpous; spores less than 100 µm usually much less. Leaves nerved   14

* In the absence of a capsule it is difficult to tell if the plant is an *Archidium*, a *Pleuridium* or *Eccremidium pulchellum*

14. Plants bulb-like; leaves nerved almost to the apex or
 beyond . .   .   .   *Acaulon* [33]
 Plants not bulb-like; leaves with nerve failing well
 below apex .   .   .   .   *Physcomitrella*
 readeri [62]

Group K
Leaves with single nerve and papillose or mamillose cells

1. Plants much branched, with persistent prostrate or
 pendulous stems bearing branches more or less at
 right angles; forming loose mats or wefts (pleurocarpous or apparently so)   2
 Plants sparingly branched with only erect stems, rarely
 pendulous; forming cushions or turfs (acrocarpous) 10

2. Leaves clearly bordered with *elongated* cells, different
 from those in the lamina, not merely paler or smooth *Dicnemoloma*
 pallidum [23]
 Leaves unbordered   3

3. Leaves usually at least 3 times as long as wide . .   .   4
 Leaves usually less than twice as long as wide .   .   .   6

4. Leaves sharply denticulate .   .   .   .   *Breutelia* [83]
 Leaves entire or crenulate   5

5. Calyptra plicate, usually with hairs. Leaves straight or
 curled and/or wound helically round the stem . .   *Macromitrium* [55]
 Calyptra smooth, hairless. Leaves wound round the
 stem .   .   .   .   *Schlotheimia* [56]

6. Stems slender, wiry. Capsules usually numerous, ±
 immersed in the perichaetial leaves .   .   .   .   .   *Cryphaea* [88]
 Stems not wiry. Capsules scarce or absent, not im-
 mersed   7

7. Leaf abruptly contracted into a long hair-point which
   the nerve usually does not enter . . . . . .   *Hymenodon pilifer*
                                                          [79]
   Leaf not ending in a hair   8

8. Stems without paraphyllia   9
   Stems with paraphyllia . . . .   *Thuidium* [114]

9. Cells in leaf base, next to the nerve, long and narrow
   parallel to nerve. Plants mainly epiphytic . . .   *Papillaria* [97]
   Cells in leaf base mostly short (2 × 1) often elongated
   obliquely. Plants mostly on rocks . . . . .   *Pseudoleskea* [115]

10. Nerve clearly papillose or toothed on the abaxial
    surface, at least in part, or covered by papillose
    cells   11
    Nerve smooth or nearly so   26

11. Upper cells elongated, at least 2 × 1   12
    Upper cells isodiametric   14

12. Rhizoids smooth . . . .   (mis-keyed
                                                     Conostomum [84])
    Rhizoids papillose   13

13. Leaf base sheathing, distinct from rest of lamina,
    smooth . . . .   *Bartramia papillata*
                                                          [81]
    Leaf base not distinct, not sheathing, plicate . . .   *Breutelia* [83]

14. Cells overlying nerve, adaxially, similar to adjacent
    cells of lamina, at least in upper half of leaf . . .   15
    Cells overlying nerve different from lamina cells,
    usually more elongated and less papillose; or nerve
    too deeply channelled to be visible   20

15. Cells similar throughout leaf . . . .   *Anomodon tasmanicus*
                                                          [49]
    Cells in basal part of leaf elongated, wide, thin-walled,
    rather hyaline   16

16. Apex of nerve toothed abaxially . . . . . .   *Tortula rubra* [39]
    Apex of nerve not toothed   17

17. Hyaline basal area bordered by narrow cells. Calyptra relatively huge, candle-shaped, covering capsule and most of seta . . . . *Encalypta vulgaris* [51]

Basal area not bordered. Calyptra not wholly enclosing capsule 18

18. Leaf margins strongly recurved . . . . . *Barbula* [46] (+ perhaps *Desmatodon* [37])

Leaf margins nearly plane 19

19. Nerve wide at apex (*c* 45 μm). Capsule cleistocarpous; seta little longer than capsule. Stems not in dense cushions. Bright pale green plants of dry sandy soils . . . . *Tetrapterum cylindricum* [42]

Nerve narrow at apex (*c* 20 μm). Capsule with peristome; seta much longer than capsule. Stems packed together to form a rather dense cushion. Dull rusty brown plants of dry clay banks . . . . . *Bryoerythrophyllum binnsii* [47]

(*Gymnostomum* may also key out here)

20. Elongated basal cells bearing single rows of papillae . *Leptodontium* [48] (also forms of *Tortella calycina* [43])

Basal cells smooth or papillose, but papillae not in a row down any cell (excluding marginal cells) . . 21

21. Stems densely packed to form a firm turf; bright pale green above . . . . *Gymnostmum calcareum* [41] or *Anoectangium bellii* [40]

Stems not densely packed, usually dull or dark green above 22

22. Cells in upper half of leaf thick-walled, especially at the corners, thus appearing rounded 23
     Cells in upper part of leaf squarish, or hexagonal, thin-walled 25

23. Leaves usually not densely papillose. Cells of basal ¼ of leaf enlarged, thinner-walled, different from cells of upper ½. Calyptra symmetrical . . . .   *Orthotrichum* [53]

     Leaves usually densely papillose. Cells scarcely different in leaf base. Calyptra side-split   24

24. Base of nerve very finely papillose abaxially; nerve reaching apex. Margins recurved in lower half of leaf . . . .   *Leptodontium* [48]

     Base of nerve smooth abaxially; nerve failing below apex. Margins usually plane . . . . . .   *Zygodon intermedius* [58]

25. Area of enlarged, thin-walled, smooth, hyaline cells extending to about ½ way up the leaf. Leaf usually widest above the middle . . . .   *Pottia* [32]

     Area of hyaline cells absent, or confined to lower quarter of leaf. Leaf usually wider below the middle   *Barbula* [46]

26. Elongated basal cells bearing single rows of papillae .   *Leptodontium* [48]

     Basal cells not seriately papillose   27

27. Marginal cells of leaf base not distinct   28

     Several rows of marginal cells at the base different from adjacent cells of lamina   31

28. Leaf contracted to an abrupt hair-point, usually not containing the nerve . . . .   *Hymenodon pilifer* [79]

     Leaf lacking a hair-point   29

29. Leaves tending to wind round the stem when dry .   *Zygodon* [58] (sometimes *Leptodontium* [48])

     Leaves crisped but not wound round the stem . .   30

30. Bog plants. Leaf margins revolute almost throughout   *Aulacomnium palustre* [75]

     Plants of wet rocks and cliffs. Leaf margins plane, or revolute only at leaf base . . . .   *Amphidium cyathicarpum* [54]

31.   Upper cells usually thick-walled, isodiametric . .   32

       Upper cells thin-walled, oblong .   .   .   .   *Breutelia affinis*
                                                        [83]

32.   Leaf with a broad sheathing base contracted suddenly
       to a long, curled subula. Cells in leaf base very
       thick-walled and porose .   .   .   .   .   .   *Holomitrium*
                                        perichaetiale [27]

       Leaf not as above or, if so, without thick-walled
       porose cells in leaf base   33

33.   Marginal cells of leaf base much longer and narrower
       than adjacent cells. .   .   .   .   .   .   *Encalypta vulgaris*
                                                        [51]

       Marginal cells of leaf base wider than adjacent cells .   34

34.   Marginal cells in leaf base with thickened transverse
       and thin longitudinal walls; other cells in leaf base
       with very thick walls. .   .   .   .   .   .   *Ulota* [57]
       Marginal cells in leaf base evenly thickened; other
       cells not heavily thickened .   .   .   .   .   *Orthotrichum* [53]

Group L
Nerved leaves, nerves not reaching beyond apex. Smooth cells

1.   Leaves strikingly silvery white, at least in upper half;
      stems very short, rather scattered. Capsule immersed   (mis-keyed
                                                              Gigaspermum
                                                               repens [59])

      Leaves not silvery; or, if so, the capsule exserted and
      the stems tightly packed together to form cushions
      (*Bryum argenteum*)   2

2.   Cells in mid-leaf (excluding nerve) isodiametric or
      short (usually not more than twice as long as wide,
      excluding perichaetial leaves)   3
      Cells in mid-leaf distinctly elongated, at least 3× as
      long as wide (seldom less than twice as long as wide
      e.g. *Cratoneuropsis* sometimes)   28

3. Lamina undulate, strongly toothed on margins and on
    the surface of the undulations . . . . . . (*Atrichum* [5]
    mis-keyed)

    Lamina not toothed on surface 4

4. Cells wide and lax, usually more than 25 μm wide,
    square-ended 5
    Cells thick-walled or lax, but usually less than 25 μm
    wide; ends either square or pointed 7

5. Capsule cleistocarpous, ovoid . . . . *Physcomitrella*
    readeri [62]

    Capsule operculate, pyriform or cylindrical . . . 6

6. Stems tomentose. Capsule with pronounced apophysis
    (longer than theca). Spores sticky, oozing from
    capsule mouth as a bright yellowish blob . . . *Tayloria* [65]
    Stems not tomentose. Capsule usually without a long
    apophysis (except *F. apophysata*) . . . . . . *Funaria* [60]
    (*Pottia truncata* may key out here)

7. Nerve very wide, occupying at least ½ width of leaf *Campylopus* [22]
    (+ perhaps *Aloina*
    [34])

    Nerve occupying less than ⅓ width of leaf . . . 8

8. Cells in mid-leaf elongated, up to 3× as long as
    wide 9
    Cells in mid-leaf and above roughly isodiametric . 16

9. Some areas of cells near mid-leaf diagonally elongated,
    aligned upwards and outwards towards the leaf
    margin (striking, in all except *Cryphaea tasmanica*) . 10
    Cells somewhat elongated, but parallel to nerve . . 12

10. Stems short, less than 1 cm; leaves slightly twisted
    round stem . . . . *Zygodon menziesii*
    [58]

    Stems longer than 1 cm; leaves not twisted round
    stem 11

11. Branches club-shaped, julaceous, tapering below
because of the smaller leaves. Plants usually on dry
limestone rock . . . .   *Pseudoleskea*
imbricata [115]

   Branches not tapering below, not julaceous. Plants
epiphytic or aquatic . . . .   *Cryphaea* [88]

12. Leaves very narrow, only *c* 6 cells wide on each side of
nerve. Outer walls of cells very thin, giving a crenulate edge . . . .   *Bartramidula pusilla*
[82]

   Leaves wider, more than 10 cells on each side of
nerve   13

13. Leaves short and wide, obtuse, very tightly appressed
to the stem, *c* 0·5 × 0·25 mm. Perichaetial bracts
ending in a long slender point, overtopped by
branches from just below . . . . . .   *Eccremidium* [16]
Leaves appressed or not, more than 0·5 mm long.
Perichaetial bracts not greatly different from
vegetative leaves. Stems not usually branched
above   14

14. Plants epiphytic. Peristome teeth densely coarsely
papillose throughout . . . .   *Daltonia* [106]
Plants mostly terrestrial; peristome teeth smooth
or finely papillose   15

15. Cells roughly rectangular, with square ends. Growing
in bogs . . . .   *Meesia muelleri*
[74]

   Cells rhomboid-hexagonal with pointed ends. Habitats various . . . .   *Bryum* [66]

16. Cells of nerve covered on the adaxial surface, at least
in parts, by the shorter cells of the lamina . . .   17
Elongated cells of nerve distinct on adaxial surface of
leaf, not covered by lamina cells   20

17. Cells at base of leaf enlarged or not, but walls clearly
thickened   18
Cells towards base of leaf enlarged, thin-walled . .   19

18. Stems creeping, often with erect branches . .   *Macromitrium* [55]
Stems erect, tufted . . . .   *Orthotrichum* [53]

19. Nerve widened above and granular on adaxial surface . . . . *Aloina* [34]
    Nerve not widened above, not granular . . . . *Ptychomitrium* [52]

20. Cells thin-walled, the cavities not rounded . . . 21
    Cells thick-walled, the cavities usually rounded by corner thickenings 24

21. Teeth on leaf-margin twinned 22
    Teeth single or absent 23

22. Leaf broad; lamellae on nerve . . . . (mis-keyed *Atrichum androgynum* [5])
    Leaf narrow, almost hair-like; without lamellae . . *Rhizogonium mnioides* [77]

23. Leaf margin more or less plane; upper cells hexagonal, rather wide (*c* 30 μm) . . . . *Pottia truncata* [32]
    Leaf margin reflexed for much of its length; upper cells square, *c* 15–20 μm wide . . . . . . *Ceratodon purpureus* [15]

    (*Zygodon menziesii* [58] may key out here)

24. Leaves bordered in lower part by elongated cells . . 25
    Leaves not at all bordered 26
    (*Grimmia apocarpa* [11] *Zygodon menziesii* [58] and *Holomitrium* [27] may key out here)

25. Upper part of leaf border intramarginal, with smaller lamina cells outside it. Alar cells not distinct. Elongated cells of the leaf base evenly thickened, not porose. Plants semi-aquatic . . . . . . *Tridontium tasmanicum* [44]

    Border confined to outermost cells. Alar cells usually enlarged and coloured. Elongated cells in the leaf base porose. Plants not aquatic . . . . . . *Holomitrium perichaetiale* [27]

26. Stems creeping, with erect branches . . . . . *Macromitrium* [55]
    Stems densely tufted, lacking prostrate stems . . . 27

44

27. Elongated cells of nerve covered abaxially by quadrate
    lamina cells . . . . *Orthotrichum rupestre* [53]

    Elongated cells of nerve fully exposed abaxially . . *Amphidium cyathicarpum* [54]

28. Cell walls strongly thickened, the walls roughly as
    wide as the cavities 29
    Cell walls in mid-leaf never as thick as the cavities . 32

29. Nerve faint, just reaching beyond mid-leaf, or shorter.
    Leaves scarcely longer than wide . . . . . *Lembophyllum divulsum* [103]

    Nerve strong, reaching above ½ way. Leaves much
    longer than wide 30

30. Longitudinal cell walls with very corrugated thicken-
    ings . . . . *Rhacomitrium* [12]
    Longitudinal walls not corrugated 31

31. Alar cells conspicuous, inflated, thick-walled and
    orange. Leaves partially bordered with hyaline
    cells, often narrowly . . . . *Dicranoloma* [25]
    Alar cells not distinct. Leaves not bordered . . . *Macromitrium* [55]

32. Leaves of shoot-tips falcate and secund . . . . *Drepanocladus* [120]
    Leaves not falcate and secund 33

33. Plants acrocarpous, forming cushions or turfs; main
    stems more or less erect and parallel, bearing cap-
    sules and gametangia at the apices 34
    Plants pleurocarpous, mostly forming loose wefts,
    the stems not parallel. Gametangia and capsules on
    special lateral branchlets 36

34. Cells rhomboid–hexagonal, usually with straight side
    walls . . . . *Bryum* [66]
    (*Funaria glabra* [60] and non-distichous forms of
    Goniobryum subbasilare [78] may key out here)
    Cells linear, usually with rather square ends; side
    walls often slightly sinuous 35

35. Seta apparently basal. Outer peristome practically
absent . . . . *Mielichhoferia* [70]
Seta clearly terminal. Outer peristome teeth conspicuous . . . . *Pohlia* [72]

36. Aquatic or semi-aquatic plants growing either in water
or in boggy habitats 37
Plants normally not at all aquatic 41

37. Leaves plicate when dry. . . . . *Brachythecium* [122]
(also sometimes
*Eurhynchium
austrinum* [123])
Leaves not plicate 38

38. Plants yellowish or brownish 39
Plants dull green . . . . *Eurhynchium
austrinum* [123]

39. Leaves (at least some) strikingly squarrose, the upper
half hooked backwards away from the stem apex
and pointing almost towards the stem base. Small
hair-like or leaf-like paraphyllia usually detectable
on stems . . . . *Cratoneuropsis
relaxa* [119]

Leaves squarrose or not, but never pointing backwards.
Paraphyllia never present. 40

40. Upper, narrow, part of the leaf flat and straight . . *Leptodictyum
riparium* [121]

Upper part of leaf U-shaped in cross section, often
somewhat twisted helically . . . . *Campylium
polygamum* [118]

41. White or hyaline tomentum in leaf axils . . . . (mis-keyed *Sauloma
tenella* [110])

Tomentum, if present, not hyaline. 42

42. Leaves, when dry, spreading widely from the stem
all round 43
Leaves, when dry, tightly pressed to stem at least in
one plane 45

43. Leaves oblong, almost parallel-sided .   .   .   .   .   *Goniobryum* [78]
     Leaves ovate, sides clearly curved   44

44. Seta smooth .   .   .   .   *Rhynchostegium*
                                                                [125]
     Seta rough   .   .   .   .   *Eurhynchium* [123]
                                                                (including
                                                                Rhynchostegiella)

45. Leaves wide, concave, ending in a short abrupt point.
        Introduced .   .   .   .   *Pseudoscleropodium*
                                                                purum [124]

     Leaves not or scarcely concave, tapering to a fine
        point   46

46. Alar cells forming a very wide, conspicuous, usually
        dark group of square or over-square cells, in regular
        rows   47
     Alar cells quadrate or not distinct, not in regular
        rows .   .   .   .   *Brachythecium* [122]

47. Cells in mid-leaf rather short, 30–40 µm long. Leaves
        denticulate or ciliate, sometimes almost entire .   .   48
     Cells in mid-leaf 50–100 µm long; leaves entire .   .   *Ischyrodon lepturus*
                                                                [127]

48. Leaves denticulate or entire .   .   .   .   *Fabronia australis*
                                                                [126]
     Leaves bordered with long, hair-like cells .   .   .   .   *Fabronia hampeana*
                                                                [126]

Group M
Mosses with no nerve, or nerve short, usually forked

1. Leaves very concave, not tapering, not or scarcely
        pointed   2
    Leaves tapering to a point; flat or channelled (i.e. U–
        shaped in TS)   7

2. Leaves at least twice as long as wide, at least at the
   shoot tips . . . .
                            3 (sometimes also
                            Eucamptodon [30])

   Leaves scarcely longer than wide . . . . . .
                            5

3. Tasmanian bog plants. Cells in mid-leaf 25–30 μm
   wide . . . .
                            Pleurophascum
                            grandiglobum [31]

   Plants rarely in bogs; epiphytes or on wet rocks.
   Cells less than 12 μm wide . . . .
                            4

4. Cells in mid-leaf usually less than 30 μm long. Plants
   usually on wet rock; almost dendroid growth
   form . . . .
                            Camptochaete
                            gracilis [104]

   Cells in mid-leaf usually more than 70 μm long.
   Plants usually hanging from twigs or trunks; prostrate or pendulous. . . . .
                            Weymouthia mollis
                            [98]

5. Leaves at shoot tips tightly imbricated to form a sharp,
   spear-like point. Alar cells thin-walled, inflated,
   empty and hyaline . . . .
                            Acrocladium
                            chlamydophyllum [117]

   Leaves at shoot apex looser, not forming a firm point.
   Alar cells thick-walled and usually porose, dark
   and granular . . . .
                            6

6. Cells at leaf apex thick-walled, *c* 30 μm long, slightly
   sigmoid, frequently porose at the cell-ends. Shoots
   2–3 mm wide, except for microphyllous shoots . .
                            Weymouthia
                            cochlearifolia [98]

   Cells at leaf apex rather shorter, *c* 15–25 μm long,
   thick-walled but not or scarcely porose, not sigmoid. Shoots *c* 1–2 mm wide . . . . . .
                            Lembophyllum
                            divulsum [103]

7. Alar cells not clearly distinct . . . .
                            8
   Alar cells conspicuous, inflated or coloured, clearly
   distinct from adjacent cells of lamina . . . .
                            14

8. Cells papillose . . . .
                            Hedwigia [89]
   Cells smooth . . . .
                            9

9. Leaves strikingly silvery white, at least in upper half. Stems very short. Capsule immersed . . . .   *Gigaspermum repens* [59]

    Leaves not usually white. Stems long. Capsules exserted   10

10. Plants with stiff, papery, squarrose leaves strongly toothed in upper half, the teeth usually multicellular . . . .   *Ptychomnion aciculare* [91]

    Leaves soft, not squarrose, entire or denticulate by projecting ends of single cells   11

11. Leaves suddenly narrowed to a long fine point. Stems tomentose . . . .   *Lepyrodon lagurus* [95]

    Leaves tapering gradually. Stems not tomentose, although sometimes bearing rhizoids   12

12. Nerve single, failing in mid-leaf or just below . .   *Brachythecium* [122]
    Nerve absent or short and double   13

13. White rhizoids in leaf axils . . . .   *Sauloma tenella* [110]
    Leaf axils lacking rhizoids . . . .   *Isopterygium* [130]

14. Alar cells (at least the outermost), inflated, balloon-like, usually empty   15
    Alar cells dark or coloured or with granular contents, not usually enlarged, nor empty   17

15. Leaves, at least on main stems, rather abruptly contracted to a very long flexuose hair; leaves never falcate. Stems reddish . . . .   *Wijkia extenuata* [132]

    Leaves gradually tapering to a point, often falcate. Stems usually green   16

16. Leaves straight, tightly imbricated at the shoot apex to form a firm, spear-shaped point . . . . .   *Acrocladium cuspidatum* [117]

    Leaves usually falcate–secund; never tightly imbricated at apex . . . .   *Sematophyllum* [131]

17. Cells clearly finely papillose or ornamented . . .   18
    Cells smooth   19

18. Leaf margin plane or incurved . . . .   *Rhacocarpus* [90]
    Leaf margin narrowly recurved . . . . . .   *Hedwigia integrifolia*
                                      [89]

19. Leaves usually clearly falcate–secund, at least at the
    shoot apex, except in very slender stems. Alar cells
    dark and granular. Very variable in shape and
    robustness . . . .   *Hypnum*
                            cupressiforme [133]

    Leaves never falcate   20

20. Stems with numerous paraphyllia . . . . . .   *Glyphothecium*
                                 sciuroides [92]

    Stems without paraphyllia   21

21. Prostrate stems green. Leaves usually 2 × 1 or broader .   *Camptochaete* [104]
    Prostrate stems brown or reddish. Leaves usually 3 × 1
    or narrower   22

22. Leaf cells thick–walled and somewhat porose through-
    out. Leaves entire . . . .   *Eucamptodon*
                              muelleri [30]

    Leaf cells thin–walled, not porose, except at the very
    base. Leaves denticulate above . . . . . .   *Hampeella pallens*
                                 [93]

Group N
Small mosses with nerves granular or with photosynthetic outgrowths (filaments or lamellae)

1. Upper part of leaf bearing distinct loose rounded
    gemmae . . . .   (mis-keyed *Tortula*
                               papillosa [39])

    Upper part of leaf without loose gemmae . . .   2

2. Adaxial surface of nerve bearing 2–4 irregular longitudinal lamellae, 3–8 cells high   3
   Nerve not lamellate   4

3. Capsule cleistocarpous, sessile, enclosed by inner convolute leaves with or without hair-points . . .   *Acaulon* sp. [33]
   Capsule dehiscent (stegocarpous), clearly exserted.
   Inner leaves concave; lamellae with or without
   protonemal filaments; leaves with or without hair-
   points . . . .   *Pterygoneurum* [38]

4. Nerve widened at apex and papillose there; never
   with long hair-point   5
   Nerve obscured by filaments on adaxial surface in
   upper part of leaf; hair-point present or absent. .   6

5. Apex of leaf usually obtuse. Capsule gymnostomous.
   Leaves short and stumpy, less than 1 mm . . .   *Pottia brevicaulis*
                                                          [32]

   Apex of leaf subacute (nerve usually excurrent in a
   short point). Capsule peristomate. Leaves usually
   more than 1 mm long . . . .   *Desmatodon* [37]

6. Leaf margin recurved; long hair-point usually present *Crossidium geheebii*
                                                          [36]
   Leaf margin widely inflexed   7

7. Upper leaves with long hyaline hair-point . . .   *Aloina sullivaniana*
                                                          [34]
   Upper leaves without hair-point . . . . . .   *Aloina ambigua* [34]

Quick guides to identification
(see pages 10–51 for complete keys)

The inclusion of a genus in any of these categories does not necessarily apply to all its species.

A. Sporophytes and reproduction

1. Calyptra large, enveloping the capsule.
   Blindia, Bruchia, Bryobartramia, Encalypta, Funaria spp., *Goniomitrium* (see also Calymperaceae).

2. Calyptra hairy.
   Daltonia, Dawsonia, Distichophyllum, Eriopus, Macromitrium spp., *Orthotrichum* spp., *Papillaria, Pogonatum, Polytrichum, Rhacopilum, Ulota.*

3. Operculum flat, or nearly so when dry.
   Bartramia, Fabronia, Funaria spp., *Gigaspermum, Goniomitrium, Hedwigia, Meesia.*

4. Spores very large, *c* 100 μm or more.
   Archidium, Eccremidium, Gigaspermum, Goniomitrium.

5. Capsule with enlarged apophysis.
   Bruchia, Funaria spp., *Polytrichum, Tayloria, Trematodon*, some Bryaceae.

6. Capsules immersed.
   See Group J. Also: *Cryphaea, Grimmia apocarpa, Hedwigia, Neckera, Orthotrichum, Ulota.*

7. Cleistocarpous capsules.
   Acaulon, Archidium, Bruchia, Bryobartramia, Ephemerum, Physcomitrella, Pleuridium, Pleurophascum, Pottia drummondii, Tetrapterum.

8. Gymnostomous capsules.
   Amphidium, Andreaea, Bartramidula, Conostomum curvirostre, Eccremidium, Encalypta, Funaria spp., *Gigaspermum, Goniomitrium, Gymnostomum, Hedwigia, Leptostomum, Macromitrium* spp., *Physcomitrium, Pottia* spp., *Pterygoneurum, Rhacocarpus, Sphagnum, Trematodon* spp., *Weissia.*

9. Seta cygneous.
   Blindia magellanica, Brachydontium, Campylopus, Eccremidium spp., *Grimmia pulvinata, G. trichophylla, Pleuridium arnoldii.*

10. Capsules spherical or nearly so, but not sessile.
    Amphidium, Bartramia, Bartramidula, Blindia, Breutelia, Conostomum, Hedwigia, Philonotis, Pleurophascum, Rhacocarpus, Sphagnum.

11. Capsules deeply grooved when ripe.
    Amphidium, Bartramia, Brachydontium, Breutelia, Campylopus, Ceratodon, Conostomum, Dicranella, Funaria spp., *Glyphothecium, Grimmia* spp., *Hampeella, Hypnodendron, Leucobryum, Meesia, Mesochaete, Orthotrichum* spp., *Philonotis, Ptychomnion, Rhacocarpus, Rhacopilum, Tetraphidopsis, Ulota, Zygodon.*

12. Polysetous.
    Atrichum, Campylopus, Dicranoloma, Eriopus, Orthotrichum tasmanicum, Ptychomitrium mittenii.

13. Conspicuous male cups.
    Breutelia, Bryum, Campylopus, Funaria spp., *Philonotis, Polytrichum.*

B. Stems and leaves

14. Stems with paraphyllia.
    Cratoneuropsis, Glyphothecium, Lepyrodon, Thuidium.

15. Densely tomentose stems.
    Amphidium, Anoectangium, Aulacomnium, Bartramia, Breutelia, Calyptopogon, Campylopus, Conostomum, Dicranoloma, Dicranum, Distichium, Gymnostomum, Holomitrium, Hypnodendron esp. *comosum, Hypopterygium, Leptostomum, Leptotheca, Lepyrodon, Macromitrium, Meesia, Philonotis, Rhizogonium* spp., *Sauloma, Tayloria, Zygodon menziesii.*

16. Leaf arrangement rope-like, twisted.
    Barbula torquata, Leptodontium, Macromitrium spp., *Schlotheimia, Tortella calycina, Zygodon.*

17. Leaves almost circular in outline.
    See Group J. Also: *Acrocladium chlamydophyllum, Camptochaete,* Hookeriaceae, *Lembophyllum, Mnium, Pleurophascum, Ptychomnion, Weymouthia.*

18. Very broad nerve.
    Aloina, Andreaea, Campylopus, Desmatodon, Dicranella, Dicranoloma, Ditrichum, Rhizogonium.

19. Plants hoary with hyaline hair-points, when dry.
    Campylopus introflexus, Crossidium geheebii, Grimmia esp. *pulvinata, laevigata, Hedwigia ciliata, Leptostomum, Rhacomitrium* esp. *lanuginosum, Tortula* esp. *muralis.*

20. Leaves (not hair-points) silver or white.
    Bryum argenteum, Gigaspermum repens, Goniomitrium enerve, Leucobryum, Sauloma, Sphagnum.

21. Leaves with ciliate margins.
    Ephemerum cristatum, Fabronia hampeana, Hedwigia ciliata (perichaetial leaves only). (See also *Syrrhopodon*).

22. Leaves with an abrupt hair-point *not* containing the nerve.
    Hedwigia, Hymenodon, Wijkia.

23. Protonema persistent.
    Ephemeropsis, Ephemerum, Mittenia.
24. Miscellaneous striking mosses.
    Andreaea, Buxbaumia, Ephemeropsis, Fissidens, Leucobryum, Pleurophascum, Sphagnum.

C. Ecological groups

1. Aquatic, wholly or partly under water.
   Blindia, Bryum blandum, Cratoneuropsis, Cryphaea tasmanica, Drepanocladus, Eurhynchium, Fissidens fontanus, F. integerrimus, F. rigidulus, F. strictus, Leptodictyum, Philonotis australis, Sphagnum esp. *falcatulum, subsecundum, Tridontium.*
2. Bogs, marshes, etc.
   Acrocladium cuspidatum, Aulacomnium, Brachythecium paradoxum, Breutelia, Bryum laevigatum, B. pseudotriquetrum, Campylium, Campylopus bicolor, C. introflexus, Cratoneuropsis, Dicranum, Ditrichum spp., *Drepanocladus, Leptodictyum, Leucobryum, Meesia, Philonotis, Pleurophascum, Polytrichum commune, Sematophyllum* spp., *Sphagnum.*
3. Wet rocks and soil etc. near streams in forest, but not aquatic.
   Amblystegium, Bartramia, Blindia, Breutelia, Camptochaete, Campylopus, Cratoneuropsis, Cyathophorum, Dicranoloma, Distichophyllum, Echinodium, Eriopus, Eurhynchium spp., *Fissidens* spp., *Gymnostomum, Hypnodendron, Mesochaete, Mnium, Pterygophyllum, Ptychomitrium mittenii, Rhacocarpus, Rhacopilum, Rhizogonium, Rhynchostegium, Thamnobryum, Thuidium laeviusculum.*
4. Epiphytes in wet forest.
   Calyptopogon mnioides, Camptochaete, Cyathophorum, Daltonia, Dicranoloma menziesii, Ephemeropsis, Glyphothecium, Hampeella, Hypnum, Hypopterygium, Lembophyllum, Leptostomum, Leptotheca, Lopidium, Macromitrium spp., *Neckera, Orthotrichum, Papillaria, Ptychomnion, Rhizogonium, Rhynchostegium, Sauloma, Sematophyllum, Tetraphidopsis, Thuidium laeviusculum, Trachyloma, Ulota, Weymouthia, Wijkia, Zygodon.*
5. Epiphytes on tree-ferns.
   Hymenodon, Leptotheca, Leucobryum, Pterygophyllum, Rhizogonium.
6. Ground cover and earth banks in wet forest.
   Acrocladium, Atrichum, Brachythecium, Breutelia, Bryum spp., *Camptochaete, Campylopus clavatus, C. introflexus, Catagonium politum, Cyathophorum, Dawsonia superba, D. polytrichoides, Dicranella dietrichiae, Dicranoloma, Ditrichum difficile, Echinodium, Eurhynchium, Fissidens* spp., *Hypnodendron, Hypnum, Hypopterygium, Lembophyllum, Leucobryum, Mesochaete, Mittenia, Philonotis, Pogonatum subulatum, Pohlia* spp., *Polytrichadelphus, Pterygophyllum, Ptychomnion, Rhacopilum, Rhizogonium, Tayloria, Thuidium, Trematodon, Weissia.*
7. Fine twigs and lianes in wet forest.
   Calyptopogon, Daltonia, Glyphothecium, Hypnum, Orthotrichum, Papillaria, Sematophyllum, Tetraphidopsis, Ulota, Weymouthia.

8. Dry mallee soils.
   Groups J and N. Also: *Barbula* spp., *Bryum, Fissidens vittatus, Tortella, Tortula* spp., *Triquetrella*.
9. Exposed siliceous rock.
   Andreaea (mountains), *Brachythecium paradoxum, Dicnemoloma, Grimmia, Hypnum, Ptychomitrium australe, Sematophyllum homomallum*.
10. Dry calcareous rock, mortar, etc.
    Bryum argenteum, Grimmia, Macromitrium, Pseudoleskea, Rhacopilum, Tortula esp. *muralis*.
11. Cities.
    Barbula, Bryum argenteum, B. dichotomum, Ceratodon purpureus, Tortula muralis.
12. Sand dunes.
    Barbula australasiae, B. pseudopilifera, B. torquata, Bryum bicolor, B. billardieri, B. capillare, B. microerythrocarpum, Hypnum, Rhacopilum, Thuidium furfurosum, Tortula princeps, Triquetrella, Weissia.
13. Dry sclerophyll forest and roadside banks.
    Brachymenium, Breutelia affinis, Bryoerythrophyllum, Bryum billardieri, B. dichotomum, Campylopus clavatus, C. introflexus, Dawsonia longiseta, Desmatodon, Dicranella dietrichiae, Ditrichum difficile, Fissidens esp. *F. humilis, F. pallidus, F. taylorii, F. vittatus, Funaria glabra, F. gracilis, Hypnum, Mielichhoferia, Pogonatum subulatum, Pohlia nutans, Polytrichum juniperinum, Rhacopilum, Tortella, Weissia,Wijkia*.

Descriptions

SPHAGNACEAE
Sphagnum L.

There is no genus of mosses more immediately and unequivocally recognizable, and few whose species are harder to delimit. This distinctiveness extends to all phases of the life history. The protonema is thallus-like, rather similar to the prothallus of a fern, instead of filamentous; rhizoids are absent from the mature plant, and are present only on protonema and juvenile shoots. The erect stems have a hard central cylinder surrounded by one or more layers of transparent empty cells and bear branches in bunches, some thicker (divergent) branches spreading widely from the stem and other (pendent) branches slimmer, inconspicuous and running down parallel to the stem. At the stem apex the branches in almost all cases are tightly packed together to form a mop-like head or *coma*. Most characteristic of all are the leaves which, although only one cell thick, consist of a regular array of cells of two quite different sorts, narrow green *chlorophyllose cells,* alternating with much wider, empty *hyaline cells* which are held rigid by internal helical thickenings (*fibrils*) round the wall inside. The hyaline cells of the outer stem tissues may also have fibrils and all hyaline cells may have, and usually do have, openings to the outside called *pores*. These are not to be confused with the pores of other moss leaves where the term "porose" indicates thin spots in otherwise thick walls, appearing to form pit-connections between adjacent cells but having no connection to the outside. The hyaline stem cells of divergent branches may also have the pore raised on the point of a mamilla projecting from the stem at the upper end of the cell, forming a *retort cell*, so called from a fancied resemblance to a chemical retort.

The pattern of chlorophyllose and hyaline cells is laid down very early in leaf development by unequal cell divisions in a diagonal criss-cross grid from which the mature structure develops by differential growth. Even a tiny fragment of a single leaf is sufficient to betray the genus. The leaves of stems are usually flat and rather different both in shape and structure from the branch leaves. A border of narrow, thick-walled hyaline cells may be present on either.

The antheridia are spherical in the axils of short fat branches in the coma, usually recognizable by a distinctive colouration, commonly orange or reddish. The capsules, which are very rare in Australia (see Willis 1952, 1955b) are also unique. Glossy, dark brown and spherical on a colourless *pseudopodium* of stem tissue instead of a seta (cf. *Andreaea*, 2) they discharge by the famous "air-gun mechanism" by compression of an air cavity under the dome-shaped spore sac until pressure blows the lid off, and the spores with it.

A solution of Toluidine Blue will be found valuable, and indeed almost essential, in revealing the presence of pores in hyaline cells. While transverse sections of the main stem are useful in identification, sections through leaves, much used to separate some northern hemisphere species, have not yet been found necessary with most of ours.

The difficulties of this large genus are as much nomenclatural as taxonomic and attributable mainly to the over-enthusiastic erection of new species by Müller and still more by Warnstorf who seemed to have allowed almost no room for variability within his concept of a species. Willis (1952, 1953d, 1955b) has thoroughly revised the Victorian species and we have adopted his treatment with gratitude. Sainsbury aptly remarks (Handbook, p. 15), "For myself I have never been able to . . . appreciate the propriety of adopting for this genus criteria for the creation of species which would not be acceptable in practically any other genus of mosses".

In the southern hemisphere *Sphagnum* is much less abundant than in the northern, and was much less widely sampled in Warnstorf's time, but the taxonomic confusion has been formidable for all that and no doubt there are still mistakes in reconciling our species with those elsewhere. Broadly speaking there are three common species in southern Australia—*cristatum*, *falcatulum* and *subsecundum* (of which the first is by far the commonest)—and at least two others which are much rarer. They are predominantly upland plants, only really common in mountain bogs (e.g. in snow-gum country) but nowhere covering the acres of ground which they can do in the northern hemisphere. The distributions which we record for Australian states are only tentative.

Key to species

1. Branch leaves ovate, cucullate (or apparently so)
    at the apex which is often jagged abaxially with
    cells exposed by the leaf's curvature. Stem and
    branch hyaline cells in 3 layers or more, some-
    times with helical fibrils   2
    Branch leaves more narrowly lanceolate, chan-
    nelled or tubular at the apex but not cucullate;
    smooth abaxially. Stem and branch hyaline
    cells in 1 or 2 layers, without fibrils   3

2. Hyaline cells of stem and branch with pores and
    helical fibrils .   .   .   .   *cristatum*
    Hyaline cells sometimes porose but without fibrils   *australe*

3. Branch leaves with numerous small pores along
    each wall of the hyaline cells. Stem hyaline cells
    predominantly in 1 layer .   .   .   .   .   .   *subsecundum*
    Branch leaves almost without pores. Stem hyaline
    cells in predominantly 2 layers .   .   .   .   .   *falcatulum*

S. cristatum Hampe

These are usually big, dull-coloured plants, coarsely leaved and with shoots
which are *c* 2 cm across in the terminal heads but more attenuate below.
The hyaline cells are 3-deep round the stem and have abundant helical
fibrils and conspicuous pores (2 to many per cell). The flat, broad stem leaves
are fibrillose throughout but with rather few large pores almost confined
to the upper ⅔ of the leaf. The branch leaves are wide and concave with
rounded cucullate (boat-shaped) apices where the cells project on the
abaxial surface to give a rough profile; hence, presumably, the specific
epithet *cristatum*. The leaves of divergent branches tend to curve backwards
and give a rough texture to the branch. The stem leaves are unbordered.
The branch leaves are fibrillose and porose throughout, averaging up to
about 6 pores per cell in mid-leaf and have the chlorophyllose cells usually
exposed on the adaxial surface.

DISTRIBUTION: TAS, ?SA, VIC, NSW, ACT, QLD; also in New Zealand.

ILLUSTRATIONS: Troughton and Sampson (1973), Plates 37, 38; Allison and
Child (1971), Plate 3.

   This is a big coarse species, recognizable in the field by the size and the blunt cucullate leaves. *S. subsecundum* occasionally matches it for size but the leaves there are slim and pointed, giving a much finer texture to the whole plant. The species most likely to be confused with it is *australe* (q.v.).

Willis (1953d) gives a long list of synonyms which he has established for this species and it is probable that the records of *palustre L.* and *magellanicum* Brid. in the Index may also be the same thing, for it is notoriously difficult to equate northern and southern hemisphere species, especially when working only with herbarium specimens.

S. australe Mitt.

This is very similar to *cristatum* in size and general appearance but much rarer, although probably overlooked. According to the Handbook the branch leaves are not cucullate but have the margins inrolled at the apex to give a similar effect; the points so produced diverge to give a jagged outline to the branch. This feature, and the sometimes rough abaxial surface of the leaf apex, it shares to some extent with *cristatum*. The principal distinguishing feature is in the hyaline stem cells which have no, or almost no, internal fibrils. There are, however, pores despite the Handbook's claim to the contrary. The stem leaves are narrowly bordered. The branch leaves have the margins slightly toothed near the apex, rather numerous small pores, up to about 10 per cell.

DISTRIBUTION: TAS, VIC, NSW; also in S. America, S. Africa, and New Zealand.

ILLUSTRATIONS: ?

What little Australian material (NSW) we have seen attributable to *S. beccarii* Hampe. (Willis, 1952) differs from *cristatum* in virtually lacking fibrils in the stem cortex, and in having only one pore per cell on the stem.

DISTRIBUTION: VIC, NSW; also (*sensu lato*) India and S.E. Asia, Africa, America.

ILLUSTRATIONS: Gangulee (1969), Fig. 3.

S. subbicolor Hampe (as used of Australian plants) is a synonym of *cristatum* according to Willis.

S. subsecundum Nees†

These may be small plants, 0·5–1·0 cm in diameter below and not much bigger at the coma, but there are also forms which are nearly as big as *cristatum* although with much finer and more feathery branches. Commonly the dry plant has a dull leathery brown colouration which is rather distinctive. The hyaline outer layer of the stem contains no helical fibrils and is only one cell thick, with trivial exceptions, and sharply distinct from the inner cells of the stem. The stem leaves are fibrillose and porose practically *throughout* and clearly bordered. On slender branches the leaves are rather flat except at the apex where the margins are inrolled to make a rather fine channelled or conical point. On bigger branches the leaves are quite concave. The margins are slightly undulate, especially above, bordered with hyaline cells, and entire except at the very apex where there are 5–6 teeth. The cells are porose throughout, characteristically with two rows of numerous small pores, one down each side of the cell (*c* 12–16/cell) particularly on the abaxial surface near the apex.

DISTRIBUTION: TAS, WA, SA, VIC, NSW, ACT; widely distributed in the northern hemisphere and also in S. America and New Zealand.

ILLUSTRATIONS: Gangulee (1969), Fig. 18.

This is a fine, feathery plant more like *falcatulum* than *cristatum* in general appearance, but is distinguished by the single hyaline layer in the stem, numerous small pores in the branch leaves, and separable from *falcatulum* by the entire leaf margins and by the stem leaves being fibrillose throughout. It occurs down to sea level in swampy country.

S. falcatulum Besch.

This is the smallest and least coarse of all the Australian species and is particularly common floating in ponds when it can be extremely delicate and feathery. It appears to be always green without brown pigmentation. The stem has 1–2 imperfect layers of hyaline cells, of which the outer has larger cells, and variably bordered leaves which are fibrillose only in the upper half. The branch leaves are very long, bordered with narrow cells, not at all cucullate but channelled and finely tapering at the tip. The margins are often undulate and are clearly *toothed* with tiny sharp teeth in the upper half and *c* 3–5-toothed at the apex; the cells are fibrillose but with rather few minute pores.

DISTRIBUTION: TAS, VIC, NSW, QLD; also in S. America and New Zealand.

ILLUSTRATIONS: ?

This is the most typical *Sphagnum* under water. Its finely tapering leaves give it a very feathery appearance matched only by *subsecundum* (q.v.). It is quite a common species.

The other species recorded by the Index are:

S. compactum Lam. et Cand.

S. cuspidatum Ehrh. ex Hoffm. (TAS, NSW, QLD).

S. dominii Kavina (QLD).

ANDREAEACEAE
Andreaea Hedw.

This very clearly demarcated genus of mosses is characteristic of siliceous rocks in montane to alpine regions, down to 2000 ft (Clifford, 1952) but usually at 4000–5000 ft.

The plants fruit quite commonly and are then unmistakable; the capsule, opening by 4 valves, is unique. There is no seta, its function being taken over by a leafless prolongation of the gametophyte, the *pseudopodium*, from which the capsule falls away when old, leaving only a flattened attachment disc behind. This pseudopodium resembles a seta in appearance as well as in function but continues directly from the vegetative stem without a join and occasionally carries unfertilized archegonia. The capsule is further remarkable in its internal structure, the spore sac being dome-shaped and arching over the columella, as in *Sphagnum*. The spores germinate to form either a thalloid or a branched ribbon-like protonema which can sometimes be found from spores germinating actually inside the capsule itself.

The leaf may or may not have a single nerve of elongated cells; the cells are thick-walled, sometimes very thick and sometimes papillose. When not fruiting, *Andreaea* would be most likely to be mistaken for a small blackish plant of *Rhacomitrium* [12] or *Grimmia* [11], but the intensely corrugated walls of *Rhacomitrium* would at once distinguish it. *Grimmia* too has distinctly sinuose walls, and the basic colour is greenish, becoming black, whereas in *Andreaea* the colour is orange-brown, becoming purplish black.

The stems are brittle and fragile when dry and the lower parts of plants are often much worn; no doubt the fragments broken off can act as propagules. Physiologically, the genus is of unusual interest; *A. rupestris*, at least, occupies habitats typical of rock-dwelling lichens and probably behaves much like one. Both the whole-plant physiology and the fine structure of cell contents would be subjects well worth investigation.

The two common species are *rupestris* and *subulata*.

Key to species

1. Leaves with a distinct nerve of elongated cells   .   2
   Leaves nerveless . . . .   *rupestris*

2. Leaves tapering to a narrow, falcate-secund subula   *subulata*
   Leaves not falcate   3

3. Leaves almost triangular above, more than twice
   as long as wide. Nerve strong, reaching near the
   leaf apex, projecting as a rib on the abaxial surface of the leaf. . . . .   *australis*
   Leaves rounded-ovate, not much longer than
   broad. Nerve rather faint, short and broad . .   *nitida*

The Victorian species are described by Clifford (1952); the first part of an excellent world monograph by Schultze-Motel (1970a) deals with the nerved species.

A. rupestris Hedw. (=*A. petrophila* (Ehrh.) Fuern.)
Most commonly this species is found on siliceous rocks, drying out completely at times during summer but protected from over-heating by occurring either in full sun only at high altitudes or sheltered by dry sclerophyll woodland at lower altitudes (Clifford, 1952).

Commonly the shoots are about 0·5–1·0 cm long, forming rather loose and open cushions, but much longer and shorter plants can be found in extreme habitats. The leaves, *c* 0·5–1·0 mm long, are concave and rather sheathing at the base and often above it; very variable in shape, commonly tapering from about mid-leaf into a rather broad point, but sometimes tapering in the upper two-thirds to a rather fine channelled point, approaching falcate and secund, but this feature may be either very pronounced or quite absent. Leaves are nerveless, appressed to the stem when dry, or with only the points spreading; entire or crenulate-denticulate below and with very thick-walled, porose cells. The cells below are long and narrow, those

PLATE I. *Andreaea rupestris* (near *A. acuminata*) TAS—Fruiting shoot × 30 with dissected leaves × 50, showing range of shape. Cells above mid-leaf (a), below mid-leaf (b), from mid leaf-base (c), all × 1000

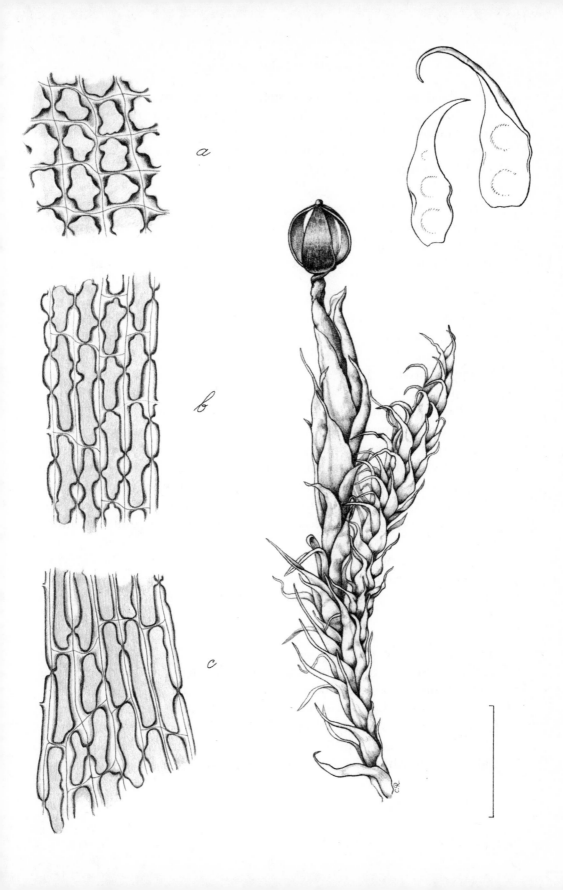

a

b

c

above quadrate with thickenings especially prominent in the angles, forming a network of rhomboidal thickenings (Plate 1a); marginal cells sometimes shorter, especially at the base, thus giving the leaf the impression of having a very wide, ill-defined nerve. Cells may be papillose (1 per cell) or smooth, but usually at least have low papillae on the abaxial surface above mid-leaf.

The perichaetial leaves have greatly enlarged sheathing bases and are either completely obtuse or have short, abrupt points. Spores measure *c* 25 μm and are smooth.

DISTRIBUTION: TAS, VIC, ACT, NSW; world-wide except in tropical Asia and Africa.

ILLUSTRATIONS: Plate 1; Grout (1928–40), I, Plate 1.

This is a very variable species containing many forms which have been described as separate species; or more probably there is a confusion between several species. "The species vary considerably and it is simply a matter of opinion into how many species the forms should be grouped" (Rodway, 1914, p. 149).

We have provisionally included in this species *A. acuminata* Mitt. in Hook.f. & Wils., from Tasmania, which has the leaves contracted slightly above the base to form a waist, and is crenulate–denticulate there. Both these features are very variable, even on the same stem, and unless more convincing features are discovered to characterize this species it is not worth retaining. The plant illustrated in Plate 1 corresponds to *A. acuminata*, and shows the two extreme types of leaf, falcate and straight, which are quite commonly found even on the same stem.

Also in this group are *A. erubescens* C.Muell., *A. eximia* C.Muell., (TAS), *A. julicaulis* C.Muell., *A. montana* Mitt. (TAS), *A. tenera* C.Muell. (TAS, NSW) which we have not seen and which may just be forms of *A. rupestris*.

A. subulata Harv. ex Hook.

This is common on rocks in wetter alpine habitats than *A. rupestris*; e.g. intermittently submerged rocks at the edge of mountain tarns, etc.

It differs from *A. rupestris* in the more slender and graceful leaves, more finely subulate, constantly falcate-secund, and not concave below; also in the nerve of long cells which is quite evident, extending well into the subula, and in the perichaetial leaves which have long, fine, usually coiled tips.

Perhaps the orange colouration shows through the black more conspicuously in this species and the stems tend to be less worn, presumably because of protection by its semi-aquatic habitat from the sand-blasting of winds at high altitudes.

DISTRIBUTION: TAS, VIC, NSW; also in New Zealand, Borneo, Africa, S. America.

ILLUSTRATIONS: Plate 2; Schultze-Motel (1970a), p. 78 (list).

A. australis F. von Muell. ex Mitt.
Like a very robust form of *A. rupestris* but with a broad rather than subulate point to the leaf, and a strong nerve which is prominent abaxially, this species has cells which are smooth, short and quadrate except in the leaf base where they reach 2 × 1, but are not usually heavily thickened. The plant is very like a *Grimmia* in all except fruit and colouration; it is rare except in parts of the Australian Alps.

DISTRIBUTION: TAS, VIC, ACT, NSW; also in New Zealand.

ILLUSTRATIONS: Schultze-Motel (1970a), Figs 3–5, p. 43 (list).

A. nitida Hook.f. & Wils.
The wide oval leaves, faintly nerved with a short broad nerve, distinguish this rare species of wet alpine rocks. Sometimes aquatic, it has been recorded up to 10 cm long when floating (Schultze-Motel, 1970a). It is even more robust than *australis* and has comose vegetative stems.

DISTRIBUTION: TAS, VIC, ACT; also in New Zealand, New Guinea, Tristan da Cunha, South America.

ILLUSTRATIONS: Schultze-Motel (1970a), Figs 10–11, p. 89–90 (list).

BUXBAUMIACEAE
Buxbaumia Hedw.

B. tasmanica Mitt. TAS and *B. colyerae* Burges NSW are both exceedingly rare endemics known only from the original collections. They would be easily recognized by the large, *Dawsonia*-like capsules produced by almost microscopic gametophytes, but are perhaps the hardest of all mosses to find. Both are forest species and no doubt occur elsewhere in Australia.

ILLUSTRATIONS: Burges (1932), Fig. 1. See also Handbook, Plate 1.

What appears to be good *B. aphylla* Hedw. has recently been found in Tasmania by Dr D. Norris. Whether it is conspecific with *tasmanica* we do not yet know.

POLYTRICHACEAE

Polytrichum Hedw.

The leaf structure is the most obviously distinctive feature of this genus and its near relatives. In all cases the leaf has a sheathing base and a narrow, lanceolate limb. Above the sheath the nerve is greatly broadened to occupy most of the width of the limb, except for a very narrow margin of lamina; on the adaxial surface this nerve is covered by densely packed green plates or lamellae running side by side down the length of the leaf like green walls a few cells high. The shapes and arrangements of these cells, especially the cap cells forming the crests of the walls, are useful guides to identification. The peristome is equally distinctive, with 32–64 short, thick teeth joining at their tips to a flat membrane or *epiphragm*, across the mouth of the capsule, the spores sifting out through the gaps between the teeth which are capable of slight hygroscopic movements. From the related genera *Psilopilum, Pogonatum* and *Polytrichadelphus* this genus is distinguished by the combination of thickly hairy calyptra, short-beaked operculum, capsule square in transverse section (except *P. alpinum*) and usually quite pronounced *apophysis* with stomata. As in *Polytrichadelphus* (q.v.) there are two kinds of paraphyses in the male cups. Because of the strong vegetative resemblance between these genera more information than usual is required to key out the species. Group A is therefore more elaborate than the rest of the main key and the accounts of species in the text correspondingly reduced.

Smith (1971), in reviewing the genera of Polytrichaceae, has created a

PLATE 2. *Andreaea subulata* TAS—Fruiting and vegetative shoots × 30, leaf × 50, cell detail from mid leaf-base showing nerve cells on the right and lamina cells on the left × 1000

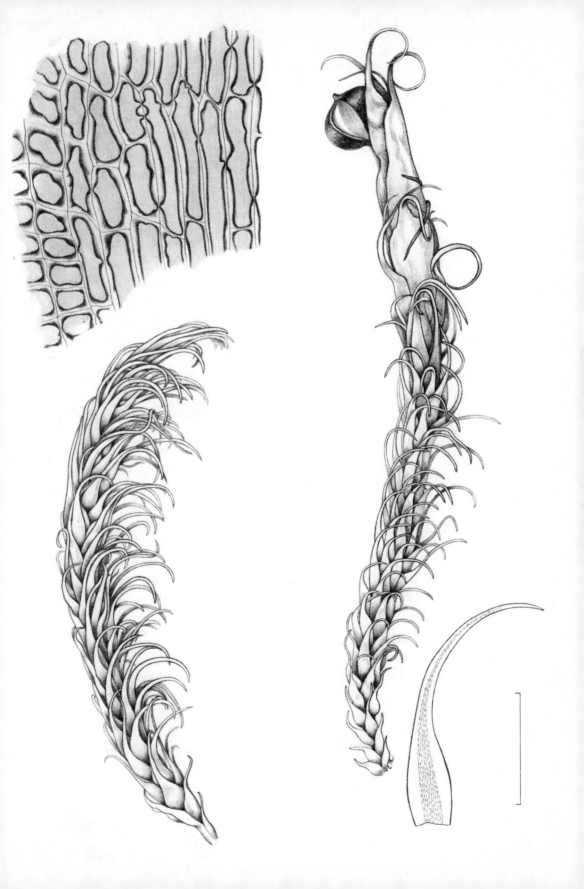

number of new genera based on characters which seem to us more appropriate to sub-genera.

As far as we have been able to ascertain no Polytrichales have ever been found in Western Australia.

Key to species (See main key, Group A)

P. alpinum Hedw.

This small to medium-sized species, commonly 1·5–6·0 cm tall in Victoria (the Handbook gives it to 15 cm in New Zealand) is almost exclusively alpine in habitat and is sometimes distinguishable in the field by having the stems 2- to 4-branched at the base. There is then a close resemblance to *Polytrichadelphus magellanicus* where, however, the branching is associated with the characteristically persistent sporophytes. When unbranched and not fruiting it is safest to rely on microscopic characters: the terminal cells of the leaf lamellae are pale (almost hyaline) and clearly papillose, unlike all other species except *Psilopilum* spp. which are also alpine mosses with somewhat papillose terminal cells but are much smaller plants with more numerous lamellae and leaves which either are almost entire or are crisped when dry, according to species. The leaves are strongly toothed throughout.

The capsule is *cylindrical*, but with distinct stomata in the apophysis region, a densely hairy calyptra, and peristome teeth irregular in both shape and number. Because of the round capsule and lack of a clearly distinct apophysis the species has often been classed as a *Pogonatum* but that genus— not one of the strongest—lacks stomata in the capsule. Smith (1971) transfers it to his new genus *Polytrichastrum*.

It is a rare plant, growing on bare soil and in wet crevices mostly above about 5000 ft.

DISTRIBUTION: TAS, VIC, NSW; also in New Zealand and mountainous parts of both hemispheres.

ILLUSTRATIONS: Plate 3; Dixon (1924a), Plate 5.

PLATE 3. *Polytrichum commune* VIC—Fruiting and vegetative shoots × 1·5 with cross section of lamellae (top right) × 500, surface view of lamella (mid left) × 500; *P. alpinum* VIC—Lamella (top left) × 500; *P. juniperinum* VIC—Lamella (bottom left) × 500

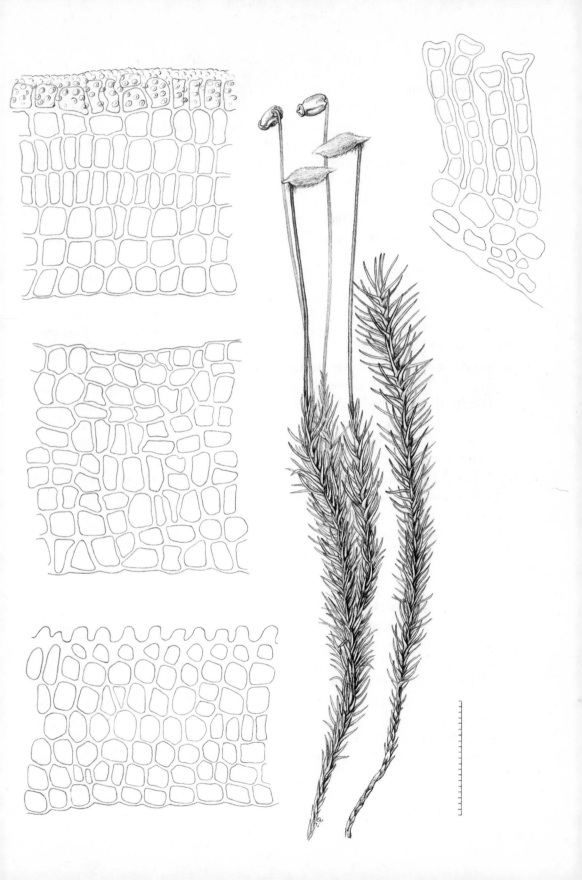

P. commune Hedw.

Commonly thought of as a robust tussock-forming species of boggy areas, this species actually occurs also quite commonly, and even fruits, as a very dwarf plant, sometimes little more than 1–3 cm high, on skeletal peat and gravelly soils in montane areas. There are all intermediates between this extreme and the tall (15–20 cm or more) hummock-forming bog-moss although large hummocks, such as occur in northern Europe, do not seem to be common in Australia. This species has some claim to being the world's largest moss—Martin (1951) records stems 6 ft long from a pool in New Zealand.

The densely tufted large form of the plant can be confused with nothing else. It has neither the glaucous colour, the very wide shoots, nor the sparse open colonies of *Dawsonia superba*, the only other species likely to match it in height. In the smaller growth forms it is most liable to confusion with *Polytrichadelphus magellanicus* (q.v.) from which it is most easily separated, when not in fruit, by the perfectly radially symmetrical shoot apices (except when dry); or with *P. alpinum* in alpine areas, which is sometimes distinguishable by the branched stems. In cases of doubt, the grooved terminal cells of the leaf lamellae in transverse section are completely diagnostic, although occasionally the grooving may be rather slight, or the terminal cells merely flattened on top.

This is a common moss especially in upland boggy areas, tolerating drier soils at higher altitudes and restricted to boggy ground at lower levels and in drier climates. The ecology is discussed by Sarafis (1971).

DISTRIBUTION: TAS, VIC, NSW, ACT, QLD; cosmopolitan.

ILLUSTRATIONS: Plate 3; Dixon (1924a), Plate 7; Watson (1955), Fig. 12, Plate 2.

P. juniperinum Hedw.

Even with a hand lens this cannot be mistaken for any other moss because of the entire membranous margins of the leaves, inflexed to cover most of the lamellae (which are also distinct in the high, domed terminal papillae). This inrolling gives the leaves a fine-pointed, sleek appearance which can easily be recognized with practice in the field. The leaf tip is prolonged into a toothed, often reddish, spine. Usually a small plant, 1–3 cm high, it may occasionally be much taller (up to 20 cm according to the Handbook). The young calyptra is a brilliant and beautiful red-brown in spring.

Cosmopolitan in distribution and ecologically widely tolerant, it is found in both shaded and exposed sites, alpine and lowland, ranging from dry to wet climates and dry to wet, but not usually boggy soils. It is very common in all kinds of forest and heathland 2–4 years after fire, following *Funaria* and *Ceratodon*. Perhaps it is at its commonest on rather sandy or gravelly peaty soils in upland areas.

DISTRIBUTION: TAS, SA, VIC, NSW, ACT, QLD, Lord Howe; cosmopolitan.
ILLUSTRATIONS: Plate 3; Handbook, Plate 2; Dixon (1924a), Plate 6; Watson [1955] Fig. 9, Plate 1.

The remaining species recorded from Australia are:
P. brachypelma C.Muell. (NSW)
P. cataractarum C.Muell. (NSW)
P. croceum Hampe (TAS)
P. lycopodioides C.Muell. (TAS)
P. novae-hollandiae Jaeg. (VIC)
P. obliquirostre C.Muell. (VIC)
P. piliferum Hedw. (Doubtful, according to Watts and Whitelegge, 1902)
P. recurvipilum C.Muell. (NSW)

Atrichum P.Beauv.
(formerly **Catharinea**)

A. androgynum (C.Muell.) Jaeg. (*A. ligulatum* of the Handbook)
The very tall (4–8 mm) narrow, slightly curved capsules with long membranous calyptras are unmistakable; the setas are *c* 4 cm long and the plants frequently polysetous. In the absence of capsules the translucent, rather drab-coloured leaves quite without sheathing bases might perhaps be mistaken for a species of *Mnium* [73] or *Rhizogonium* [77] if the few low lamellae on the upper surface of the nerve were missed. The stems are quite tall, commonly 2–5 cm, with long strap-shaped leaves which are distinctly undulate, clearly bordered and carry teeth, often in pairs, on both the margins and the leaf surface on the crests of the undulations. We have seen one colony of plants lacking teeth and lamellae on almost all leaves, but this is very exceptional. There are 32 peristome teeth, engaging on an epiphragm, as in *Polytrichum*.

CHROMOSOME NUMBER: n = 14 (VIC).

DISTRIBUTION: TAS, VIC, NSW, ACT, QLD; also in New Zealand, S. America, S. Africa.

ILLUSTRATIONS: Plate 4; Handbook, Plate 3; Allison and Child (1971), Plate 6.

This is a plant of moist banks and forest floors in wet sclerophyll forest, probably intolerant of drying out since it turns yellowish and brittle on exposure to sun; quite common.

The Australian and New Zealand *A. ligulatum* has been reduced by Nyholm (1971) to synonymy with the widely distributed southern hemisphere *androgynum*.

A. angustatum (Brid.) B.S.G. has been recorded from Australia presumably on the basis of var. *polysetum* Watts & Whitel., which is likewise reduced to *androgynum* by Nyholm.

A. pusillum (C.Muell.) Par. from TAS we do not know.

Pogonatum P.Beauv.

P. subulatum (Brid.) Brid.

From all except *Dawsonia longiseta* this is usually distinguished by the small size and the colour—a rather distinctive combination of yellowish or glaucous green young growth, bright deep green mature leaves, and dull brown old leaves. A disproportionately long seta (3–4 cm), densely hairy calyptra and narrow erect capsule, smoothly cylindrical except for inconspicuous ridges like seams down its length, together make fruiting plants immediately recognizable. Because of the erect capsules the white epiphragms are very conspicuous, especially when old. There are approximately 32 peristome teeth.

The archegonia have strikingly curled long necks when mature, emerging fully from the perichaetial leaves. There are no stomata in the capsule wall.

PLATE 4. *Atrichum androgynum* VIC—Fruiting shoot × 4, leaf cross section × 500, cells of lamina × 1000

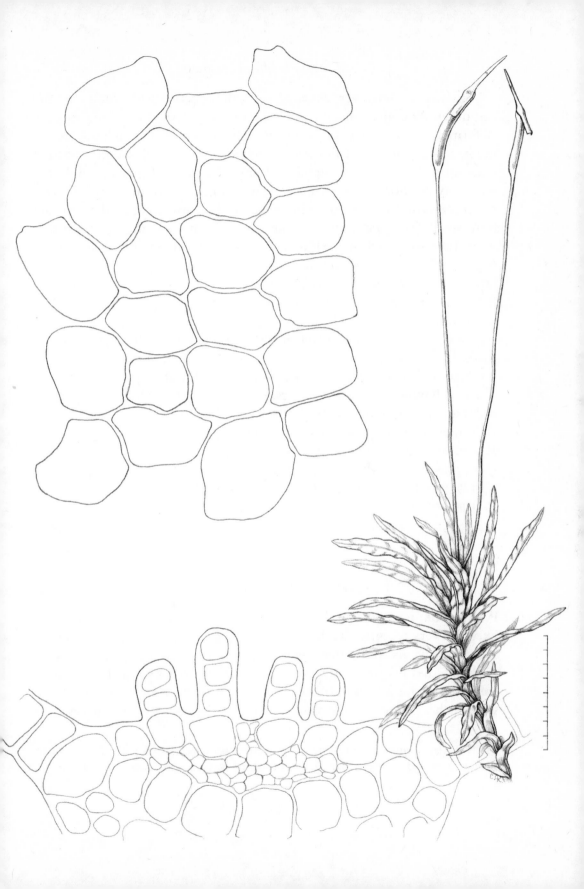

Vegetatively, however, *P. subulatum* can sometimes be confusing; the half dozen Australian endemic species admitted by the Index are probably all forms of the very variable *subulatum*. The commonest form has a narrow shoot, 1·5–2·5 cm tall, with leaves which are greatly crisped and contorted when dry, rather like the alpine *Psilopilum crispulum*. Less commonly the stem may be shorter, less than 1 cm, with thicker and much stiffer leaves which, at most, have the points incurved. Such forms can be hard to separate from small *Dawsonia longiseta* if there are no capsules; the most constant distinctions seem to lie in the leafiness and in the toothing of the leaf margins (see key, Group A, couplet 13).

CHROMOSOME NUMBER: n = 7 (NSW).

DISTRIBUTION: TAS, VIC, NSW, QLD; also in New Zealand.

ILLUSTRATIONS: Plate 5; Handbook, Plate 3.

This is a very common plant in lowland to montane forest regions, on roadside cuttings, usually coming in at a more advanced stage of colonization than *Polytrichadelphus* and also where the substrate is particularly hard.

The following Australian species, accepted by the Index, are reputedly endemic. Most, if not all, are probably *subulatum* in one of its guises:

P. brachypodium (C.Muell.) Watts & Whitel. (NSW)

P. camarae (C.Muell.) Par. (NSW, QLD)

P. gippslandiae (C.Muell.) Par. (VIC)

P. gulliveri (Hampe) Jaeg. (TAS, VIC)

P. nanocarpum (C.Muell.) Par. (VIC)

Polytrichadelphus (C.Muell.) Mitt.

P. magellanicus (Hedw.) Mitt.

Except in very recently established colonies sporophytes are usually plentiful and immediately distinguish female plants of this species from other medium-sized relatives. The calyptra, glabrous except for a few terminal erect bristles, long-beaked operculum, asymmetrical concave–convex capsule which is almost crescent-shaped in transverse section, and the persistence

PLATE 5. *Pogonatum subulatum* VIC—Small form with young sporophyte; tall form male, wet and dry × 11, lamella and toothed leaf margin × 500

74

of old sporophytes are all reliable diagnostic features. Commonly 3 or sometimes 4 generations of sporophytes are present at once on the same plant: old setas from which the capsules have dropped off, mature capsules, full-sized but still closed capsules, and full length setas (3–5 cm tall) with no signs of expansion in the capsule region. The peristome is similar to that of *Polytrichum*.

Even vegetatively it is usually distinct in having many stems (commonly 2–6 cm tall), branched 2–3 times, unlike most other related species except sometimes *Polytrichum alpinum*; female plants, when moist, can almost always be distinguished by the slightly secund toothed leaves which give the stem apex a drooping, one-sided appearance from above. Male plants, often rather narrower than the female, are much more difficult to identify with certainty in the field. They resemble *Polytrichum* in having two kinds of paraphyses in the antheridial cups—uniseriate filaments and slightly paddle-shaped filaments, expanding to 4 or more cells broad in the upper one-third. The leaves, usually somewhat glaucous, have narrow toothed margins only 1–2 cells wide, outside the lamellae, which are usually 6 cells high, with the top cells domed to give them a strongly crenulate appearance in side view.

CHROMOSOME NUMBER: n = 7 (TAS).

DISTRIBUTION: TAS, VIC, NSW, QLD; also in New Zealand and S. America.

ILLUSTRATIONS: Plate 6; Handbook, Plate 2.

This is a very common species on roadside cuttings in heavy soils where it is one of the earliest colonisers and is often present in large carpets, either in sun or shade, except where the banks dry out completely in summer. It is not confined to forest regions and occurs in lowland and montane to alpine habitats.

P. arnoldii (Hampe) Jaeg. (VIC) we know nothing of.

PLATE 6. *Dawsonia superba* VIC—Fruiting shoot (centre) × 1·5, lamella (bottom right) × 500; *D. longiseta* VIC—Fruiting shoot (right) × 1·5, lamella (top right) × 500; *D. polytrichoides* NSW—Lamella (mid right) × 500; *Polytrichadelphus magellanicus* VIC—Fruiting shoot × 1·5, section through capsule × 10, paraphysis × 50 (bottom left), lamella (top left) × 500

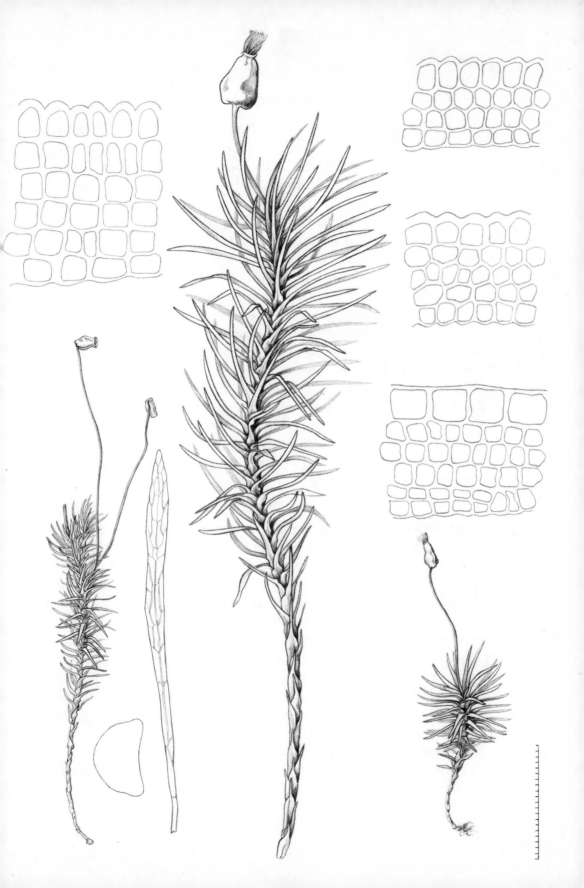

Psilopilum Brid.

Plants of this genus are distinct, when in fruit, in the glabrous calyptra, rough at the apex, in the biconvex, narrow-mouthed capsule flattened from the sides (not unlike *Dawsonia*) and in the down-curved, horn-shaped beak on the operculum. They are distinguished from all other Australian Poly-trichales by having a portion of the lamina bistratose but still bearing lamellae, and a stem with a central strand composed of only one type of cell. Both species are exclusively high-country plants with short stems and crisped leaves, unmistakable in fruit but liable to confusion vegetatively with *Pogonatum subulatum,* from which they are separable by having clearly papillose cells along the crests of the lamellae, as well as by the leaf and stem structure and the habitat. In the papillose terminal cells, these species re-semble *Polytrichum alpinum* but the capsule, calyptra and stem and leaf structure are otherwise quite different; moreover the leaves are either entire or strongly crisped in *Psilopilum*, and the leaf base is proportionately much longer, sometimes exceeding the tapering, lamellate limb.

The two species are most easily separated by the height of the lamellae, 2–4 cells in *crispulum* and 5–10 in *australe*. Both are transferred by Smith (1971) to his new genus *Notoligotrichum* based chiefly on the shape and structure of the peristome teeth. The argument is plausible but the evidence not yet sufficiently compelling.

P. crispulum (Hook.f. & Wils.) Mitt.

Of the species in Australia this appears to be the commonest. The leaves are normally strongly crisped when dry, the lower leaves spreading widely; sometimes the apical leaves are not crisped but rigid and with incurved apices, deceptively like those of *australe*. The sheathing base varies from narrower than the limb to slightly wider than it. In our material there are usually 40–50 lamellae, as in *australe*, rarely as many as the 60–70 claimed in the Handbook, so that this is not a useful distinguishing character. But the lamellae are somewhat sinuous, rather loose and far apart, and only 1–4 cells high. The nerve is only 70–250 μm wide, with an adjacent bistratose lamina 14–20 cells wide on either side, bearing lamellae above. The unistratose margin is 4–10 cells wide, usually clearly denticulate. The marginal cells of the sheath are shortly rectangular and scarcely different from the inner cells; those of the limb are squarish or slightly elongated transversely. The leaves and stems are commonly, but not always, finely papillose all over. There are 32 peristome teeth (alternate ones often reduced), with an epiphragm as in *Polytrichum*.

DISTRIBUTION: TAS, VIC, NSW; also in New Zealand.

ILLUSTRATIONS: Handbook, Plate 4.

This is not uncommon on wet earth banks along streams in the Alps. *P. pyriforme* (Hampe) Jaeg. is probably conspecific (Clifford and Willis, 1952).

P. australe (Hook.f. & Wils.) Mitt.

This seems to be much rarer than *crispulum*, from which it is distinguished by the deeper lamellae, 5–10 cells high, the nerve *c* 600 μm in width with a narrower bistratose portion only 6–7 cells wide. The unistratose portion is up to 10 cells wide, only apparently narrower than in *crispulum*, and usually entire. The sheathing leaf base is clearly wider than the limb and has a margin of very thin-walled hyaline elongated cells forming a distinct although flimsy border. The marginal cells of the limb are shortly rectangular and tend to be elongated longitudinally. There are only 16 relatively long peristome teeth, sometimes with rudimentary short teeth between.

We have seen little Australian material of this species.

DISTRIBUTION: TAS, VIC, NSW; also in New Zealand and (?) S. Africa.

ILLUSTRATIONS: Wilson (1854), Plate 87.

DAWSONIACEAE

Dawsonia R.Br.

While vegetatively there is little to separate this genus from *Polytrichum*, without recourse to the detailed internal anatomy of the stem, the capsule is quite different—and indeed unique—in having a bunch of bristles for a peristome.

The capsule is ovate in outline and flattened, rather like an antique gunpowder flask in shape; the seta is quite short and the calyptra densely felted with bright reddish-chestnut hairs. The stem is thick and stiff and the leaves large and glaucous—commonly more robust and densely leafy than in *Polytrichum* and its near relatives. Van Zanten (1973) in his recent revision,

79

treated this genus as a member of the Polytrichaceae but we believe the sporophyte is so distinctive that the family Dawsoniaceae is well founded.

D. superba Grev.

When fully grown its combined height and girth make this, one of the most robust of all mosses, quite unmistakable. Most commonly it measures 10–20 × 2–4 cm when fresh and the leaves have a characteristic glaucous bloom, but in poor conditions the stems may be stunted and éven occasionally branched, and the leaves yellowish. The colonies tend to be open, the individual stems scarcely touching. *D. intermedia* C.Muell. we believe to be merely a small form of *superba*; it has been the subject of physiological experiments by Thrower (1964). The number, size and structure of the leaf lamellae and the shape of the leaf are very variable characters. In general, deep lamellae, supposedly characteristic of *intermedia*, tend to occur on fruiting plants in exposed situations; long leaves usually have shallow lamellae and shorter leaves may have either shallow or deep lamellae. *D. pulchra* Wijk, often held to be the correct name for the Australian plant, seems to be merely *superba* (isotype checked). Van Zanten (1973) has reached the same conclusion.

CHROMOSOME NUMBER: $n = 7$ (NSW. *D. pulchra*).

DISTRIBUTION: TAS, VIC, NSW, QLD; also in New Zealand.

ILLUSTRATIONS: Plate 6; Handbook, Plate 5; Brotherus (1924–5), Fig. 796.

It is not uncommon on outcrops of heavy soil, such as red clay-loam, usually in shade, e.g. upturned soil mounds of fallen trees, road cuttings etc. in wet sclerophyll forest, especially in montane areas, e.g. *Eucalyptus regnans* forest in Victoria.

D. polytrichoides R.Br.

This species is distinguished chiefly by the size of the shoot; 1–12 or more cm tall and rather wider than either *Polytrichum commune* or *Polytrichadelphus magellanicus* but narrower and shorter and more commonly branched than even stunted *D. superba*. In the absence of capsules the most characteristic feature is in the lamellae; when viewed from the side, i.e. mounted flat, these have some of the terminal cells *asymmetrically convex*, the biggest bulge being somewhat towards one end of the cell instead of in the middle.

CHROMOSOME NUMBER: n=7 (NSW).
DISTRIBUTION: TAS, VIC, NSW, ACT, QLD; endemic. Recorded for New Zealand in the Index, but apparently in error.
ILLUSTRATIONS: Plate 6; Brown (1811), Fig. 23; Brotherus (1924–5), Fig. 796.
This is far from common and rarely fruits in VIC where it is apparently almost confined to eastern districts extending as far west as Mornington Peninsula, but the commonest *Dawsonia* in NSW. It is found on exposed soils of cart ruts, mounds etc. in dry and wet sclerophyll forest.

D. longiseta Hampe

This species closely resembles small Polytrichaceae, especially *Pogonatum subulatum* [6, q.v.] but is distinguished from that, when not fruiting, by its rather wide leafy shoot, bigger leaves, and by the position of the leaves when dry; erect and appressed, neither contorted nor with the tips incurved to the stem.

The stem is quite short, usually not more than 1 cm tall, with a relatively long seta, commonly 2 cm tall. The proportionately long seta distinguishes the species from *polytrichoides*; the leaves too are wider than there, and less tapering.
CHROMOSOME NUMBER: n=7 (NSW).
DISTRIBUTION: TAS, SA, VIC, NSW, ACT, QLD; endemic.
ILLUSTRATIONS: Plate 6; Brotherus (1924–5), Fig. 796; Mueller, F. (1864), Plate 9 (as *D. longisetacea*), Plate 10 (as *D. appressa*).

It is rather a common plant on exposed soil, e.g. of road cuttings, mainly in dry sclerophyll forest, especially in montane regions of VIC and NSW.

FISSIDENTACEAE
Fissidens Hedw.†

Once it is known, this genus can be mistaken for no other. The leaves are always clearly distichous and usually lie flat when moist like a miniature fern frond, and have a unique structure, appearing to have a quarter of the lamina (basal and adaxial) duplicated. Morphologically this double portion (the *vaginant lamina*), which clasps the stem, is the true leaf which is folded along the line of the nerve and prolonged as a single layer, the *upper lamina*. The junction of the two halves of the vaginant lamina may be confined to the line of the nerve, in which case the lamina is termed *open* and the two halves often gape apart. Where the two halves are joined right out to the margin, the lamina is said to be *closed*. Intermediate conditions also occur and the same species sometimes varies in this feature. The back of the nerve carries a wing, the *dorsal lamina*, which is usually similar in size to the vaginant and upper laminas combined. The laminas may be wholly or partially bordered or unbordered. In *F. dealbatus* the nerve is absent. Capsules are frequent in many of the smaller species and are usually carried at the tips of the stems; the slender peristome teeth are often red and quite conspicuous, usually split to half-way. The habitat is usually on rocks or soil, but *F. tenellus* and *F. humilis* can be epiphytic; little, however, is known about the comparative ecology of the different species although it would be a useful and interesting field for investigation.

Key to species

1. Leaves nerveless . . . .   *dealbatus*
   Leaves nerved   2

2. Leaves completely without a border (except sometimes on perichaetial leaves)   3
   Leaves bordered, at least in part, by elongated cells   11

3. Cells with multiple or single papillae, or mamillose with highly convex walls   4
   Cells smooth and level, without papillae . . .   6

4. Cells densely papillose . . . .   *humilis*
   Cells with single low papillae, or mamillose . .   5

5. Stems less than 5 mm high. Vaginant lamina usually clearly toothed. Cells sharply uni-papillose; leaves nerved right to the apex . .   *tenellus*
   Stems usually more than 1 cm high. Vaginant lamina only crenulate. Cells mamillose; nerve failing below apex . . . .   *oblongifolius* (see also *asplenioides*)

6. Leaves extremely widely spaced and reaching 5 mm or more in length. Lax aquatic plants   .   *fontanus*
   Leaves mostly close together, rarely exceeding 3 mm. If aquatic, stiff and erect   7

7. Leaf tips rolled up (circinate) when dry   .   .   .   8
   Leaf tips sometimes crisped or twisted when dry but not rolled   9

8. Leaf apex acute; margins entire or nearly so.   .   *pallidus*
   Leaf apex widely pointed (90°) or rounded; margins crenulate with projecting cells   .   .   *asplenioides*

9. Marginal cells clearly reduced in size. Leaves more than 1 cell thick. Plants aquatic   .   .   .   *strictus*
   Marginal cells similar to adjacent cells. Leaves 1 cell thick. Plants usually not aquatic .   .   .   10

10. Very robust (over 2 cm). Margins lightly toothed at apex   .   .   .   .   *adianthoides*
    Small (less than 1 cm). Margins entire   .   .   .   *taylorii*

11. Border intramarginal in the vaginant lamina, constantly with at least one row of lamina cells outside the border   12
    Border all marginal (rarely with one row outside)   13

12. Cells above papillose, obscure.   .   .   .   .   .   *vittatus*
    Cells smooth and clear .   .   .   .   *hunteri*+
    integerrimus

13. Border usually only on vaginant lamina, often indistinct. Plants minute.   .   .   .   .   .   *taylorii*
    Distinct border on all leaf margins. Plants minute or large   14

14. Plants aquatic or semi-aquatic. Border bistratose or thicker.   15
    Plants on damp earth, not aquatic. Border not thick   16

15.   Cells large and thin-walled, 12–18 μm. Plants
small, less than 1·5 cm high (rare) .   .   .   .   *crassipes*
Cells small and dark, 6–9 μm. Plants large,
usually more than 2 cm high (common) . .   *rigidulus*

16.   Leaf cells isodiametric and obscure throughout,
c 6 μm diam..   .   .   .   .   *leptocladus*
Leaf cells clear throughout, *c* 10 μm diam.,
larger and wider in the vaginant lamina .   .   .   *pungens*

F. humilis Dix. & Watts

This minute plant is at once distinct from the other Australian species in the leaf cells which are covered with hobnail papillae rather like those of *Tortula* and its relatives. The stems are usually only 1–2 (–4) mm high with deep green leaves, *c* 1·0 × 0·2 mm, slightly curled and even rolled up when dry. The leaves are acute, usually with a hyaline tip caused by the projecting nerve. The nerve actually appears to fail lower down the leaf where it becomes covered and concealed by a layer of lamina cells. The cells of the stem, nerve and rhizoids are all smooth.

The peristome teeth (actually half-teeth) are disputedly either spirally thickened or papillose. Our material shows oblique bars on both faces of the transparent teeth which thus appear to have crossed thickenings; in the upper halves of the teeth, which are extremely narrow and thread-like, the thickenings overlap both sides of the tooth giving a distinctly papillose appearance.

DISTRIBUTION: VIC, NSW, QLD; also in New Zealand.

ILLUSTRATIONS: Plates 7–9; Watts (1916), Plate 2.

This is not a common plant. It usually grows on earth banks but we have a specimen from charcoal tree stumps in dry sclerophyll forest near Sydney.

There is no species in Australia with which this can be confused.

PLATE 7. *Fissidens rigidulus* VIC (centre)
Top row (left to right): *F. leptocladus* VIC, *F. oblongifolius* NSW, *F. pungens* VIC
Middle row (left to right): *F. asplenioides* VIC, *F. pallidus* VIC (two shoots)
Bottom row (left to right): *F. vittatus* VIC, *F. humilis* NSW, *F. taylorii* VIC (vegetative and fruiting shoots), *F. tenellus* VIC
All ×7

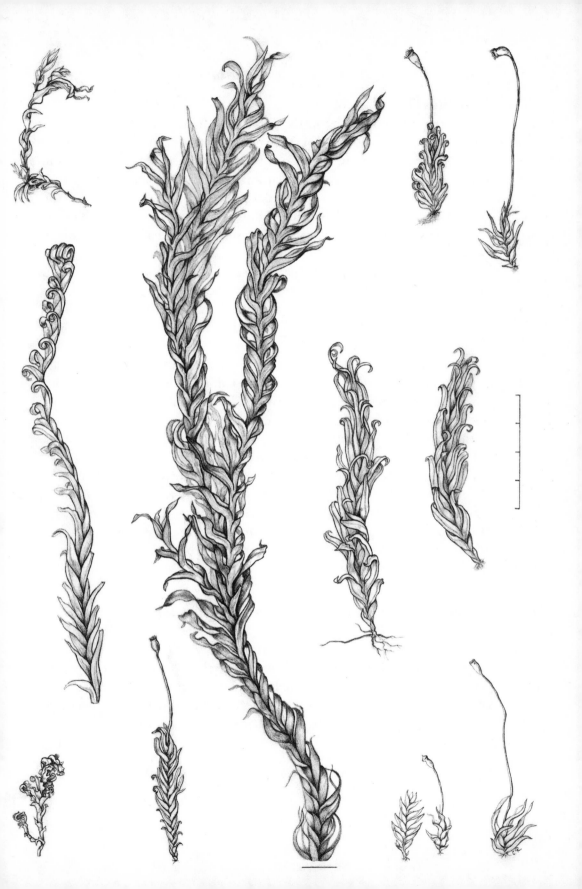

F. tenellus Hook.f. & Wils.

One of the smallest species in the genus, this has stems about 1·5–2·0 mm high with only 4–6 pairs of narrowly pointed leaves which measure 0·5–1·5 × 0·1–0·2 mm. It forms open colonies on damp earth, or, more rarely, is epiphytic. The vaginant lamina, about ½ the length of the leaf or less, is pointed and joins up at the margin. There is a strong nerve, failing in the apex or shortly excurrent. The cells are hexagonal or squarish, *c* 9 μm, sharply *uni-papillose* (sometimes surprisingly hard to see), making the leaf margin crenulate. The cuticle may be minutely roughened, especially when dry. In the *vaginant lamina the cells are enlarged* to *c* 8 × 12–15 μm and the *margin* there is usually irregularly and *distinctly toothed*. The dorsal lamina is sometimes very narrow and often disappears towards the base of the leaf. Sometimes the marginal row of cells is paler but never forms a distinct border.

There is a relatively long seta, *c* 5–7 mm, with a tiny capsule 0·5 mm long or less, oblong and tapering slightly into the seta and tightly constricted below the bright red/orange peristome. The operculum, also red, is as long as the capsule and is abruptly contracted to a fine curved beak.

DISTRIBUTION: TAS, SA, VIC, NSW, QLD, Lord Howe; also in New Zealand.

ILLUSTRATIONS: Plates 7–9; Wilson (1854), Plate 83.

The denticulation of the vaginant lamina, which is usually obvious, and the papillose cells easily distinguish this species. It is a common plant but easily overlooked because of its small size. We have found it in great abundance on the base of *Melaleuca* in swamps. There are plants in which the rhizoids are papillose, instead of smooth, but the taxonomic significance of this requires investigation.

F. fontanus (Pyl.) Steud. (=*F. muelleri* of the Handbook=*Octodiceras muelleri* (Hampe) Jaeg.)

Often put into the separate genus *Octodiceras*, which is more conservatively treated as a subgenus of *Fissidens*, this plant has a very different appearance from the other Australian *Fissidens*. It is a floating aquatic attached at the base, soft and with rather widely spaced, alternate leaves with weak nerves and large cells, up to *c* 20 × 30 μm. It is either rare or overlooked.

PLATE 8. *Fissidens* leaves (same arrangement as Plate 7). All × 50

DISTRIBUTION: SA, VIC, NSW, QLD, Lord Howe; also in New Zealand.
ILLUSTRATIONS: Plate 10.

Except in the youngest leaves the cell walls may be sinuose in dry material but this appears to be an artefact on drying. The cells get noticeably smaller towards the margins but do not form a distinct border. Rhizoids, small leafy branches, and separate male and female branches arise in the leaf axils.

F. pallidus Hook.f. & Wils.

An aptly named species which forms pale and usually yellowish patches on bare earth banks, sometimes carpeting large areas. The stems are mostly short, 0·5–1·5 cm long, bearing pale yellowish green leaves which are curved or even rolled up at the tips when dry. The long slender leaves are commonly 2–3 mm long × 0·25–0·3 mm wide and have the vaginant lamina ½–⅓ as long, rounded at the tip and with the junction usually mid-way between the nerve and the margin, but sometimes on the nerve. The nerve is strong, failing just below the apex, flexuose in the upper half and often bent sharply at the beginning of the upper lamina.

The cells are characteristic, pale and very clear and transparent; they are irregular in size, varying from 10 to 18 μm but averaging about 13 μm. There is no border but sometimes the cells of the marginal row are distinctly smaller, c 7 μm, and almost square. The margins are entire almost throughout but sometimes slightly crenulate at the acute apex.

CHROMOSOME NUMBER: n=12 (NSW).
DISTRIBUTION: TAS, VIC, NSW, QLD; also in New Zealand.
ILLUSTRATIONS: Plates 7–9; Wilson (1854), Plate 83.

Perhaps most likely to be mistaken for *F. asplenioides* which differs in the blunt apex, bulging convex cells, and crenulate margins. *F. oblongifolius* has mamillose cells, not transparent as in *F. pallidus*, giving a crenulate margin.

F. strictus Hook.f. & Wils.

This is a rare plant of rock crevices in rivers and waterfalls, where the short fronds (c 0·5–1·0 cm × 1·5 mm) form black silky tufts. Characteristically the fronds remain stiff and straight even when dry, possibly because the dorsal and upper laminas are 2 or more cells thick, except for a broad

PLATE 9. *Fissidens* cells (same arrangement as Plate 7). Cells all × 1000. *F. tenellus* cells (bottom right): vaginant lamina below, dorsal lamina above

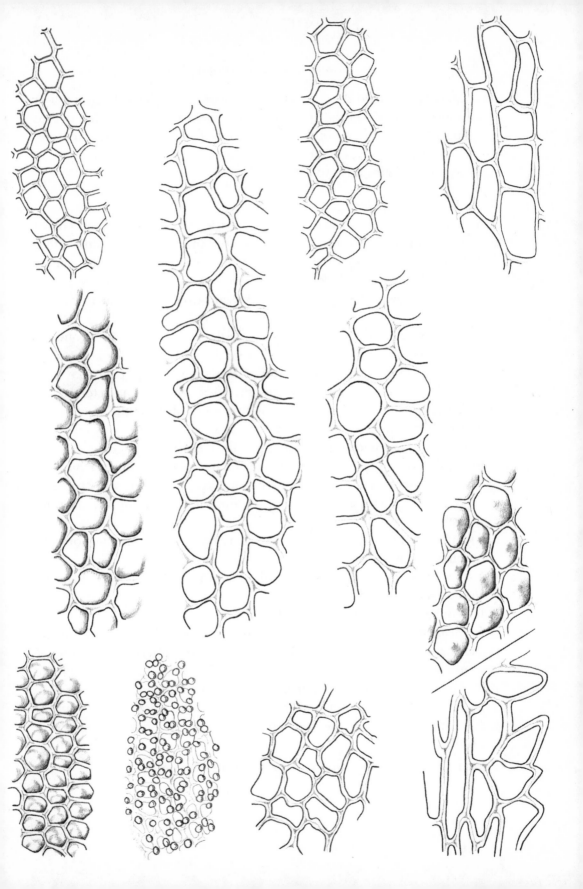

unistratose band round the edge. The long acute leaves have a yellowish or reddish nerve, failing in the apex. The marginal row of cells is reduced to 6–8 µm, about half the size of those in mid-leaf, forming an evident margin.

DISTRIBUTION: TAS, VIC; endemic.

ILLUSTRATIONS: Wilson (1859), Plate 171.

F. asplenioides Hedw.

The shoots of this species grow in rather thick turfs *c* (0·5–)2–3 cm tall, of which the uppermost 3 mm or more is growing, coloured a dull rather yellowish green but sometimes glistening when fresh because of the convex cells. Leaves are *c* 2 mm long ×0·4 mm (0·3 mm above the vaginant lamina), rolled up at the tip when dry. A year's growth is about 10 pairs of leaves.

The vaginant lamina occupies ⅔ to ¾ of the leaf length, and is rounded at the end where the blades join together on the nerve. The dorsal lamina is prone to end above the base of the leaf. The nerve is strong, bent above the vaginant lamina and failing just below the apex which is characteristically rounded or broadly pointed (*c* 90°). There is no border and the margins are crenulate throughout, especially at the apex where they are sometimes almost denticulate with the projecting cells. Cells are hexagonal, *c* 8 µm, convex; slightly larger, *c* 10–12 µm, and squarer in the vaginant lamina.

The seta is short, *c* 4 mm, and the capsule very short and urn-shaped, *c* 0·5–1·0 mm long with a very long operculum of *c* 1 mm, slightly slanting.

CHROMOSOME NUMBER: $n = 12$ (NSW).

DISTRIBUTION: TAS, WA, SA, VIC, NSW, ACT, QLD, NT; from the northern hemisphere tropics southwards, widely distributed.

ILLUSTRATIONS: Plates 7–9; Handbook, Plate 6; Wilson (1854), Plate 84 (as *ligulatus*).

It may be confused with *F. pallidus* (q.v.).

F. adianthoides Hedw. is commonly confused with this species by name rather than appearance; it is a larger and rarer moss differing in the bigger cells and the acute leaves serrate at the apices. The marginal cells tend to be pale, forming a contrasting margin. It occurs in TAS and, doubtfully, SA, as well as in New Zealand, but is predominantly a northern hemisphere species. It is said to bear the sporophytes laterally. We have not seen it in Australia.

PLATE 10. *Fissidens fontanus* SA—Shoot ×7, leaf ×50, cells ×1000

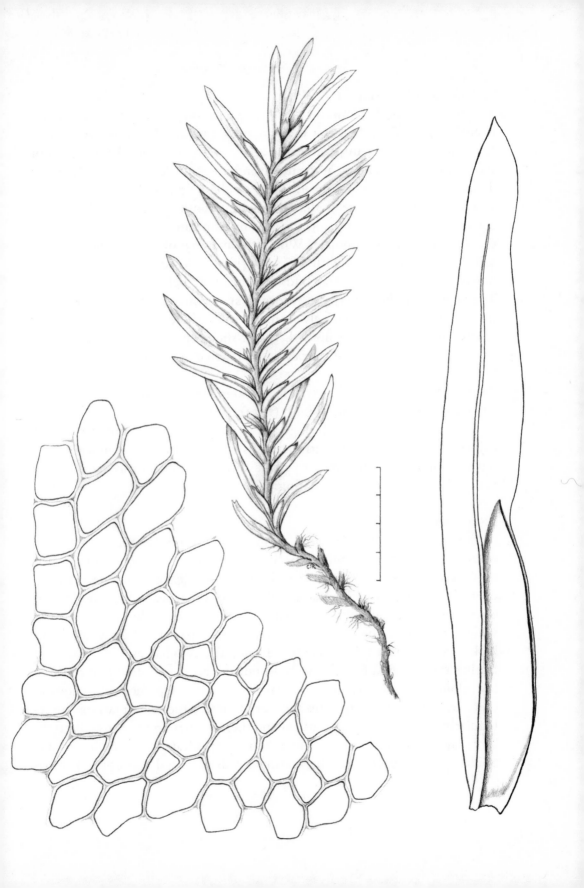

F. oblongifolius Hook.f. & Wils.

This species is said by Sainsbury (1955d, pp. 26–27) to be a different species from *F. asplenioides*; the latter is distinguished from it by "the leaves when dry being circinately inrolled at the apex, in the junction of the blades of the vaginant lamina being at the nerve and not at the leaf margin, and in the dioicous inflorescence". The junction of the vaginant lamina, however, varies even on the same stem, the leaves of both species may be almost equally strongly rolled up and the cells in both can be equally convex and the margins equally crenulate. What little material of *oblongifolius* we have seen has been autoicous with the antheridia enclosed by small perigonial bracts, forming little lateral buds on the stem below the inflorescence. The Handbook, however, also records a synoicous variety. While the two species are distinct vegetatively in many cases, there are clearly specimens where the inflorescence is needed. If that should prove to be an unreliable character then the separation between the two species would no longer be easy to maintain.

DISTRIBUTION: TAS, SA, VIC, NSW, QLD; also in New Zealand.

ILLUSTRATIONS: Plates 7–9; Wilson (1854), Plate 83.

This appears to be rare throughout its range and to occupy a range of habitats from epiphytic to terrestrial. Its ecological and anatomical variability combine with its rarity to make this plant suspiciously hard to circumscribe. It is worth investigating.

F. taylorii C.Muell.

Another of the minute *Fissidens,* this is common on damp earth either as open colonies or frequently mixed with other species. The unbranched stems, 1·5–2·5 mm high or even less, bear very tiny leaves, 0·4–0·6 × 0·1 mm, of which the vaginant lamina usually makes up at least ⅔ of the length, often much more, and the dorsal lamina is frequently reduced to a narrow wing. The leaves as a consequence seem widely spaced and the whole stem looks like a miniature *Blechnum* fern frond. The two halves of the vaginant lamina are bordered by narrow cells, but the border is often reduced to a few elongated cells or, in some leaves, missing altogether. Occasionally there are traces of a border on the dorsal lamina too. The margins are irregularly crenulate or even toothed and there is a strong nerve usually right to the apex. The cells are clear and smooth, quadrate to somewhat rhomboidal, c 8–12 × 9–16 µm. The marginal cells of dorsal and upper laminas are sometimes paler but do not form a distinct border. Some specimens show a fine intense stippling over all stem and leaf surfaces, like very minute papillae (*c* 1 μm). They do not seem to be of taxonomic importance.

It is often found fruiting in abundance, with masses of tiny red-tipped capsules on short setas (2–3 mm). Fertile shoots are distinctly shorter, with fewer leaves than vegetative shoots and with the upper leaves enlarged.

DISTRIBUTION: TAS, WA, SA, VIC, NSW, ACT; also in New Zealand.

ILLUSTRATIONS: Plates 7–9; Wilson (1854), Plate 83 (as *brevifolius*).

The weak border and frequent reduction of the dorsal lamina often separate this common species from the other tiny *Fissidens*: *pungens*, *tenellus*, and *humilis*. The relationship between *taylorii* and the Victorian species *elamellosus* Hampe and C.Muell., *macrodus* Hampe, and *semilimbatus* Hampe and C.Muell. (all illustrated in Mueller (1864) and all reduced to *taylorii* by Sainsbury, 1955d) requires further investigation. There are, at least, some curious forms in Victoria. There seem, however, to be intergrading forms with *pungens* which require further study.

F. epiphytus Allison (1960) is recorded from Victoria and New Zealand. It needs research in connection with the *taylorii* group.

F. vittatus Hook.f. & Wils.

This is another species in which the leaves are curled at the tips when dry, like *asplenioides* and *pallidus* although the whole stem tends to curl as well. It is at once recognizable under the microscope by the border which is more or less throughout the leaf and is stronger and conspicuously intramarginal in the vaginant lamina, with 3 or more rows of lamina cells, often forming a lightly toothed margin, outside it. The stems, although 6–7 mm high, can withstand burial by sand and are often found in dry sandy soil with only a few, tightly rolled-up leaves showing at the soil surface. The leaves are *c* 0·3–0·5 × 0·1 mm, the vaginant lamina occupying *c* ¾ of the leaf length, with the 2 halves joining at the nerve and usually gaping wide apart. The nerve is strong, reaching the apex or excurrent as a short point. The cells are rather thick-walled, small, 5–8 μm, and singly papillose.

Sporophytes are very rare; the perichaetial leaves are noticeably long and narrow, giving fertile stems a quite different appearance (as in *taylorii*).

DISTRIBUTION: TAS, WA, SA, VIC, NSW, ACT; also in New Zealand and S. Africa.

ILLUSTRATIONS: Plates 7–9; Brotherus (1924–5), Fig. 122; Wilson (1859), Plate 171.

The intramarginal border, papillose cells and gaping vaginant lamina usually make this an easy species to recognize. *F. leptocladus* occasionally has a single row of cells outside the border but this is not conspicuous and is infrequent.

F. hunteri Willis (Willis, 1951, p. 83), so far known only from far eastern Victoria, differs in having smooth cells. It is perhaps conspecific with *integerrimus*.

F. integerrimus Mitt. (TAS) also has an intramarginal border sometimes present (Sainsbury, 1955d). It is illustrated in Wilson (1859), Plate 171.

F. variolimbatus Allison (1963), a New Zealand species, ought to be borne in mind when investigating Australian material. With its variable border it could give endless trouble. It is not in the Index.

F. rigidulus Hook.f. & Wils.

This species, the most robust in our region, is also one of the easiest to identify in the field and is common in suitable habitats. The stiff, dark green, large fronds are closely packed together to form soft turfs on dripping rocks, stones in rivers and other aquatic and semi-aquatic habitats. We have recorded it from a depth of 35 ft in clear water. The stems vary from 2–6 cm and may be branched or unbranched. The leaves, 2·5–3·0 mm long × 0·6–0·7 mm wide, are usually turned towards one side of the stem (secund) even when moist but especially when dry, and are curled and twisted when dry. They are bordered by a stiff yellow border of elongated and thick-walled cells throughout, except right at the apex and at the base of the dorsal lamina, where the border normally fades out. This border is several cells thick and can persist in eroded specimens, together with the nerve, after the lamina has disintegrated. The vaginant lamina is pointed, the two halves joining together either at the margin or between the margin and the nerve, and occupies about ⅔ of the length of the leaf. The nerve is very strong and yellowish or reddish, often bent on leaving the vaginant lamina, and fading out, like the border, just at the apex. Cells are very small and dark,

6–9 μm, irregular in size and shape but mostly quadrate or hexagonal; the cell walls in old leaves are quite commonly reddish.

DISTRIBUTION: TAS, VIC, NSW, ACT; apparently not known from SA, WA and QLD. Also in S. America and New Zealand.

ILLUSTRATIONS: Plates 7–9; Handbook, Plate 6; Wilson (1854) Plate 83.

The only other well known species with completely bordered leaves, in our region, are *F. pungens*, *F. crassipes* and *F. leptocladus*. They are all considerably smaller than most specimens of *F. rigidulus*, since they rarely exceed 2 cm, and are plants of damp soil and rocks, not aquatic; in *F. pungens*, and *F. leptocladus* the border is only unistratose.

F. leptocladus C.Muell. & Rodw.

The stems of this variable species are brown or green, commonly 0·5–1·5 cm tall, simple or branched, and forming rather open turfs, sometimes very sparse. We have found it on limestone and basalt, but whether it is consistently calcicolous remains to be discovered. The leaves, when dry, are slightly secund, bent downwards in one direction and sometimes even inrolled at the tips; they measure 1·0–1·5 × 0·3 mm, half the length being taken up by the vaginant lamina which is pointed, with the blades joining at the leaf margin. The dorsal lamina is contracted at the base. There is a border on all leaf margins, evident and often very strong at the base but thinning out above and usually failing completely just before the apex. Occasionally there is one row of marginal cells outside the border in the vaginant lamina but the intramarginal condition is never pronounced. The nerve is strong, often flexuose above, and reaches right to the apex. The cells are very small, *c* 5 μm, hexagonal or squarish, and rather irregular in shape. The leaf apex is acute and slightly crenulate or denticulate by projecting cells.

DISTRIBUTION: TAS, WA, SA, VIC, NSW, ACT; also in New Zealand.

ILLUSTRATIONS: Plates 7–9; Studies, Plate 7.

This species is separable from *F. pungens* principally by the small cells and failing border and from *F. rigidulus*, which is a much more robust plant, by the thin, not bistratose border. It is not uncommonly associated with *F. taylorii* from which it differs in the almost complete border.

F. pungens C.Muell. & Hampe

These minute plants form dense carpets on moist earth. The stems are only 2–5 mm long and tend to be of two sizes; very short with 3–4 pairs of leaves and bearing capsules, and taller ones with about 10 pairs of leaves and apparently barren. However, archegonia do occur on the tall plants but it seems as if they are rarely or never fertilized; continued close observation of marked plants might reveal the reasons.

The leaves, 1·0–1·5 mm×0·25 mm are usually contorted when dry and are distinctive in the strong border on all margins, joining together with the nerve at the apex to form a stout excurrent point. The junction of the halves of the vaginant lamina is at the margin and the nerve is usually bent at the start of the upper lamina. The dorsal lamina narrows and sometimes disappears at the base. The cells are clear and quite wide, *c* 10–12 μm in mid-leaf, elongated to 10–30 μm in the vaginant lamina.

CHROMOSOME NUMBER: n=13 (NSW).

DISTRIBUTION: TAS, WA, SA, VIC, NSW, QLD, NT; probably in ACT; also in New Zealand.

ILLUSTRATIONS: Plates 7–9; Studies, Plate 7 (as *inclinabilis* and *campyloneurus*); Handbook, Plate 6; Mueller (1864), Plate 17.

This species is closest to *F. leptocladus* but differs in the larger cells, in the enlargement of the cells of the vaginant lamina, and in the border which is typically strong above and fused together with the nerve at the apex, whereas in *F. leptocladus* and *F. rigidulus* the border tends to fade out and disappear just below the apex. There are, however, forms of *pungens* with weak or perhaps absent borders and these may give difficulties.

F. dealbatus Hook.f. & Wils. is also a minute plant which is bordered throughout, but is immediately distinguished from all other species in our area by having no nerve. It is very rare or overlooked, on clay. soils in forest and perhaps elsewhere.

DISTRIBUTION: TAS, VIC, NSW; perhaps in Queensland; also in New Zealand and Fiji.

ILLUSTRATIONS: Wilson (1854), Plate 84.

F. crassipes B.S.G., only recently discovered by Willis, turns out to be quite widely distributed in Australia. It is like a very small *F. rigidulus*.

DISTRIBUTIONS: SA, VIC, NSW, ?QLD. It is widely distributed in the northern hemisphere.

ILLUSTRATIONS: Dixon (1924a), Plate 16.

Of the 38 other species given by the Index Muscorum for Australia, *F. splachnoides* Broth. may be *dealbatus* according to Willis (1955a).

The others are:

F. amblyothallioides Broth. & Watts (Lord Howe)

F. ampliretis (C.Muell.) Broth. (NSW)

F. arboreus Broth. (NSW, QLD, Lord Howe)

F. arcuatulus Broth. & Watts (Lord Howe)

F. aristatus Broth. (NSW)

F. australiensis Jaeg. (VIC, NSW) (*F. perpusillus*, Mueller (1864), Plate 16).

F. basilaris C.Muell. & Hampe (SA, NSW)

F. bryoides (Hedw.) Bertsch.

F. bryoidioides Broth. (NSW)

F. cairnensis Broth. & Watts (QLD)

F. calodictyon Broth. (NSW, QLD)

F. cambewarrae Dix. (NSW, QLD)

F. delicatulus Aongstr. evidently not Australian (Watts & Whitelegge, 1902).

F. forsythii Broth. (NSW)

F. homomallulus Dix. (NSW)

F. howeanus Broth. ex Whitelegge (Lord Howe)

F. hyophilus Mitt. (QLD)

F. kerianus Broth. (QLD)

F. kurandae Broth. & Watts (QLD)

F. linearis Brid.

F. longiligulatus Broth. & Watts (Lord Howe)

F. maceratus Mitt. (QLD)

F. micro-humilis Dix. (QLD)

F. montecolli Broth. & Watts (NSW)

F. patulifolius Dix. (QLD)

F. pauperrimus C.Muell.

F. perangustus Broth. (NSW)

F. serrato-marginatus C.Muell. (SA)

F. sordiderirens Broth.

F. subkurandae Bartr. (QLD)

F. subtenellus Broth. & Watts (Lord Howe)

F. tenelliformis Broth. & Watts (Lord Howe)

F. terrae-reginae Bartr. (QLD)

F. tortuosus Geh. & Hampe (TAS)
F. victorialis Mitt. (NT)
F. wattsii Broth. (NSW, Lord Howe)
F. wildii Broth. (QLD)

GRIMMIACEAE

Grimmia Ehrh. ex Hedw.

This is a well defined but troublesome genus much in need of a world-wide revision. Like many another genus it shows great variability under alpine conditions and a plethora of species has been described, probably mainly based on environmentally produced modifications. At least four of the Australian species are well known overseas and have stood up to critical scrutiny. The leaves of most species have hyaline points but these may be greatly reduced or even absent and there is then a risk of confusion with species of *Orthotrichum* [53] and even of *Andreaea* [2]. Characteristically the dry shoots absorb water very rapidly and the leaves flex backwards excessively when soaking out, before relaxing forwards to their normal moist position. A diagnostic feature of the genus is usually held to be the cell walls which, in most species, are sinuose, but this is often expressed only as a slight bulging of the walls of cells in mid-leaf, or of a few cells overlying the base of the nerve adaxially, and is sometimes quite un-detectable even to the eye of faith. Capsules are common and may be either erect or pointing downwards from the end of a curved seta or, in *apocarpa*, immersed. The single peristome consists of 16 teeth which, in our species, are variously split and perforated; the calyptra is smooth and membranous.

There are only four common species.

Key to species

1. Abaxial surface of dry leaf ribbed with projecting
    nerve and recurved margins. Leaves V-shaped
    in TS   2
    Abaxial surface of leaf almost smooth, slightly
    convex, with the nerve not much protruding.
    Leaves widely U-shaped in TS. Margin of leaf
    plane .   .   .   .   *laevigata*

2. Hair-point very short or absent. Capsules im-
    mersed . . . . *apocarpa*
    Hair-point usually long and hyaline. Capsules
    exserted 3

3. Leaves scarcely twisted when dry. Plants greenish
    grey, very hoary. Most of cells in mid leaf-base
    short and wide, c 12 × 24 μm . . . . . . *pulvinata*
    Leaves somewhat twisted round stem when dry.
    Hair-points smooth. Plants yellowish not hoary.
    Most of cells in mid leaf-base long and narrow,
    c 35–55 × 9 μm. Upland to alpine moss . . . *trichophylla*

G. laevigata (Brid.) Brid.

Almost invariably this species has a very strong and conspicuous hyaline
point, broad and flat at the base and forming a striking contrast with the
blackish leaves and stems so that the patches, which are often quite extensive, have a piebald appearance when dry.

The shoots vary in length, commonly 0·5–1·5 cm tall, and usually form
extensive turfs, but sometimes small dense cushions. The most distinctive
feature is the shape of the leaf which is wide and smooth-surfaced, with
the nerve projecting almost imperceptibly abaxially as a broad flat rib.
The leaves are densely arranged and closely imbricate when dry. The nerve
is rather thin and inconspicuous, reaching to the apex where it runs into
a very pronounced hair-point which is strongly toothed and broadened
at the point of contact with the lamina.

The leaves are bistratose above, not just at the margins, and the cells
there are square and thick-walled with rounded cavities, c 7 μm. In the
leaf-base the cells next to the nerve are very elongate, reaching 45 × 10 μm;
those outside them are broadly hexagonal, c 21 μm across, and the outermost 1–3 rows are sometimes more or less hyaline forming an indistinct
border. Above and to the side of the patch of hexagonal cells in the leaf-
base there is generally a large area of over-square cells, wider than long,
measuring typically 8 μm in length and 13–15 μm wide. There is rarely
much sign of sinuose or corrugated wall thickenings in this species, although
there is often a curious, rather spiky, lining to the cell cavities.

The capsules are smooth, just emergent on a short straight seta, overtopped by the tips of the longest leaves.

DISTRIBUTION: TAS, WA, SA, VIC, NSW, ACT, QLD, Lord Howe; widely distributed in both hemispheres.

ILLUSTRATIONS: Plate 11; Handbook, Plate 7.

Grimmia campestris and *G. leucophaea* are common synonyms.

This species is distinct in the bistratose smooth leaf, widely U-shaped in transverse section, and in the rough hair-point, widened at the base, although the hair-point can be narrow in alpine forms. The leaf shape distinguishes it from other Grimmias and the obvious nerve will separate it from Hedwigias [89] which resemble it in leaf shape. It is probably the commonest of all the Grimmias in Australia.

G. apocarpa (Hedw.) B.S.G.

By no means the commonest of the Grimmias, this is the most unmistakable when fruiting, with the capsules immersed in the dark perichaetial leaves and showing not much more than the red mouth and peristome. This feature has led to the genus *Schistidium* but that seems a rather weak genus, since the immersion of the capsule is very much a matter of degree. It is found on the two extremes of dry rock and intermittently submerged (var. *rivularis*) on rocks in streams, lakes and waterfalls.

The shoots form rather open olive or blackish green turfs or cushions 0·3–2·5 cm tall, with leaves usually having no more than a hyaline apiculus and often not even that; more rarely there is a short hair-point. The nerve is very strong and forms a projecting rib abaxially, with the leaf lamina folded forwards from it, V-shaped in cross section. The margins are entire and two or sometimes more cells thick, forming a stiffened border and apex. The cells are very irregular in size and shape. The uppermost cells vary from square, $c\,6\times6$ μm, to 7×4 μm; the cells in the leaf-base are rectangular $c\,10\times8$ μm to 36×10 μm. The lamina cells, somewhat below mid-leaf but above the basal cells, tend to have quite strongly sinuose or corrugated walls and the shorter cells above them are often irregularly thickened to give an hour-glass shape of cell cavity. The basal marginal cells tend to be square.

The capsule is wide, $c\,2$ mm long, usually with a red mouth and peristome.

PLATE 11. *Grimmia laevigata* VIC—Shoot × 22, leaf × 50, cells × 1000. Note: the capsule is much more bent to one side than is customary for this species

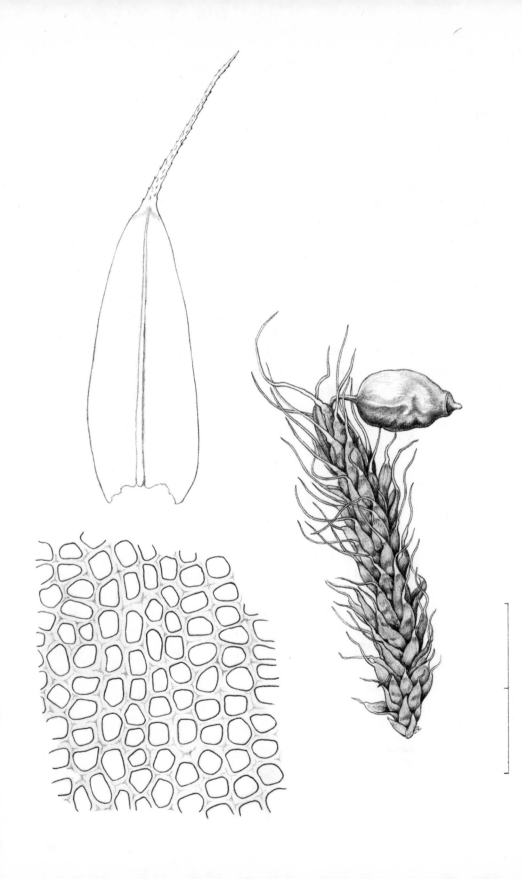

DISTRIBUTION: TAS, ?WA, SA, VIC, NSW, ACT, QLD; widely distributed almost throughout the world.

ILLUSTRATIONS: Plate 12; Watson (1955), Fig. 60.

G. pulvinata (Hedw.) Sm.

The dense, rounded greenish cushions of this common moss are hoary with the long hair-points of the leaves. The shoots are seldom much more than 1 cm tall of which only the upper 3 mm or so is green. The leaves average 1–2 × 0·3–0·6 mm, with a hair-point smooth or slightly denticulate at the tip, sometimes as long as the lamina but more commonly ½–¾ as long, more rarely very short. The leaf is folded about the prominent nerve, as in *apocarpa*, with usually slightly recurved margins. The cells in the upper part of the leaf are rather irregular in shape, roughly quadrate or slightly elongated, *c* 6–9 μm, thick-walled and rounded-off by corner thickenings, not infrequently dumb-bell-shaped. The cells in the leaf base are oblong, often with corrugated longitudinal walls, mostly 2–3 times as long as wide (24–36 × 7–12 μm) or rarely longer. The marginal cells are rectangular at the very base of the leaf, becoming square higher up with a tendency to thickenings on the transverse walls only.

The capsule is wide-mouthed, and *c* 8-ribbed both wet and dry, *c* 1 mm long on a curved swan's-neck seta *c* 1·5 mm long. The operculum is reddish and shortly pointed.

DISTRIBUTION: TAS, WA, SA, VIC, NSW, ACT, QLD, Lord Howe; practically world-wide in distribution.

ILLUSTRATIONS: Handbook, Plate 7; Watson (1955), Plate 5, Fig. 62.

This species is likely to be mistaken for moist *laevigata* (when moist, the black colouration is lost) but is easily distinguished from it by the recurved margins and prominent nerve. More troublesome is the distinction from *trichophylla* which, however, is mostly a subalpine to alpine moss in Australia. Where the hair-points are reduced it becomes difficult to separate from *apocarpa* in the absence of fruit but usually, at least, lacks the thickened border of that species.

PLATE 12. *Grimmia apocarpa* VIC—Fruiting shoot × 15, leaf × 50, basal cells (below) and upper cells × 1000

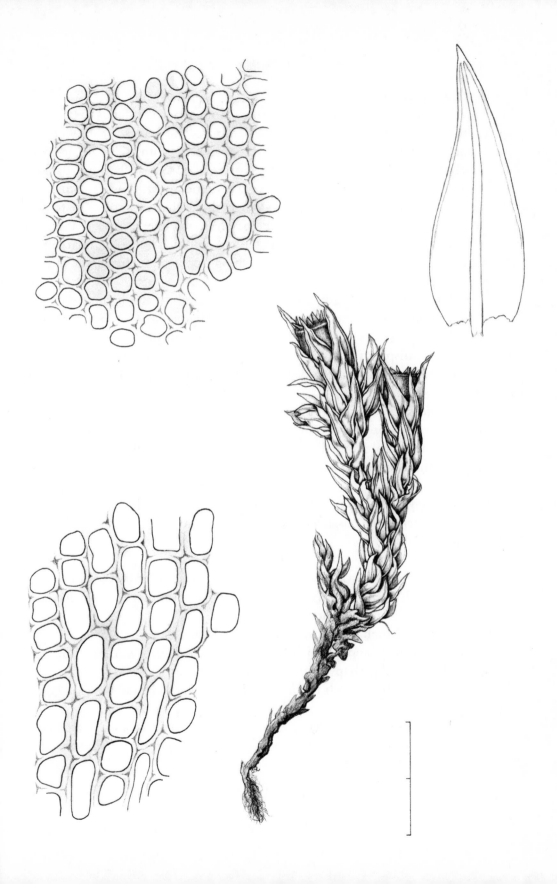

G. trichophylla Grev.

This species can be very hard to separate from *pulvinata*. It tends to be taller, up to 2 cm, and yellowish brown in colour. The leaves are longer and proportionately narrower (typically *c* 2–4 × 0·4–0·7 mm) with a hair-point which is virtually smooth and much shorter or absent so that the cushions are not at all hoary. When dry the leaves tend to be twisted round the stem, especially at the stem apex. When these characteristics fail to distinguish this species from *pulvinata* the most reliable feature is the cells which, in the middle of the leaf base, tend to be much longer and narrower, mostly *c* 35–60 × 9 μm, sometimes reaching 100 × 9 μm. In *pulvinata* they never reach this length.

In Australia the best guide of all, in most cases, is the habitat which is almost exclusively alpine or subalpine, replacing *pulvinata* at high altitudes. DISTRIBUTION: TAS, WA, SA, VIC, NSW, ACT, QLD; throughout much of the northern hemisphere and also in S. America and New Zealand. ILLUSTRATIONS: Watson (1955), Fig. 63; Grout (1928–40), II, Plate 14.

G. austro-funalis C.Muell. (VIC) is very similar but has strongly corrugated cell walls except in the extreme leaf base. It may be conspecific with *trichophylla* but is probably a distinct species. The cells look like a cross between *Grimmia* and *Rhacomitrium*.

G. cylindropyxis C.Muell. in herb. MEL 1002777, probably the type, is mainly *Rhacomitrium crispulum* with an admixture of *G. austro-funalis* from the same locality.

The other species recorded from Australia, and accepted by the Index, are: *G. funalis* (Schwaegr.) B.S.G., which may be a misnomer for *trichophylla*. *G. cyathocarpa* (Hampe) Mitt. (VIC, NSW) *G. obtusata* (C.Muell. & Hampe) Jaeg. (VIC, NSW) *G. ovalis* (Hedw.) Lindb. which has been recorded from N. QLD as *G. commutata* Hueb. and perhaps from TAS (Sainsbury, 1955c). *G. procumbens* Mitt. which is *Rhacomitrium crispulum* (Handbook, p. 63). *G. readeri* Broth. (VIC) and the following Muellerian species: *antipodum, callosa, compactula, crispatula, parramattensis, pygmaea, stenophylla, stirlingii, subcallosa, tasmanica, truncato-apocarpa, woollsiana*.

Rhacomitrium Brid.

Microscopically this genus shares with *Sphagnum*, *Leucobryum* and a few
others the distinction of being recognizable even from tiny leaf fragments.
The longitudinal walls of the cells have heavily corrugated thickenings
unlike any other genus except, to a much lesser extent, its close relative
Grimmia. It shares with *Grimmia* the feature of over-expansion of the leaves
on soaking out (see above). In habitat too it resembles *Grimmia*, normally
occurring on rocks although also found on soil. We have never seen it
growing epiphytically. The two principal species in Australia are very
readily separated. *R. crispulum* is dark green or brown, with the hyaline
hair-point smooth and very short or absent. *R. lanuginosum*, which seems
to be almost confined to Tasmania, is usually very hoary with wide,
long, papillose hyaline points. The plants are usually branched and the
sporophytes are terminal on short lateral branches. Despite its relationship
to *Grimmia* and other acrocarpous mosses, it is hard to regard it as other
than pleurocarpous. In accordance with Crundwell's recommendation
(1970, p. 137) we have retained the customary spelling, not *Racomitrium* as
used in the Index.

R. crispulum (Hook.f. & Wils.) Dix.

This common moss grows on siliceous rocks in dry or wet regions and is
sometimes aquatic. The dark green or olive-yellow stems, up to 6 cm long
or more, form quite extensive mats on rocks, the shoot tips and branches
tending to curve upwards to an erect position from the series of prostrate
stems. The lanceolate leaves, *c* 2–3 × 0·6–0·8 mm, are tightly pressed to the
stem when dry, sometimes slightly curly, arching backwards over-far when
moistened and then relaxing forwards to a spreading position. They are
1-plicate on one side, below, and with the margins recurved on one or
both sides and slightly decurrent at the angles. The strong single nerve is
sunk in a channel and fails below the apex. This is slightly cucullate and
has a very short terminal hyaline spicule occasionally prolonged as a quite
long smooth hair, or is quite commonly absent altogether. The cells in the
upper part of the leaf are *isodiametric*, *c* 9 μm, sometimes slightly elongated
and dumb-bell-shaped, those in the lower half of the leaf long and narrow,
20–50 × 7–9 μm, with thin transverse walls but very thick, strongly corrugated longitudinal walls, wider than the cell cavities. From one to three

marginal rows at the leaf base are often sinuose-walled but not corrugated; a few alar cells may be different—oval, thick-walled and yellow. The surfaces of all cells are quite smooth.

The erect cylindrical capsule, *c* 2–3 mm long, is borne on a short seta of similar length, terminating a short lateral branch. There is a single peristome of red, deeply split teeth, arched in over the capsule mouth when moist, completely closing it, and erect when dry. The operculum has a slim erect beak, *c* 0·5 mm long, with a symmetrical calyptra covering only the uppermost part of the capsule, split at the base into a fringe.

CHROMOSOME NUMBER: n=12 (TAS).

DISTRIBUTION: TAS, WA, VIC, NSW, ACT, apparently missing from SA (Willis 1955d, p. 73); widely distributed in the southern hemisphere.

ILLUSTRATIONS: Plate 13; Handbook, Plate 8.

R. ptychophyllum Lindsay has been collected by Willis on alpine rocks in the Bogong district of VIC. It resembles rather big *crispulum* but lacks a hair-point and has strongly plicate leaves with 2 plications on either side of the nerve. It is not uncommon in similar habitats in New Zealand and also occurs in S. America. The authorship of the name is attributed usually to Mitten in 1866 but he seems to have been anticipated by Lindsay in *Trans. bot. Soc. Edinburgh* Vol. 8, 1865.

R. amoenum (Broth.) Par. has been recorded for Kosciusko, NSW, last century and from Bogong High Plains, VIC, by Willis in 1946. It is "a more densely tufted plant, having slender branches, smaller leaves and very much smaller capsules" (Clifford and Willis, 1952, p. 156), but is otherwise like *crispulum*.

R. lanuginosum (Hedw.) Brid. var. *pruinosum* Wils.

This is a much rarer plant than *crispulum* recognizable by the extremely hoary, long, curled hyaline leaf-points which are toothed and densely papillose, the hyaline area extending some distance along both margins below the apex. The elaborate fretting and frosting of the leaf tip make it one of the most spectacularly beautiful of all microscopical objects. Apart from these hyaline cells, all cells are said to be smooth, but cf. *canescens*.

PLATE 13. *Rhacomitrium crispulum* VIC—Shoot × 7, leaf × 50, cells × 1000

DISTRIBUTION: TAS, VIC, NSW, QLD (?); apparently confined to Tasmania and the Australian Alps.

ILLUSTRATIONS: Handbook, Plate 8; Watson (1955), Plate 6.

It is a very widespread species in the northern hemisphere, but with the var. *pruinosum* restricted to S. Africa, New Zealand and Australia.

R. canescens (Hedw.) Brid. has been reported from VIC. It differs principally in the papillae which are present on all cells and not restricted to the hyaline parts; but the papillae in typical *canescens* are strongly marked, whereas those Australian and New Zealand plants that we have seen have had rather low rounded papillae and seem to us to be probably no more than a slightly papillose form of *lanuginosum*, with which they agree in other respects.

Also recorded for Australia by the Index are:

R. heterostichum (Hedw.) Brid. (TAS)

R. pseudo-patens (C.Muell.) Par. (NSW)

R. striatipilum Card.

ARCHIDIACEAE

Archidium Brid.

Australian species of this genus have recently been described and illustrated (Stone, 1973a). The sporophyte is unique among mosses because there is no columella penetrating the spore-bearing region; there is a dome-shaped air space between the capsule wall and the spore sac (cf. *Sphagnum*) and the few large spores completely fill the central region.

A. stellatum I. G. Stone

These plants, which fruit freely, form a very low yellowish-green perennial turf on bare earth or are scattered amongst other bryophytes.

The short erect stem, 1–3 mm long, has a terminal perichaetium from

PLATE 14. *Archidium stellatum* VIC—Fertile plant × 30 (top), cells (top right) from margin of upper part of perichaetial leaf × 1000; *A. clavatum* VIC (type)—Corresponding cells of perichaetial leaf (top left) × 1000; *Bruchia brevipes* NSW—Fruiting plant × 30, cells (bottom right) × 1000, portion of spore surface from scanning electron microscope photograph (bottom left) × *c* 3825

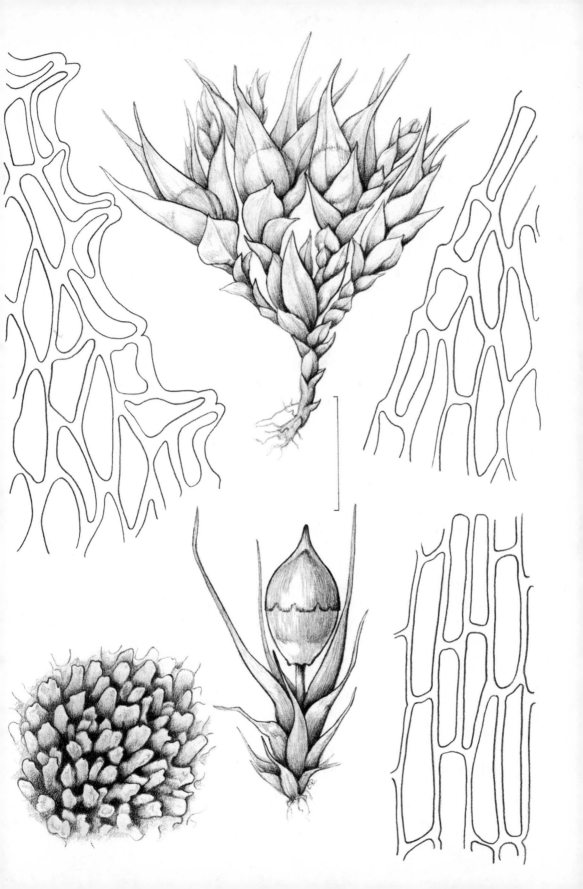

within which arise new branches (innovations) which are short and erect at first and more or less julaceous. When the capsule is mature the new branches spread to give a star-like arrangement; eventually they too become fertile at the apex and repeat the same pattern of branching as the primary stem. The leaves are about 0·35–0·5 mm long, closely overlapping below the comal leaves and on new branches. They are almost triangular or broadly ovate with a small apiculus; the margin practically entire and the nerve failing short of the apex or percurrent. The cells are mostly quadrate or rhomboidal or very shortly rectangular, *c* 8 μm wide in the upper half of the leaf, slightly wider below.

The perichaetial leaves, 5–8 in number and 1·0–1·5 (–2) mm long, are ovate and suddenly narrowed to a short subula; the margin practically entire and the nerve percurrent (rarely slightly excurrent). The capsule is globose, 340–450 μm, completely sessile with a short thick foot in a cup-shaped vaginula, cleistocarpous with no sign of a dehiscence line or an apiculus; often dark brown to black and shining on top when mature, and readily dislodged from the plant. The calyptra is small and soon shrivels away. The spores are few, variable in number (16–48) angular and very large, to 150 μm in the longest diameter, almost smooth and very yellow because of the oil content. Their outline can be readily seen through the capsule wall.

CHROMOSOME NUMBER: n=13 (VIC, det. I.G.S.)

DISTRIBUTION: SA, VIC; ?endemic.

ILLUSTRATIONS: Plate 14; Stone (1973a), Plates 23, 24, Figs 55–57.

A. clavatum I. G. Stone, a second species, appears to be very rare, growing amongst other bryophytes, lichens and blue-green algae on hard gravelly soil in depressions in rock. It differs from *stellatum*: in the more strongly julaceous and fewer club-shaped innovations, the leaves wider than long, the apices obtuse and sometimes almost cucullate, the margins cristate-denticulate.

At least some of the perichaetial leaves have an obtuse or truncate apex and the margins are cristate-denticulate. The capsule measures *c* 550 μm in diameter, the finely granular spores 160 μm.

DISTRIBUTION: VIC; endemic.

ILLUSTRATIONS: Plate 14; Stone (1973a), Plate 25, Figs 58–60.

A. stolonaceum C.Muell., on examination proves to be *Eccremidium pulchellum* [16, q.v.] (Stone, 1973a). *A. rothii* Watts ex Roth. is a northern species (QLD).

DITRICHACEAE†

Includes: *Garckea flexuosa* (Griff.) Marg. & Nork. (QLD) = *G. comosa* of the Index *fide* Margadant and Norkett (1973). *J. Bryol.* **7**, 439–441.

Ditrichum Hampe

This is not altogether an easy genus taxonomically although there are several well defined and easily recognizable species. It is liable to confusion with *Dicranella* [24] and sometimes with *Distichium*. The leaves of *Distichium* are usually clearly distichous and the abaxial surface of the nerve is clearly papillose, but *Dicranella* cannot always be separated without capsules. The capsules are erect, rather wide and short and deeply grooved in most Dicranellas, with a long operculum. In *Ditrichum* the operculum is relatively short, the capsule smooth, long and narrow. Typically, the ditrichoid peristome has the peristome teeth split to the base into two papillose, hair-like segments, whereas the dicranoid peristome of *Dicranella* has them divided only to half-way, often incompletely, i.e. with perforations rather than complete splits from the tip, and the teeth have transverse bars. But the distinction is a weak one and intermediate conditions are common. A thorough study of ditrichoid and dicranoid peristomes with the scanning electron microscope to reveal the fine detail is much needed.

Key to species

1. Leaf base suddenly contracted to the subula, with specialized small dark cells at the shoulder. Cells of subula all short, from 1×1 to 2×1 2

   Leaf base not abruptly contracted, not forming distinct shoulders, without specialized cells. Subula cells, except for the marginal rows, long and narrow (6×1), at least in parts 3

2. Leaf subula usually tightly corkscrewed at the tip when dry. Capsule relatively short and wide, 3–5 × 1, brown . . . . *punctulatum*

 Leaf subula only lightly twisted. Capsule very long and narrow, 7–10 × 1, pale straw-coloured except at base and apex . . . . *cylindricarpum*

3. Capsule narrowly ovate in profile, with curved sides, flat when dry, tapering to the mouth. Leaves usually very long and silky, 3–5 mm. Lowland to montane plant; very common . . *difficile*

 Capsule shortly cylindrical, with straight sides, scarcely narrowed to the mouth, not flattened when dry. Leaves shorter and stiffer, *c* 2·0–2·5 mm. Rare alpine plant . . . . *rufo–aureum*

D. difficile (Dub.) Fleisch. (=*flexifolium* of the Handbook)

By far the commonest species, this plant is to be found on damp roadside clay banks and damp exposed soil generally. It fruits abundantly and is one of the mosses most likely to be collected by beginners. The stems form dense soft greenish or yellow-green patches with spreading, rather wavy leaves which usually have a very long flexuose subula from a wide base. On the whole the base is narrowed to the subula quickly but not abruptly enough to create the rounded shoulders of *cylindricarpum*. Apart from the marginal row, the cells of the subula are long and narrow, *c* 30 × 5 μm, although this is sometimes very hard to see. The cells in the leaf-base are narrowly rectangular and there is no hyaline area at the margins.

Sporophytes are usually abundant and characteristic: very long setas, 2·5–3·0 cm or more, asymmetric ovate capsules (carried obliquely) which are flattened when empty, often with a slight spiral twist. In profile the lower side of the capsule tends to be almost straight and the upper one strongly curved. The capsule is 2·0–3·5 × 1 mm, and is not unlike that of *Ceratodon* in outline, but not deeply grooved. The 16 peristome teeth are long and slender, *c* 0·5 mm or more, and divided to a short basal membrane into two papillose hair-like divisions. They are very fragile and soon break off. Spores are 12–15 μm, finely warted on the surface. The inflorescence

PLATE 15. *Ditrichum difficile* VIC—Vegetative and fruiting shoots × 11, cells of subula (above) and sheathing base (below) × 1000

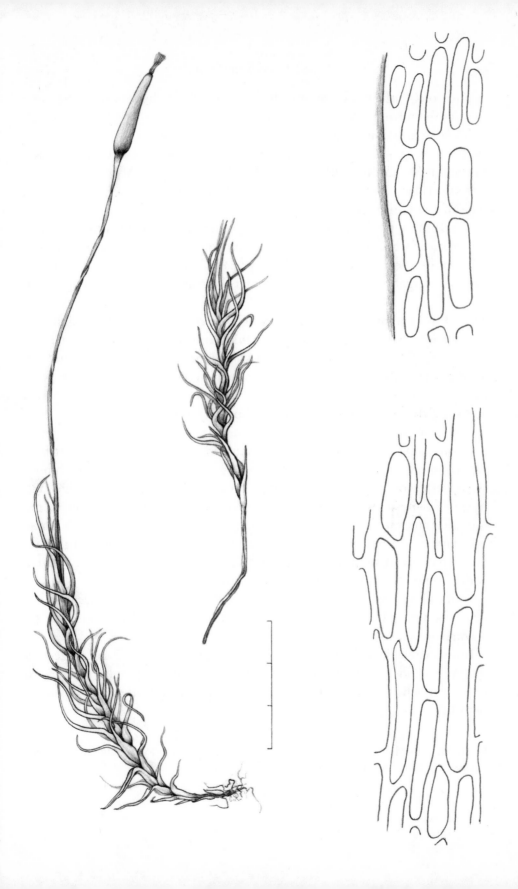

is autoicous with tiny male branches in the axils of the leaves on the same stem as the terminal female perichaetia.

CHROMOSOME NUMBER: n=13, 13+m (NSW).

DISTRIBUTION: TAS, WA, SA, VIC, NSW, ACT, QLD; not yet known from NT, but otherwise throughout. Also in New Zealand, Asia, Africa.

ILLUSTRATIONS: Plate 15; Wilson (1854), Plate 84 (as *Trichostomum setosum*); Allison and Child (1971), Plate 9.

While the typical flattened and asymmetric capsule is unmistakable there are many forms which have larger, more symmetrical and cylindrical capsules on shorter setas. These are especially common in upland areas and need investigating. It is not unlikely that there are several species involved.

D. cylindricarpum (C.Muell.) F.Muell. (=*elongatum* of the Handbook)
When fruiting this is easily distinguishable from *difficile* by the very narrow capsule (2–3 × 0·3–0·4 mm), pale straw-coloured except at the blackish mouth and neck. The seta, 1–2 cm, is straw-coloured above and reddish below. Vegetatively it differs from *difficile* in the shape of the leaves which are abruptly contracted to the long subula (3–6 mm long), making conspicuous shoulders at the junction. The subula cells are short, mostly about 2 × 1 but sometimes quadrate, occasionally rough. The cells in the leaf-base are narrowly rectangular with thick walls next to the nerve, and thinner walls farther out; towards the margin they are often very thin-walled and hyaline, sometimes forming a very wide hyaline band. The cells at the shoulder are very thick-walled, isodiametric or over-square, with dense and dark contents. The leaf is channelled throughout and the margins are denticulate from shoulder to leaf tip. When dry, the leaves usually have a gentle spiral curl, but are not tightly twisted.

DISTRIBUTION: TAS, VIC, ACT; also in New Zealand and S. America.

ILLUSTRATIONS: Plate 16; Studies, Plate 5; Wilson (1859), Plate 173 (as *Trichostomum elongatum*).

Immediately identifiable when fruiting, this species is characterized by the very slender capsule and by the short cells of the subula. It seems to occupy much the same sorts of habitat as *difficile* and often grows mixed with it.

PLATE 16. *D. cylindricarpum* VIC—Fruiting shoot × 7, leaf × 50, cells of sheathing base (top left) and subula (bottom right) × 1000

D. punctulatum Mitt. is characterized by leaves like *cylindricarpum* but with the tip of the subula golden and corkscrewed. The capsule is short, erect and symmetrical, $1 \cdot 0 - 1 \cdot 5 \times 0 \cdot 3 - 0 \cdot 4$ mm. This species, or a very similar one, occurs in upland areas in Victoria.

DISTRIBUTION: TAS, VIC; also in New Zealand.

ILLUSTRATIONS: Studies, Plate 5; Handbook, Plate 10.

The corkscrew twist of the subula is said to be diagnostic (Handbook) but seems to be shared to some extent by other species.

D. rufo-aureum (Hampe) Willis was re-discovered by Willis in the Victorian Alps. It is characterized by the small, almost symmetrical brown capsules, $1 \cdot 0 - 1 \cdot 5$ (–2) $\times 0 \cdot 4 - 0 \cdot 5$ mm or even shorter, with short papillose, incompletely perforate peristome teeth. The leaf subula has elongate cells like *difficile*. The spores are 13–20 µm.

DISTRIBUTION: VIC, ?NSW; also in New Zealand (included with *D. calcareum* in the Handbook).

ILLUSTRATIONS: Willis (1955c), p. 10.

The relationship of this species to the smaller forms with erect capsules, apparently referable to *difficile*, is not yet clear. This may turn out to be quite a widespread species in alpine regions. Some of the taxonomic problems could be explained by hybridization but proof of this is lacking.

D. strictum (Hook.f. & Wils.) Hampe is a very dark, straight-leaved species confined to alpine areas of TAS, ?VIC, and NSW. It would not be easy to recognize as a *Ditrichum* in the absence of capsules, but it is apparently very rare. It is sparsely distributed throughout the southern hemisphere.

ILLUSTRATIONS: Wilson and Hooker (1845), Plate 59 (as *Lophiodon*).

D. calcareum (R. Brown ter.) Broth., similar to *D. rufo-aureum*, may be conspecific with *D. subbrachycarpum* (C.Muell.) Par., but Willis (1957a, p. 24) makes a good case for treating the identity with caution.

D. brevirostre (R. Brown ter.) Broth. has recently been found in subalpine Victoria. It is distinguished by the small size (2–5 mm) and regularly falcate leaves.

The remaining Australian species given by the Index are:

D. brachycarpum Hampe (NSW)

D. muelleri (Hampe) Hampe (VIC, NSW, QLD)

D. semi-lunare (C.Muell.) Par. (VIC)

D. strictiusculum (C.Muell.) Par. (TAS)

D. subcapillaceum (C.Muell.) Watts & Whitelegge (TAS) one would expect
to be *Distichium capillaceum* (Watts and Whitelegge, 1902, p. 38)

D. viride (C.Muell.) Par. (NSW, QLD)

Distichium capillaceum (Hedw.) B.S.G. is an exclusively alpine moss
in our region, forming dense soft cushions rather like *Ditrichum* spp. but
the leaves are truly distichous, the stems tomentose and the subula cells
short. The back of the nerve is papillose in the subula which is a diagnostic
feature according to the Handbook, although *Ditrichum punctulatum* and
D. cylindricarpum can show the same feature.

DISTRIBUTION: TAS, NSW, possibly also in VIC; almost world-wide. (See
Willis 1955d, p. 76).

ILLUSTRATIONS: Handbook, Plate 11; Studies, Plate 5.

Ceratodon Brid.

C. purpureus (Hedw.) Brid.

This ubiquitous but rather nondescript plant is never very easy to recognize
in the field except when fruiting. Under the microscope the leaf and cell
characteristics usually make it quite easy to identify. It can grow in a wide
variety of habitats, mostly on rather dry gravelly or sandy soils but not
uncommonly on rotting wood in dry or wet sclerophyll forest. It is particularly common on a wide range of substrates, about 1–2 years after fire.

The stems vary in height from 0·5 to 2 cm, of which roughly the upper
0·5 cm is current season's growth, forming open or dense turfs or sometimes
cushions. The leaves when dry usually have a slight twist, rather like a
Barbula, and frequently have a marked reddish tint although the colour is
very variable. Under the microscope the leaf shows a consistent set of
characteristics: rather narrowly triangular shape, 1·5–2·0 mm long, contracted at the base; strong nerve reaching the apex or slightly excurrent;
margins revolute for almost their whole length and usually lightly toothed
just at the apex; cells all smooth, almost square and arranged in regular

longitudinal files in the lower half or more of the leaf. The cells are about 10 μm across in mid-leaf but towards the apex are often slightly *larger* and more elongated, up to 13–15 μm long, and much less regular in shape and arrangement. Those at the base are distinctly elongated, 12–50 × 10 μm, but very varied.

Capsules are highly distinctive, 2–3 mm long, held horizontally on the end of a long (2–4 cm) seta, curved downwards and with deep longitudinal grooves. The seta varies from yellow to dark red (commonly burgundy coloured, hence the name) and the capsule from brown to very deep red, sometimes almost black, becoming paler and faded with age; sheathed at the base by a tube of large broad perichaetial leaves. Usually the setas are present in great numbers in the colony; with changes of humidity they twist and untwist hygroscopically. This movement helps to scrape off the sharply conical opercula and to jerk the capsules, helping in spore discharge. The peristome teeth are incurved at the tip, like a lobster trap, the spores sifting out between the sides of the teeth which part and close-up hygroscopically. Possibly the contraction of the grooves in the capsule at maturity also helps to squeeze out the spores.

CHROMOSOME NUMBER: n=13 (NSW)

DISTRIBUTION: Apparently throughout all parts of Australia; virtually in all other parts of the world.

ILLUSTRATIONS: Plate 17; Handbook, Plate 12; Watson (1955), Fig. 21.

Barbula spp. [46] which perhaps resemble this species most closely, differ in the entire apex, papillose cells and of course in the capsules.

C. crassinervis Lorentz (TAS) and *C. stenocarpus* Bruch & Schimp. ex C.Muell. (TAS, VIC, NSW, QLD) have both been recorded for Australia. In view of the striking variability of *C. purpureus* they should be considered tentative.

Eccremidium Wils.

The plants of this widely misunderstood genus are mostly tiny, usually forming a low open or compact turf on bare ground; the stems varying

PLATE 17. *Ceratodon purpureus* ACT—Fruiting shoot × 15, leaf × 50, cells from top, middle and base of leaf (top to bottom) × 1000

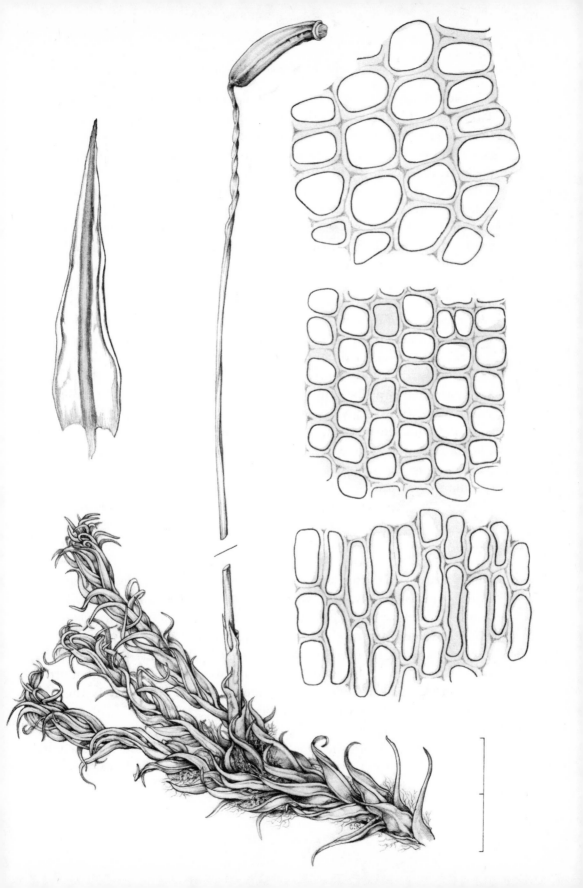

from less than 1 mm to a few mm tall, simple or with a few branches. The leaves, which are mostly single-nerved, fall broadly into two categories: small, ovate, appressed and overlapping on the stem (julaceous) or else lanceolate or narrowly lanceolate extending into a long narrow subula; the margins are entire to variously crenulate or serrulate by projecting cells. The cells are smooth, more or less isodiametric in julaceous species, oblong or irregularly hexagonal or rhomboidal in the long-leaved species and in the perichaetial leaves. The perichaetial leaves in the julaceous species are much larger than the vegetative leaves and contracted to a subula which is either falcate or straight. There is a short, usually thick seta, arcuate to straight, emerging from a cylindrical or elliptic vaginula. The capsule is globose or pear-shaped with an equatorial or nearly equatorial dehiscence line of one or two rows of narrow, elongated, thin-walled cells and with stomata confined to the lower half; peristome absent; operculum domed, with a usually small thick apiculus; exothecial cells and opercular cells usually thickened and often very thick-walled with only a small unthickened area in the centre of the outer wall which may be distended as a shallow papilla; calyptra conic or widely bell-shaped and lobed at the base, and with remains of the archegonial neck often persisting at the top. The spores are large to extremely large and when mature have a tessellated pattern composed of polygonal areas separated by grooves (the negative reticulum of Erdtman, 1943, pp. 51–2).

There are six species in the genus: three in the subgenus *Eccremidium*, which are readily recognized when fruiting by the short, thick, arcuate seta and more or less rounded, obovate, pear-shaped capsule which is operculate and pendulous. The outermost cells of capsule wall and operculum usually have characteristic collenchymatous thickenings. *Eccremidium arcuatum* and *pulchellum* have very julaceous stems with small, very broad, concave, closely appressed leaves and could be mistaken for *Pleuridium* [17] which however has a much broader, longer and stronger nerve and usually longer pointed leaves. *Eccremidium exiguum* is like a tiny *Ephemerum* [64] with long fine leaves but without the copious protonema, and often grows in damp clay pans covered with filamentous algae. Fortunately there are usually capsules to indicate its affinities.

The other three species, in the subgenus *Pseudo-pleuridium* Broth., have very short setas slightly bent near the apex when dry, and upright when moist, but this is not a very well defined character and is variable. The operculate capsule is globose with a small apiculus, but is not pendulous as in the other species. The calyptra is always widely bell-shaped and lobed. Although Brotherus describes the capsule wall as thin in *whiteleggei* some of the type material in Herb. BM shows somewhat thickened walls and, in specimens from WA, which are attributable to this species, the characteristic collenchymatous thickening is present. The stems are not julaceous; those of *brisbanicum* and *whiteleggei* are a few mm long while *minutum* is practically stemless and is distinguished from *exiguum* by the erect capsule without collenchymatous thickenings. The upper leaves are mostly long and subulate, with or without a nerve.

Key to species

1. Sporophytes present 3
   Sporophytes absent 2

2. Stems julaceous 5
   Stems not julaceous . . . . (not certainly identifiable)

3. Seta strongly arcuate; capsule pendulous. Stems sometimes julaceous. Leaves very short or long and narrow 4
   Seta not arcuate; capsule erect. Stems never julaceous. Leaves long and narrow 6

4. Stems less than 1 mm, not julaceous. Leaves with a long fine nerve-filled subula . . . . . *exiguum*
   Stems 1 mm or more, julaceous, with tiny, appressed, closely overlapping leaves with nerve usually ending short of apex 5

5. Stems 1–3 mm, unbranched, sterile stems markedly triquetrous. Leaves obtuse and serrulate; perichaetial leaves shortly subulate, often truncate at apex and falcate. No propagules . . . . *arcuatum*
   Stems 3–5 (10) mm, often with 2–3 erect branches, not triquetrous. Leaves usually apiculate, entire or minutely denticulate; perichaetial leaves long subulate, pointed (rarely falcate), rarely fruiting but often with specialized deciduous branchlets . . . . *pulchellum*

6.   Stems less than 1 mm. Spores very variable in size   *minutum*
     Stems usually 1–5 mm. Spores 100 μm, often much
     more   7

7.   Stems lax, often tortuous. Leaves distant, nerveless
     or almost so, concave, ovate–lanceolate to
     narrowly acuminate, spreading .   .   .   .   .   *whiteleggei*
     Stems straight. Leaves spreading, not distant, with
     overlapping bases, nerved into the long point.   *brisbanicum*

E. arcuatum (Hook.f. & Wils.) C.Muell.

These plants usually have a distinctive greenish gold sheen and are red–gold
at the base of the stem. The triquetrous appearance is particularly evident
in sterile plants and the fertile plants are markedly arcuate, a feature further
emphasized by the curved seta and pendulous capsule. The stems are
typically simple with concave leaves increasing in size up the stem from the
base, 0·2–0·7 mm long, broadly oval, obtuse to cucullate, and slightly
keeled in the upper part. The larger comal leaves, *c* 0·9 mm, have a drawn-
out apex which is often truncate. The margin has a peculiar cristate ser-
rulation with the cells extended into curled, finger-like projections, but
sometimes inflexed so much that the margin appears entire. The narrow
nerve usually finishes short of the apex. The cells are shortly rectangular
below, 20–40 × 10 (–15) μm and thin-walled; those above, 8–12 μm wide
and irregularly hexagonal or rhomboidal, usually with thicker walls.

The perichaetial leaves are usually falcate, and curved all to one side, up
to 1·4 mm long, broadly lanceolate and concave at the base and then
suddenly narrowed to a flat subula less than one-third the length of the
leaf and mostly truncate at the apex (occasionally acute) and serrate.

The red-brown capsule, after losing the operculum, is shallow with a
wide mouth. The yellowish calyptra is bell-shaped, flared and usually lobed
at the base, covering the capsule down to the line of dehiscence. The spores
are large, *c* 100 μm, brown.

PLATE 18. *Eccremidium pulchellum*—(a) (TYPE) WA. Fruiting plant with 2 immature
capsules × 22; (b) VIC. Vegetative shoot × 22 and cells × 1000; (c) WA. Plant
with mature capsule × 22, section through leaf nerve × 500; (d) *E. arcuatum*
VIC—Fertile and vegetative shoots × 22, cells from leaf margin of vegetative
shoot × 1000; (e) *E. exiguum* NSW—Fruiting plant × 22

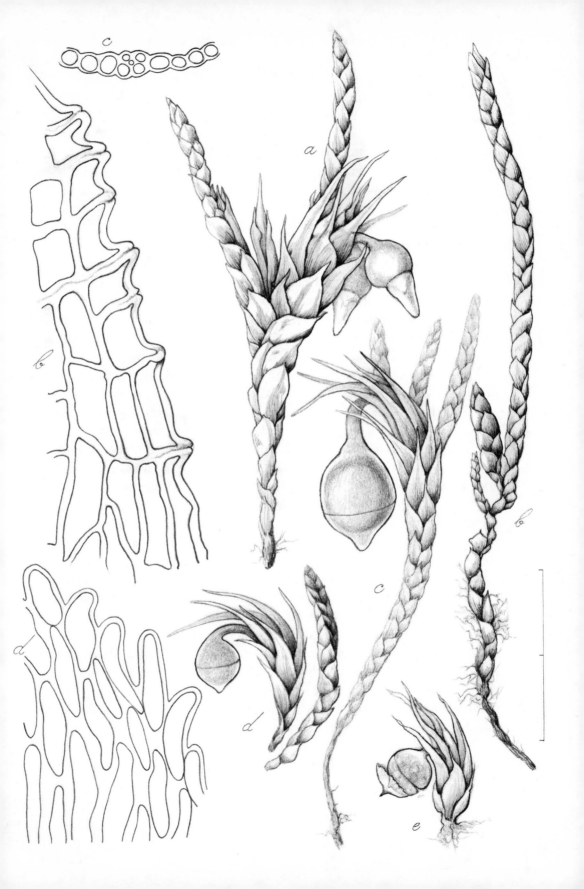

DISTRIBUTION: ?TAS, WA, SA, VIC, NSW, ?ACT, QLD, ?NT; ?endemic.

ILLUSTRATIONS: Plate 18; Hooker, W. J. (1840), Plate 738 (as *Schistidium*).

This species is commonest on sloping ground in dry sclerophyll forest, especially in Ordovician gold-mining areas of Victoria, and on shallow soils on rock outcrops in Western Australia.

Sterile plants may be distinguished from other julaceous mosses by the peculiar serrated margins of the leaves.

E. pulchellum (Hook.f. & Wils.) C.Muell.

Unlike *arcuatum*, the stems branch, usually by 2 or 3 erect innovations from below the perichaetium or from the base. The leaves are concave, broadly ovate, apiculate or rarely rounded, with the margins entire to serrate. The serrations, when present, are formed—unlike those of *arcuatum*—more by the collapse or erosion of the outer cell wall giving a similar appearance to the margin of some leaves of *Pleuridium nervosum*, *Archidium clavatum* and other species. The nerve is weak, not reaching the apex, and the cells in mid-leaf are irregularly rhomboidal with thick or thin walls.

The perichaetial leaves are ovate–lanceolate, long acuminate or subulate, sometimes straight, sometimes arched; the margin is entire and the nerve fills the subula and is occasionally excurrent. The capsule differs from that of *arcuatum* as the dehiscence line is nearer the apex giving a narrower mouth and smaller operculum. The deep-yellow calyptra is firmer, conical, and not flared or crenate at the base. The spores are not as large, 60–80 μm, and sometimes appear aborted.

DISTRIBUTION: TAS, WA, SA, VIC, NSW, ?ACT, QLD, ?NT; probably also in New Zealand.

ILLUSTRATIONS: Plate 18; Hooker, W. J. (1840), Plate 738 (as *Schistidium*).

This is a common and very widely distributed species often mixed with *Pleuridium*, *Archidium* and other *Eccremidium* spp.

Unfortunately the capsules are extremely rare and have only been found in WA; the perichaetia are common but the archegonia abort. Sterile plants are very variable and have sometimes caused considerable trouble to bryologists. A form with long slender filiform branches in WA was named *Anomobryum filescens* by Bartram (1951), but sections of the leaves of part of the type specimen reveal its true identity.

The sterile branches often bear deciduous specialized side or apical shoots which are short and thick, with oil-filled cells, and bear reduced closely appressed often serrated leaves. These shoots may grow out while attached or fall and resume growth on the ground. The plant in this form was suspected by Sainsbury (1935) to be a sterile state of *Pleuridium nervosum*.

The type material of *Archidium stolonaceum* C.Muell., which was examined in the NSW herbarium and at MEL herbarium, also belongs to this taxon (Stone, 1973a).

The safest way to make a diagnosis of sterile material is to section stem leaves and examine the nerve which, in *E. pulchellum*, consists of a pair of cells, each with a large lumen, on both the upper and lower surface enclosing one or two very small central cells (Plate 18, top left).

E. exiguum (Hook.f. & Wils.) Wils. in Salmon

The stems are extremely short—almost immeasurably so—with 2–3 basal branches so that the plants are often in small clusters. The leaves, except for a few very tiny and nerveless ones at the base of the stem, are concave and ovate–lanceolate narrowing to a long to very long, almost setaceous subula, sometimes flexuose and slightly secund. The margin is entire or lightly serrulate by irregular projection of cell ends. The nerve is often weak below but is stronger above, filling the subula. The cells are oblong below, narrower, more rhomboidal and thick-walled above.

The line of dehiscence is about the mid-region of the globular to obovate capsule which is therefore wide-mouthed when the operculum falls. The calyptra is brownish, lobed at the base. The spores vary from 60–120 μm. DISTRIBUTION: WA, ?SA, VIC, NSW; recently found in S. Africa (Drakensberg Mts.; Stone and Schelpe, 1973).
ILLUSTRATIONS: Plate 18; Hooker, W. J. (1840), Plate 737 (as *Phascum*).

This seems to be a rarer plant than the two preceding species. It is found on clay pans in mallee and ironbark forests on occasionally inundated soils. Like *E. whiteleggei*, it shows considerable variation in the grades of thickening of the capsule wall cells; the spores, which are often abortive, are also very variable in size from plant to plant.

E. whiteleggei Broth. (= *Sporledera minutissima* of the Handbook)

This species is most like *exiguum* but with a much longer stem, the seta not distinctly arcuate and the capsule not pendulous. The stems are 3–5 mm high, often ending in a male inflorescence from which one or two female branches arise; lax, and slightly zig-zag with erecto-patent, often very distantly spaced leaves. Green protonema often arises from the male inflorescence and leaf axils. The leaves are concave, ovate–lanceolate to acuminate; sometimes the upper few have a long subula and margins with very small serrations. The nerve is absent or practically so. The cells are rectangular to irregularly 6-sided or rhomboidal and the cell walls not thick.

There is a very short seta, and a vaginula as long as the capsule. The outer capsule walls are thin to firm or even collenchymatous. As with the other species the columella is resorbed during development and at maturity the few extremely large spores fill the capsule. The spores are polyhedral, round, brown and fewer in number than in the other species varying from 4 or 8 to 24 or more. Sometimes there are two capsules from the one perichaetium.

DISTRIBUTION: WA (new record), NSW; also in New Zealand?

ILLUSTRATIONS: Brotherus (1924–5), Fig. 134.

Sporledera minutissima from NZ appears to be the same taxon or at least very close to it, and is certainly neither a *Sporledera* nor a *Bruchia* as its capsule has a line of dehiscence. We include it provisionally with *E. white-leggei*.

There are at least four recorded species in three different genera which have a closer affinity with *Eccremidium* (subgenus *Pseudo-pleuridium*) than with the genera in which they have been placed. *Ephemerum whiteleggei* Broth, & Geh., *Bruchia* (*Sporledera*) *whiteleggei* C.Muell., *Bruchia minuta* Mitt. and *Micromitrium* (*Nanomitrium*) *brisbanicum* (Broth.) Crosby have sporophyte and gametophyte features which clearly indicate that they belong to the same genus; the first three of them, indeed, are probably conspecific (*E. minutum* (Mitt.) Stone & Scott). They have had several changes of generic name indicating their troublesome nature but evidently the differentiated dehiscence line on the capsule has not been observed except in *M. brisbanicum* (=*E. brisbanicum* (Broth.) Stone & Scott). All these show the following characteristics: the stem is almost absent or a few millimetres long; the leaves are very variable in length, usually very small, concave and ovate, at the base of the stem, much longer upwards with a concave–lanceolate base and a subula of variable length, often flexuose and subsecund and variably serrate through the projection of cell angles.

The cells are usually more or less rectangular below, irregularly rhomboidal to 6-sided above, 12–20 μm wide, mostly 4–5× as long as wide, sometimes much longer; usually thin-walled below and thin to firm or thick-walled in the upper lamina.

The sporophyte is of the typical *Pseudo-pleuridium* form.

We prefer to place these two species or species groups in *Eccremidium* until a closer anatomical study and estimation of the degree of variability from a larger number of populations have been considered.

E. brisbanicum (Broth.) Stone & Scott

The stems, usually 2–3 mm long, may be bare below except for a few distant tiny ovate leaves or have leaves increasing in length up the stem, in both cases with an upper tuft of long narrow subulate finely pointed leaves which are straight or slightly curved back, the longest 2·0–2·5 mm. These upper leaves are more or less concave–lanceolate in their basal part, *c* 250 μm wide at the broadest region, and taper gradually to the tip which is usually only *c* 25 μm wide. The nerve is mostly quite clear but may be very weak near the base. The margins are plane and entire or almost so but the leaf tips sometimes slightly rough by projecting cell ends. The spores measure 100–140 μm.

DISTRIBUTION: NSW, QLD; ?endemic. *Micromitrium neocaledonicum* may be conspecific.

ILLUSTRATIONS: Roth (1911), Plate 11 (as *Archidium brisbanicum*). It occurs in sandy coastal districts from Sydney northwards.

E. minutum (Mitt.) Stone & Scott

This species differs from *brisbanicum* in the extremely short stem, usually much less than 1 mm long, the leaves less finely pointed and rarely as long, 1·5–2·0 mm × 250 μm tapering to about 50 μm at the apex, sometimes straight but more often flexuose and subsecund. The leaf tips are variably serrated, often conspicuously so, by projecting cell ends and the margins may be somewhat crenulate even well down the leaf; the nerve is often obscure.

The calyptra is occasionally rough with projecting cell ends and the spores are extremely variable in size—from 50 to more than 100 μm—and often aborted.

DISTRIBUTION: TAS, WA, SA, VIC, NSW, QLD; also New Zealand.

ILLUSTRATIONS: Wilson (1859), Plate 171 (as *Bruchia minuta*); Roth (1911), Plate 13 (as *Bruchia (Sporledera) whiteleggei*), Plate 23 (as *Ephemerum whiteleggei*); Roth (1914), Plate 10 (as *Bruchia (Sporledera) minuta*).

This plant could readily be mistaken for an *Ephemerum* [64] but careful observation will reveal the narrow line of dehiscence on the capsule. It grows on bare ground in similar situations to *Eccremidium exiguum*, but commoner, often mixed with other small earth mosses like *Acaulon, Bryobartramia* etc., on clay pans and clay banks and in soil-filled, periodically waterlogged depressions in rock, coastal or inland.

Pleuridium Brid.

These tiny mosses, with stems usually less than 5 mm long, grow densely crowded on bare earth. The leaves are single-nerved and have smooth elongate cells which are larger in the leaf base. The leaves, which are unaltered when dry, increase in length up the stem, with the lanceolate–subulate perichaetial leaves the longest. The capsule, on a very short seta, has stomata at the base and is cleistocarpous with no dehiscence line, partly covered by a side-split calyptra.

P. nervosum (Hook.) Mitt.

These plants tend to form a dense turf on bare ground in a wide range of climates, the stems 2–5 mm high, julaceous and erect. The leaves are erect, concave, closely imbricated, ovate to ovate–lanceolate with a wide insertion and usually acute apex, *c* 0·5 mm long increasing in size up the stem, the upper more spreading. The margin is entire or irregularly crenate–denticulate, the nerve wide and strong except in the lowest leaves, usually percurrent to excurrent. The perichaetial leaves are much larger, 2–3 mm long, lanceolate–subulate, concave at the base, the nerve filling the subula. Cells in mid-leaf are *c* 10–12 μm wide and 3–5 × 1.

The capsule is erect, *c* 1 mm, ovoid to almost globose, apiculate with a short obtuse beak, shining red-brown or orange when mature. The calyptra is hood-like, covering about half the capsule; the spores papillose, 25–35 μm. DISTRIBUTION: TAS, WA, SA, VIC, NSW, ACT, QLD, NT, ?Lord Howe; also New Zealand, S. Africa.
ILLUSTRATIONS: Plate 19; Handbook, Plate 9; Roth (1911), Plates 14, 21.

PLATE 19. *Pleuridium nervosum* VIC—Fruiting plant × 30, cells × 1000; *P. arnoldii* VIC—Fruiting plant × 30 (top)

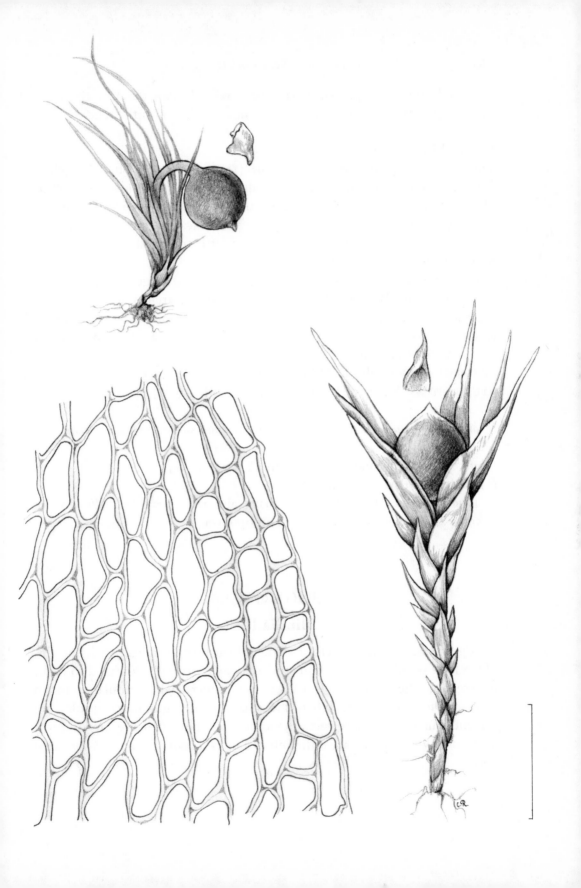

Several other similar species have been recorded for the area and more study is necessary to determine the extent of variation in the species.

P. arnoldii (R. Brown ter.) Par.
These are tiny plants with very short (1–2 mm) stems, often branched, forming dense silky tufts on earth in alpine regions. The lower leaves are the shortest, the upper lanceolate–subulate narrowing from the concave base to a long fine subula, straight or slightly falcate in any direction. The margin is entire except in the subula where it is frequently serrate and the nerve almost fills the subula.

The perichaetial leaves are similar but with a longer often flexuose subula to 2 mm. Because of the short and strongly arched seta the capsule protrudes from between the leaves near the base of the plant. The capsule is almost globose with a rather long often curved beak, and granular papillose spores, *c* 30 μm.
DISTRIBUTION: VIC; also in New Zealand.
ILLUSTRATIONS: Plate 19; Studies, Plate 5; Roth (1911), Plate 16.

P. austro–subulatum Broth. ex Roth. is *Eccremidium brisbanicum* (Stone and Scott, 1973).
P. brachycaulon (C.Muell.) Kindb. (NSW)
P. krauseanum Par. (SA, VIC, NSW)
P. viride (C.Muell.) Kindb. (NSW) have also been recorded. We do not know them well enough to form an opinion.

SELIGERIACEAE

Blindia B.S.G.

This genus of greenish black or dark golden brown mosses with the tips of the long subulate leaves often eroded away is virtually confined to rocks in streams, often submerged. There is a superficial resemblance to an aquatic *Dicranum* except when the small usually globular capsules are present, but in transverse section the cells of the leaf nerve are uniform in *Blindia*, whereas they are differentiated in *Dicranum* [20] and *Dicranoloma* [25].

Key to species

1. Plants small, usually less than 3 cm. Leaves 1·5–
   3·0 mm long   2
   Plants robust, usually more than 3 cm (often 6–12).
   Leaves 4–9 mm long   3

2. Alar cells distinct. Seta cygneous when wet . .   *magellanica*
   Alar cells not distinct. Seta straight. . . . .   *ferruginea*

3. Alar cells distinct. Operculum usually separate
   from columella; spores *c* 20 μm . . . . .   *robusta*
   Alar cells not distinct. Operculum adherent to
   columella; spores 28–32 μm . . . . . .   *tasmanica*

B. robusta Hampe (=*tenuifolia* of the Handbook)

This is often an abundant plant in sub-alpine to alpine areas, submerged in streams and pools or sometimes exposed on wet rocks, forming large golden brown to blackish green patches. The flexuose stems are 5–12 cm long, usually not much branched but densely leafy. The leaves, especially the apical ones, are normally strongly falcate to circinate *c* 5(–10) mm long, gradually tapering from a concave lanceolate base to a long, very fine, channelled, entire subula with an acute apex. The cells have a very similar appearance to those of *Dicranum* and *Dicranoloma*, very long and very narrow, (20–) 30–50 × *c* 7 μm at the top of the subula, 35–70 × 6–10 μm in the mid-subula; those below longer (50–) 100–150 × 8–10 μm and often very thick-walled, the wall thickenings sometimes slightly sinuose with the cavity reduced to a mere slit. A group of rather square, coloured cells at the angles form a conspicuous alar group, thickish-walled but often collapsing when dry. The nerve is wide, but sometimes indistinct at the base, *c* 60–80 μm, and extends into the subula.

The seta is straight or nearly so, yellowish brown, rather thick, 1–2 cm long. The capsule is highly characteristic, blackish and more or less globular or turnip-shaped, wide-mouthed when empty, with 16 short red broadly lanceolate peristome teeth, split above. The operculum is low-conic with a very long oblique beak; the calyptra large, covering the capsule and side-split.

DISTRIBUTION: TAS, VIC, NSW, ACT; S. America and New Zealand.

ILLUSTRATIONS: Studies, Plate 6; Mueller, F. (1864), Plate 7.

When fruiting this species is charming; the calyptra brown above and white below, these colours reversed on the operculum. When not fruiting it can be recognized most easily by the colour and habitat. Its closest resemblance is to some forms of *Dicranoloma billardieri* but that species is usually distinctly toothed in the subula and has a margin of hyaline cells in the sheathing base. A more frequent source of confusion is with [120] *Drepanocladus fluitans* (and perhaps *aduncus*) which has a similar colour, habitat and falcate leaves, but differs in being usually pinnately branched, and in the much less thick-walled cells and of course in the lateral instead of terminal fruit.

B. magellanica Schimp.

This small species was recently found for the first time in Victoria in a mountain stream on rocks either submerged or exposed (according to the Handbook not submerged in New Zealand). The brownish green leaves in our specimens are slightly falcate or secund at the stem tips, mostly straighter and darker below. The subula is not much longer than the sheathing base. The cells which may be thick-walled are distinctly *shorter* than those of *robusta*; those at the top of the obtusely tipped subula 10 (–15) × 10 μm, sometimes oblique, in mid-subula 10–20 × 8 μm and in the sheath 30–50 × 6–8 μm. The alar cells are inflated and conspicuous, and thin- or firm-walled, more or less square. The lamina may be 2 cells thick in the subula. The subula cells are shorter and thicker-walled than those described in the Handbook but the areolation agrees very well with New Zealand specimens in Herb. NSW. Whether the leaves are straight or slightly falcate is a variable feature, as is the golden or blackish colouration.

The perichaetial leaves are broader than vegetative leaves at the base and narrow more suddenly to the long subula, and the extreme apex is denticulate with short projections of the cells. The seta, *c* 3 mm long, is *cygneous* when wet, flexuose and more or less erect when dry; the tiny dark or light brown capsule, subglobose, occasionally flaring at the mouth when empty; teeth 16, fragile; operculum with a curved beak. Dioicous.

DISTRIBUTION: TAS, VIC; also New Zealand.
ILLUSTRATIONS: Studies, Plate 6.

The other recorded species are:

B. ferruginea (Mitt. in Wils.) Broth., (TAS, VIC, NSW), which is similar to *magellanica* but the alar cells are not differentiated and the seta is straight. It is illustrated in Wilson (1859), Plate 172 (as *Dicranum*).

B. tasmanica Sainsb., TAS endemic, apparently differs from *B. robusta* in the nerve being much stronger, alars not differentiated and the operculum adherent to the columella.

Brachydontium B.S.G.

B. intermedium I. G. Stone

These tiny plants grow in crevices and attached to the upper surface of cavities in non-calcareous rocks in alpine country, so far only found on Bogong High Plains.

The stems are less than 1 mm long, occasionally with one or two short branches at the base; the fine leaves are crowded, a few below very small and oval, the longest at the top 1·5–1·8 mm, lanceolate, somewhat sheathing at the base, narrowing above to a long fine terete subula 30–60 μm wide, with a rounded apex. The nerve almost fills the subula and is obscured by lamina cells on both surfaces. The cells of the leaf base are 10–12 μm wide and 2–4 × 1, those of the subula above are more or less quadrate, thick-walled with irregularly rounded cavities, 6–8 μm wide.

There is a yellow seta, 2·0–2·5 (–3) mm long, cygneous when moist. The capsule is symmetrical, pale brown, oval, ridged, 300–500 μm long; the operculum yellow, low-convex with a straight subulate beak *c* 250 μm long, truncate at the apex and often falling with the calyptra; the annulus is large and red, curling off at maturity; the peristome imperfect, usually of 16, occasionally bifurcated often truncate teeth. Spores are 10–12 (–15) μm with a brain-like pattern of convoluted ridges (under oil immersion lens). The calyptra is symmetrical, narrowly conical above, broadened and with splits at the base.

DISTRIBUTION: VIC; endemic.

ILLUSTRATIONS: Stone (1973b).

The only record of the genus in the southern hemisphere, this species is obviously very closely related to the other two species, one in Granada, S. America and in central Africa, the other more widely spread in mountains of Europe and North America. When moist and fruiting, it is rather like a tiny *Ptychomitrium* [52].

DICRANACEAE

Includes: *Campylopodium euphorocladum* (C.Muell.) Besch. (TAS)

C. lineare (Mitt.) Dix. (TAS)

Dicranodontium tapes (C.Muell.) Par. (TAS)

?*Kiaeria starkei* (Web. & Mohr) Hag.

Leucoloma bauerae (C.Muell.) Par. (QLD)

L. circinatulum Bartr. (QLD)

L. leichhardtii (Hampe) Jaeg. (VIC, NSW, QLD)

Microdus tenax (C.Muell.) Par. (QLD)

Pseudephemerum nitidum (Hedw.) Reim. (NT. Recorded by Catcheside (1958) as *P. axillare* but omitted in the Index.)

Wilsoniella karsteniana C.Muell. (QLD)

Dicranum Hedw.

The differences between this genus and *Dicranoloma* [25] are as much those of general appearance and stature as any clearly defined characteristics. Most Dicranums are rather small plants with leaves which are usually unbordered, whereas Dicranolomas tend to be robust with usually bordered leaves, but the distinction between the two genera is neither logical nor fully satisfactory. There appears to be a difference in the colour and texture of the cell contents so that electron microscope studies of cell structure might be expected to reveal more convincing differences. The peristomes of the two genera also are slightly different and so is the general size and morphology of the capsule. The leaves of *Dicranum* are slender, tapering, usually falcate–secund, nerved to the apex and with distinct alar cells. The peristome has 16 teeth, divided to almost ½ way.

D. trichopodum Mitt. (=*Dicranoloma* in the Index)
The soft, rather tall cushions of this species reach 4 cm or more in height, of which 0·5–1·0 cm is current yellow-green growth; they are rhizoidal below but scarcely tomentose. The leaves are long and very slender, slightly falcate, *c* 2–4 mm long from a base 1·0 × 0·6 mm broad, tapering to a long and very fine subula most of which is nerve; it is sometimes hyaline at the tip. The nerve is broad below, *c* 100 µm, and both leaf base and base of subula are U–channelled in transverse section. The subula cells are more or less isodiametric, thick-walled and rounded, *c* 10 µm; those in the leaf bases 30–50 × 9 µm, narrowly rectangular, very thick-walled, the walls slightly porose at the extreme base. The alar cells are in a conspicuous group, quadrate, orange and very thick-walled. In the leaf base, there are usually 2–3 rows of marginal cells, which are narrow, thick-walled and colourless, forming a fairly distinct border.

We have not seen sporophytes in Australian material, but the New Zealand plants have erect cylindrical capsules and a short peristome of irregular teeth split at the tips but not deeply.
DISTRIBUTION: TAS; also in New Zealand.
ILLUSTRATIONS: Handbook, Plate 21.

It seems to be confined to subalpine scrub in central Tasmania where it occurs on both ground and small trees. The isodiametric cells separate it from all species of *Dicranoloma* except *menziesii* which has far finer and longer subulate leaves, usually of a much darker green. The genus into which this species should go is a matter of personal judgement on rather inadequate grounds. The general stature of the plant, and shape and size of the capsule suggest *Dicranum* while the border suggests *Dicranoloma*. Sainsbury himself vacillated between the two (1955b, p. 10; Handbook, p. 121). The Index chooses *Dicranoloma* but the cell structure is perhaps more in keeping with *Dicranum* and we have opted for that, without strong convictions either way. The rhizoids are smooth and curly, much as in *Holomitrium* which also has a similar short peristome and long operculum as noted by Sainsbury (Handbook).

D. aucklandicum Dix., treated by the Index as in the genus *Holodontium* (otherwise a subgenus of *Dicranum*) occurs as an epiphyte in the subalpine scrub in the Cradle Mountain area of TAS. It is a tiny plant, the size of *Dicranoweisia* but with very narrow, not curled, leaves and long, narrow cells in the subula. The epithet *falcatum* appears to have priority.

The other species given by the Index are:

D. contortifolium Bartr. (WA, SA)

D. novae-hollandiae Hornsch. ex Card. (Locality only given as "Nova Hollandia".)

D. subviride Mitt. (NSW) is apparently a *nom. nud.*

D. antipodum Hampe, given in the Index Supplement, is a *nom. nud.* probably a synonym of *Campylopus pallidus.*

Bruchia Schwaegr.

B. brevipes Harvey ex Hook.

These small plants of bare earth in low rainfall areas are characterized by the large truncate apophysis, about one third of the capsule length, and by the very large calyptra, lobed at the base and covering most of the capsule. These features distinguish it from *Ephemerum* [64], with which it might be confused. The plants fruit freely and are usually in groups with a slight persistent green protonema.

The stems are short, c 1 mm high, with erecto–patent, straight, stiff leaves increasing in size from below upwards. The leaves have a transparent ovate sheathing base contracted to a long narrow concave subula, rough on the surface with projecting cell ends which may be reflexed. The margin is entire to irregularly denticulate by projecting angles of cells. There is a strong nerve often filling the apex of the long subula which tapers to a single cell. The cells are more or less rectangular, those in the sheath region c 10 μm wide and $3–4 \times 1$, in the subula gradually narrower c 5 μm wide and $6–8 \times 1$.

The seta is pale and much shorter than the capsule which in turn is overtopped by the perichaetial leaves. The capsule is obovoid, shortly apiculate, truncate at the base of the apophysis which is green and about one-third of the capsule length. The outermost cells of the capsule wall contain an orange pigment at maturity and contrast vividly with the green apophysis. The calyptra is translucent, large, bell-shaped and lobed at the base. The spores are 35–45 μm and ornamented with long flattened spines.

Specimens are attributed to *B. drummondii* Hampe in MEL herbarium but the spores are not correct for this species and appear essentially the same as the S. African *B. brevipes*. There is a specimen named *B. brevipes*, collected in WA by J. Drummond (1845), in the British Museum.

DISTRIBUTION: WA, SA, VIC, NSW; also in S. Africa.

ILLUSTRATIONS: Plate 14; Roth (1913), Plate 1; Hooker, W. J. (1840), Vol. 3, Plate 231.

B. amoena C.Muell. has now been transferred to *Trematodon* [29, q.v.].

B. minuta Mitt. in Hook. is *Eccremidium minutum* [16, q.v.].

Campylopus Brid.

This distinctive genus is characterized by the thick and very wide nerve which in some cases almost completely fills the leaf base and is always at least 200 μm wide. The leaf base, which is long and in some species comprises the whole leaf (e.g. *kirkii*), is held stiffly erect, nearly parallel to the stem. The capsules are often abundant, commonly with several capsules together (polysetous), with the seta usually curved like the neck of a feeding swan (cygneous) so that the capsule, at least when immature and when moist, is buried among the leaves. The side-split calyptra is elegantly fringed at the base.

It is a perplexing group of plants in which characteristics usually distinctive of one species may turn up, sporadically at least, in others. Hair-points, alar cells, brittle stem apices, stem tomentum, nerve width and structure, distribution of cell types in the leaf base, are all crucial characteristics upon which species have been based but which are liable to fail and even mislead the taxonomist at the critical moment. Hence the long list of synonyms in the literature; undoubtedly it should be even longer.

Key to species

1. Leaves with a slender subula, sometimes tubular but never boat-shaped at the apex; often with hyaline hair-points 2
   Leaves blunt and cucullate at the apex; no hair-points 5

2. Leaf tips hyaline, squarrose, reflexed strongly out at right angles to the stem . . . . . . *introflexus* p.p.
   Leaf tips hyaline or not, not squarrose 3

3. Leaf tips (except *arboricola*) long and very slender, flexuose when dry. Plants pale green, matted together with red tomentum. Alar cells absent. *pallidus* (+*arboricola*)
   Leaf tips rather stiff. Plants olive or khaki in colour, rather dark. Tomentum usually dark brown or whitish, or absent. Alar cells usually distinct . 4

4. Stem apices fragile, breaking off when rubbed. Alar cells usually conspicuously inflated and orange-red. Nerve moderately wide, not much more than ½ width of leaf base, smooth abaxially. Leaves all erect . . . . *clavatus*
   Stem apices usually not fragile. Alar cells little inflated and not coloured. Nerve very wide, more than ¾ of the leaf base, deeply grooved abaxially. Usually at least a few of the leaves are somewhat squarrose . . . . *introflexus* p.p.

5. Alar cells inflated and coloured. Leaves straight, wide, obtuse. Nerve with lateral branches above *kirkii*
   Alar cells not distinct. Leaves narrowed to a stiff slender point, cucullate at the tip; some leaf tips bent inwards. Nerve not branched . . . . *bicolor*

PLATE 20. *Campylopus introflexus* VIC—Fruiting shoot × 11, leaf × 50, cells × 1000

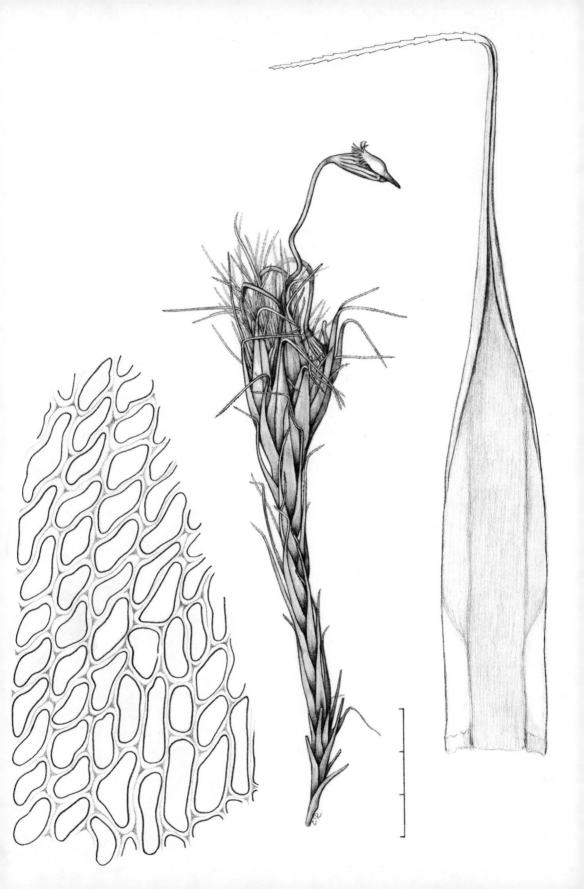

C. introflexus (Hedw.) Brid.

At its most typical this species is completely unmistakable with its spreading bottle-brush of hyaline leaf-tips reflexed stiffly outwards. No other species in Australasia ever develops this characteristic. However the hair-point can be almost or totally lacking and such forms are very troublesome. It is a common plant of roadside banks, dry and wet sclerophyll forests and even of subalpine bogs; almost always found in habitats rich in raw humus such as decaying wood, peat, etc. and on charcoal but also found in small patches of litter on otherwise bare rock.

The stems are commonly 1–2 cm long, dark (almost blackish) below, greenish above. The leaves are about 2–4 mm long not including a hair-point of up to 1–2 mm; the nerve usually very wide, c ½–¾ or even more of the leaf-base in width; the alar cells not usually inflated but sometimes distinct. The hyaline cells of the leaf base extend obliquely upwards and outwards to the margin. On the abaxial side the nerve is deeply grooved above so as to have low lamellae while the adaxial surface of the nerve is covered with thin-walled cells.

The capsules are usually conspicuously ribbed, oblique-mouthed and asymmetrical, on a cygneous seta; the peristome teeth 16, split in two to half-way, longitudinally ribbed below, papillose above. The spores are almost colourless, nearly smooth, c 12 µm diam., but aborted spores are common.

CHROMOSOME NUMBER: n=12 (NSW)

DISTRIBUTION: TAS, WA, SA, VIC, NSW, ACT, QLD, ?NT, Lord Howe; southern hemisphere, recently introduced and spreading in Europe where it has been confused with *C. polytrichoides* De Not.

ILLUSTRATIONS: Plate 20; Sim (1926), p. 170; Richards (1963), Plate 1, Fig. 1.

C. subappressifolius Broth. & Geh. is probably conspecific (Sainsbury 1955b, p. 8).

C. clavatus (R.Br.) Wils.

Apparently this is a very common species of roadside banks, disused paths, fallen logs etc., in wet or dry sclerophyll forests, open heathland, woods and scrub, usually forming quite dense turfs and in some cases covering large areas. Probably, however, many specimens are really forms of *introflexus*.

The stems are about as long as in *introflexus*, sometimes matted together below with whitish, brown, or even red tomentum but this is often missing. The leaves are rather stiffly erect or slightly spreading; dull yellowish green above, brown below; the shoot tips fragile and easily detached.

The leaves are usually shorter than in typical *introflexus*, 3–4 mm long, straight, denticulate at the apex and sometimes hyaline in the upper ¼ or less; the nerve occupying usually ⅓–½ of the leaf base and all of the subula. Alar cells are usually conspicuous, inflated, reddish or orange in colour, sometimes bulging out from the leaf to form distinct auricles, but sometimes scarcely distinct at all. In the upper part of the lamina the cells are chlorophyllose, variable in shape, usually thick-walled; those below more elongated, thin-walled and often hyaline; the area of hyaline cells sometimes extending obliquely upwards and outwards towards the margin.

The capsule is not or scarcely grooved. There are 16 peristome teeth, split into two filiform papillose segments right to the base. The spores are 9–11 µm.

DISTRIBUTION: TAS, WA, SA, VIC, NSW, ACT, ?QLD, Lord Howe; also in New Zealand and Africa.

ILLUSTRATIONS: Handbook, Plate 19; Wilson (1859), Plate 172 (as *C. insititius*); Allison and Child (1971), Plate 10.

This very variable species sometimes has quite long erect hyaline points on the leaves, more commonly none at all. The erect straight leaves, neither squarrose nor flexuose, and the fragile shoot tips are useful field characteristics. A pencil or small twig scored lightly across a patch of this species releases a row of shoot tips which spring to the surface. The greatest difficulties in identification occur in separating this species from *pallidus* and from forms of *introflexus* where the hair-point is missing. From the former it can be separated in the field by the stiff, not flexuose leaf subula, by the brownish green colour and usually by a lack of conspicuous red tomentum on the stem. The oblique boundary between hyaline and green cells in the leaf base, not uncommonly found, is more typical of *introflexus* and also occurs in *pallidus*.

We can find no consistent set of distinguishing characters to separate sterile *clavatus* and *introflexus*. At its most typical the fragile shoot tips, straight leaves with smooth abaxial surface, well developed alar cells and relatively narrow nerve are very distinct from typical *introflexus* and we have separated typical plants in the key using them, but any or all of these features can be present in *introflexus*. The only (compatible) characteristics which rarely seem to occur together are fragile shoot apices and reflexed hyaline points so that these features have become the hallmarks of "typical" *clavatus* and *introflexus* respectively. Even the nerve structure is not as helpful as the literature suggests: *introflexus* supposedly has large thin-walled cells on the adaxial surface of the nerve and *clavatus* small thick-walled cells. However, both cell types can occur on the same leaf in either species, so it seems of little use as a discriminating character. The areolation of the lamina in both species is extraordinarily variable; the size, shape, thickness and orientation of the cells seem to vary almost indiscriminately.

This is an interesting taxonomic puzzle which will only surrender to an extensive and careful biometrical investigation.

C. arboricola Card. & Dix.

This looks like a bright yellow epiphytic *clavatus* with red tomentum, scarcely distinct alar cells, and a dicranoid peristome of the *introflexus* type. It differs from *clavatus* also in the fragile apical buds which have broad cucullate leaves that are rough abaxially, and in the adaxial surface of the main leaves which have a strip of narrow, thick-walled stereid cells adaxially over the nerve right to the leaf base; in *clavatus* and *introflexus* the corresponding cells are relatively thin-walled and wide-lumened.

DISTRIBUTION: TAS; also in New Zealand. The Index omits this species from the Australian list but Sainsbury (1955b) records it and we have found good material. The colouration is closest to *pallidus* but the leaf shape is quite different.

ILLUSTRATIONS: Studies, Plate 7.

To the same group also belong *C. "angustilimbata"* Bartr. (an illegitimate name), *C. perauriculatus* Broth. (NSW), and *C. wattsii* Broth. (QLD).

C. pallidus Hook.f. & Wils. (=*torquatus* of the Handbook)

Probably this species is not uncommon in dry and wet sclerophyll forest, especially on old stumps, and on tussocks in boggy ground—habitats typical of *introflexus*. It is distinguished by the pale green colour, with red-brown tomentum on the stem, by the long flexuose subula which is

PLATE 21. *Campylopus bicolor* VIC—Vegetative shoot × 11, cells × 1000

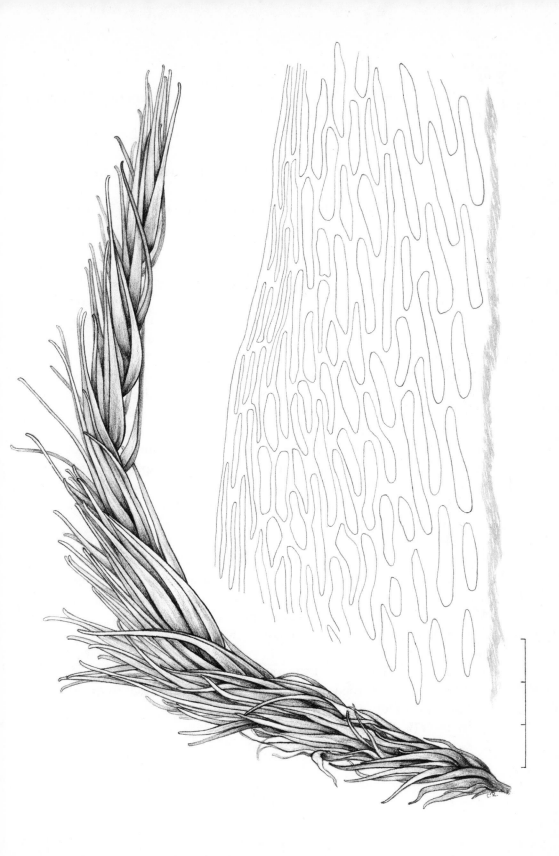

rarely hyaline and by the usually complete absence of alar cells. The lamina cells are very similar to those of *introflexus*.

DISTRIBUTION: TAS, ?WA, SA, VIC, NSW, ACT, QLD; also in New Zealand.

ILLUSTRATIONS: Wilson (1854), Plate 84.

It could easily be passed over in the field for a vivid, pale, silky form of *clavatus*.

C. woollsii (C.Muell.) Par., recorded from VIC to QLD and Lord Howe, is either conspecific or closely related.

C. bicolor (C.Muell.) Wils. is a plant of wet rock and ground, very similar to and sometimes mixed with tall forms of *introflexus* with neither hair-points nor alar cells. The stiff, straight, rather narrow leaves with cucullate apices are the most characteristic feature; some leaves can always be found which are bent in at the extreme tip. Specimens of var. *ericeticola*, with short hyaline points, seem to us often to be *introflexus*, but Dixon (Studies) considered that *bicolor* typically has hair-points here and there.

Capsules, which are rare, are nearly symmetrical, deeply ribbed, on a cygneous seta; the peristome similar to that of *introflexus* but the spores larger, 16–22 μm, pale brown and finely papillose.

DISTRIBUTION: markedly western; seems to be common in WA, extending through SA into TAS, western VIC, NSW and QLD; also in S. Africa, but extending eastwards to New Zealand.

ILLUSTRATIONS: Plate 21; Studies, Plate 7; Sim (1926), p. 180.

C. kirkii Mitt. is a very distinctive bog plant with wide cucullate leaves and a very broad nerve with faint lateral branches from it in the upper part of the leaf. The shoot tips here, as in *clavatus*, can be deciduous. The alar cells are inflated and coloured, the other cells of the leaf base thin-walled. All other cells are very thick-walled, porose, and quite long.

DISTRIBUTION: TAS, WA, SA, VIC; also in New Zealand. A rare species confined to Australasia.

ILLUSTRATIONS: Handbook, Plate 1a; Studies, Plate 7.

Other names, recorded for Australia, of plants unknown to us are:

C. australiensis Dub. (VIC)

C. brunneus (C.Muell.) Par. (TAS)

C. cockaynii R. Brown ter.

C. erythropoma Dub. (VIC)

C. homalobolax (C.Muell.) Par. (WA)
C. lenormandii Thér. (VIC)
C. nigro-flavus (C.Muell.) Par. (WA)
C. novae-valesiae Broth. (NSW, QLD)
C. senex (C.Muell.) Par. (VIC, NSW)
C. viridicatus (C.Muell.) Par. (NSW, QLD)

Dicnemoloma Ren.

D. pallidum (Hook.) Wijk et Marg. (=*sieberianum* of the Handbook)
This species is common in its most typical habitat of rock outcrops and boulders, especially siliceous rocks, in regions of dry sclerophyll forest but can also occur on soil and on calcareous rock. Usually there are dense mats of a rather distinctive greyish green, formed by prostrate main shoots, 2–4 cm or more long, with erect branches. Characteristically the prostrate shoot tips are turned up like the tips of skis.

The leaves are arched when dry, with the tips pointing in to the stem, and spreading when wet; the margins rather undulate and, in the upper half, rolled in to form an almost tubular apex; the nerve strong and single, reaching almost to the apex (sometimes excurrent in a short hyaline point) and covered abaxially in the uppermost part by a layer of lamina cells; otherwise exposed on both surfaces. Cells over most of the leaf are square, very small, *c* 9 μm across, thick-walled and porose, each crowned by a single papilla; towards the base of the leaf the papilla is branched on top, in cells nearer the leaf apex spicule-like and slightly curved, especially prominent at the leaf apex. There is a clear margin of long, smooth, hyaline cells in the lower half of the leaf or throughout. The basal cells are elongated, up to *c* 4 × 1, and smooth, with a conspicuous group of squarish brown-walled alar cells, the brown colouration extending in a band right across the leaf base.

The capsule, exserted on a seta *c* 5 mm long, is funnel-shaped and asymmetric; the operculum with a very long beak equal to or longer than the theca; peristome teeth deep red-brown, split into two in the upper half. The perichaetial bracts are conspicuous, sheathing, with hyaline points.

CHROMOSOME NUMBER: n = 11 (NSW).

DISTRIBUTION: TAS, SA, VIC, NSW, ACT, QLD, Lord Howe; also in New Zealand.

ILLUSTRATIONS: Plate 22; Handbook, Plate 18; Brotherus (1924–5), Fig. 167 (as *sieberianum*).

Under the microscope the leaf, with its spiculose–papillate cells especially at the apex, resembles no other species in the region. In the field its growth habit is very similar to a *Rhacomitrium* but the incurved leaf tips and slightly undulate leaf margins are characteristic. *D. clavinerve* (C.Muell.) Ren. is confined to QLD, *fraseri* (Mitt.) Ren. to NSW, and *imbricatum* (Broth. & Geh.) Ren. to NSW. We have seen none of these.

Dicranella (C.Muell.) Schimp.

D. dietrichiae (C.Muell.) Jaeg.

The total height above ground of these little plants, including fruit, is only about 1 cm, but they form quite noticeable dense turfs on wet clay banks and roadsides in lowland forests. The leaves increase from *c* 2·5 mm at the base of the stem to 4 mm long in the terminal, perichaetial, leaves; the lower leaves suddenly contract from a wide sheathing base, *c* 0·8–1·6 mm long, to a fine subula; the perichaetial leaves more gradually contracted with the leaves tending to be curved to one side and rather falcate at the stem apex. The cells in the leaf base are generally rectangular but rather variable, commonly square at one end and pointed at the other, 50–80 × 9 μm, those towards the margins narrower but never forming a conspicuous border. There are no alar cells. In the subula the cells tend to be rather short, *c* 18 × 7 μm, except in the nerve which is rather wide at the base, *c* 80 μm, and extends almost to the apex, taking up most of the subula. There is a cluster of small teeth right at the very tip and sometimes the margins are crenulate–denticulate at the shoulders, but are otherwise entire.

The capsules are usually very abundant, erect and oval to cylindrical

PLATE 22. *Dicnemoloma pallidum* VIC—Vegetative shoot × 11, basal cells (left) and upper cells (centre) × 1000, leaf × 50

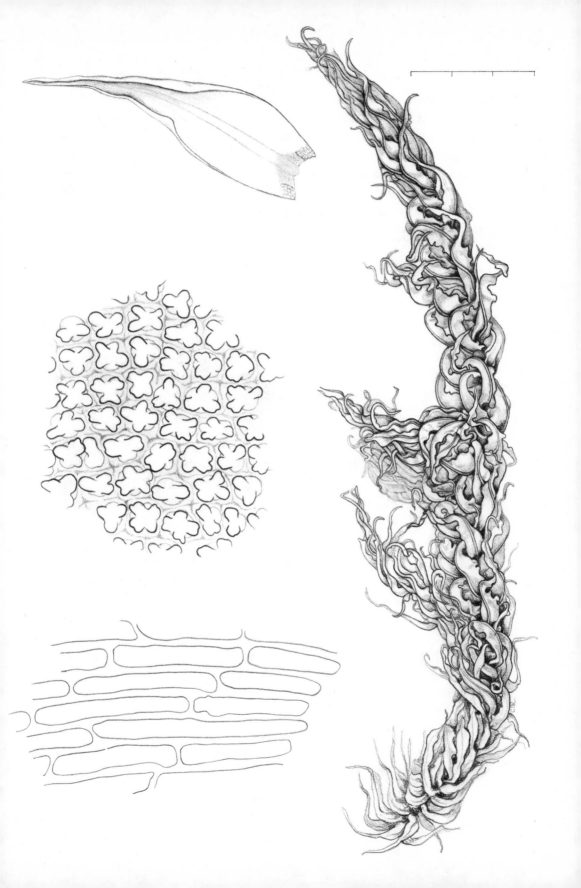

c 1·4 × 0·7 mm, golden-brown with a red mouth; dark brown when old, and grooved but not deeply. The operculum has a long slanting beak of *c* 1·2 mm, and a membranous, long and side-split calyptra. The peristome teeth are *c* 230 μm long, split to halfway into two, and rather irregularly partially split elsewhere; the outer face deep orange-red with strong transverse bars and slightly vertically striolate. The teeth are hyaline at the tips and there is a strong, conspicuous annulus.

CHROMOSOME NUMBER: n=12 (NSW).

DISTRIBUTION: VIC, NSW, QLD; endemic.

ILLUSTRATIONS: Plate 23.

This is a common species in the coastal forests of NSW, where it is a soil-binder on the fresh crumbly clay-loams. Old capsules can be found almost buried at soil level, overtopped by subfloral branches forming the current year's growth. It has only recently been found in Victoria and may well have been carried into the State on vehicles. It is rather like a *Ditrichum* but differs in the grooved capsules and in the dicranoid peristomes, with teeth split only to half way and finely longitudinally striolate.

D. cardotii (R.Br.ter.) Dix. differs in having a very squarrose, papillose subula and smooth top-shaped dark capsules. It is dioicous; spores are 20–30 μm.

DISTRIBUTION: TAS, NSW; also in New Zealand.

ILLUSTRATIONS: Studies, Plate 6 (as *D. wairarapensis*).

D. jamesonii (Mitt.) Broth. (=*Anisothecium hookeri* of the Index) has recently been found in Victoria. The shoots are 0·5–1·0 cm tall, with leaves gradually tapering from a scarcely sheathing base to a spreading or slightly falcate subula; the uppermost, perichaetial, leaves contracting more abruptly. The cells are smooth throughout. The capsule is short and wide, not grooved but slightly curved and also inclined, *c* 0·6 × 0·4 mm, with a wide mouth and a dicranoid peristome as in *dietrichiae*.

DISTRIBUTION: TAS, VIC; also in New Zealand.

ILLUSTRATIONS: Handbook, Plate 17.

PLATE 23. *Dicranella dietrichiae* VIC—Fruiting plant ×11, leaf ×50, cells from sheathing leaf-base ×1000

This differs from *cardotii* in the smooth leaf-cells, and from *dietrichiae* both in the short, smooth capsule and in the gradually tapering leaves.

Of the remaining species allowed by the Index *D. tricruris* (C.Muell.) Mitt. is considered the same as *dietrichiae* by Burges (1952), *D. paucifolia* (C.Muell.) Par., from N.E. Victoria is unknown to us, and the rest are all tropical or sub-tropical plants:
D. apophysatula (C.Muell.) Broth. (NSW)
D. euryphylla Dix. (QLD)
D.stackhousiana (C.Muell.) Broth. (NSW)

As at present delimited *Anisothecium* Mitt. does not seem a very satisfactory genus; pending a complete revision of the whole group of genera it seems best to retain in *Dicranella* the following species which the Index treats in *Anisothecium*:
clathratum (Hook.f. & Wils.) (=*D. vaginata* var. *clathrata* of the Handbook)
hookeri (C.Muell.)
schreberianum (Hedw.) (TAS)

Dicranoloma Ren.

On the whole these are big plants with erect little-branched stems, forming tall turfs or cushions in wet forests. In most species the leaves are falcate with very thick-walled porose cells, conspicuous alar cells, and hyaline borders. The capsules are also large, erect and usually curved and frequently with conspicuous perichaetia; polysety is frequent. The cells are very rich in oil-globules and are fairly distinctive. As Sainsbury has pointed out (Handbook, p. 126) the structure of the nerve, although widely used in the past, is not a very reliable taxonomic feature, especially in critical cases where it is most needed. The differences from *Dicranum* are discussed under that genus. Most of the species can be found either on the ground or as epiphytes.

PLATE 24. *Dicranoloma eucamptodontoides* TAS—Vegetative shoot × 4, cells × 1000

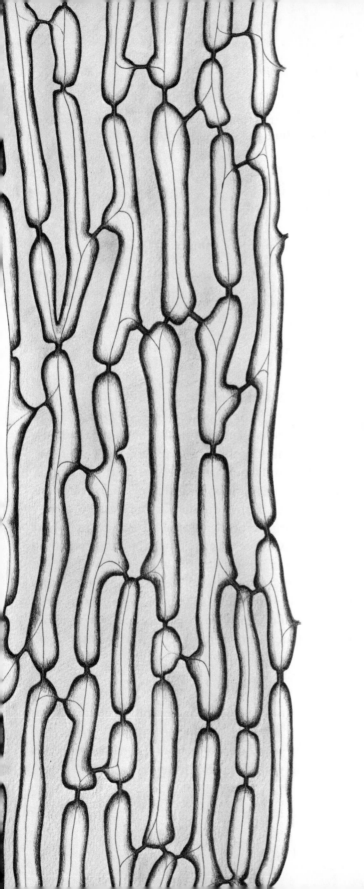

Key to species

1. Leaves ovate, concave, very shortly pointed . .   *eucamptodontoides*
   Leaves narrow, channelled, tapering to a long fine
   point   2

2. Leaves neither plicate nor undulate   3
   Leaves plicate or undulate in mid-leaf   4

3. Cells in subula and uppermost part of lamina
   7–9 µm, isodiametric or very short . . . .   *menziesii*
   Cells in upper part of lamina clearly elongate,
   c 45–60×9 µm . . . .   *billardieri+robustum*

4. Leaves longitudinally plicate . . . . . .   *dicarpum*
   Leaves undulate about mid-leaf . . . . .   *platycaulon*

D. eucamptodontoides (Broth. & Geh.) Par.

This handsome Tasmanian endemic looks nothing like a *Dicranoloma* in
general appearance, although the cells are quite typical. The short, fat shoots
are erect, occasionally branched, and loosely packed together, c 5 cm tall ×
4–5 mm diam., dark brown below and tapering above to a strikingly conical,
golden-olive apex. The leaves are very glossy and densely imbricate, undulate, concave and ovate, c 6×2·5 mm, rounded at the base to a very
narrow insertion and with the margins inrolled at the apex to form a short
tubular cusp. There is a very narrow nerve, c 25–35 µm at the base, extending
right to the apex. The cells are long and narrow, very thick-walled and
porose throughout the leaf, c 60–90 × 10 µm in mid-leaf, those at the angles
shorter and wider, forming a conspicuous alar group; cells towards the
margins are narrower, with several marginal rows hyaline and very narrow
forming a conspicuous border which sometimes reaches nearly to the apex.

Sporophytes are unknown.

DISTRIBUTION: TAS; endemic.

ILLUSTRATIONS: Plate 24.

This is a rare plant of wet button-grass plains in central Tasmania and is
exclusively terrestrial.

It is a comparable plant to the New Zealand endemic, *obesifolium*, which
differs in the obtuse apex and the variable, shorter, nerve.

D. menziesii (Tayl.) Par.

Despite the claims of the Handbook (p. 127) the commonest colour for this species is not dark yellow but dark green. This colour, and the very long narrow leaves, usually make it an easy plant to recognize in the field. The shoots are commonly 2–5 cm tall, although they can exceed 8 cm, usually very dark green above and blackish below (not unlike *Echinodium*), contrasting strongly with the pale sheathing leaf bases and whitish tomentum. In fertile stems the colouration tends to be dull yellowish above and brown below, with much less white showing.

The leaves are very long and narrow, reaching 13 mm or more, of which only *c* 2 mm is leaf base; the subula minutely and irregularly toothed in the uppermost portion. Usually the leaves are falcate and secund but not as clearly as in other species. There is a very broad nerve, to *c* 200 µm at the base, narrower above and occupying most of the subula which is fragile and tends to break off easily. The cells of the subula and upper part of the lamina are small, isodiametric and thick-walled, basically square, triangular, or oval, 7–9 µm or less across. The cells of the upper ½ of the leaf base are also short and very irregular in shape, those below larger, *c* 40–50 × 8 µm, thick-walled and porose with a very narrow border of hyaline cells sometimes present in the leaf base and with square or oblong, deep yellow alar cells forming conspicuous rounded auricles.

There is a short seta, *c* 1·0–1·5 cm, and a short erect or inclined slightly curved capsule, *c* 2–3 mm long, lightly grooved or striated, with a wide red mouth. The single peristome is of 16 deep red teeth split to ½ way. The perichaetial leaves usually have a very fine hair-like point, but are commonly overtopped by vegetative leaves or branches from below and are not conspicuous.

CHROMOSOME NUMBER: $n = 8$ (NSW).

DISTRIBUTION: TAS, VIC, NSW, QLD, Lord Howe; also in New Zealand, S. America, and Oceania.

ILLUSTRATIONS: Plate 25; Studies, Plate 4, Plate 7 (as *Dicranodontium australe*); Wilson and Hooker (1845), Plate 58.

This is normally the only deep green *Dicranoloma* in our area and that, together with the very long narrow leaves, will separate it from other species on most occasions. Some forms of *billardieri* can approach it in leaf shape and sometimes colour but the small isodiametric cells are then adequate to distinguish it. It is predominantly an epiphyte although sometimes terrestrial.

D. bartramioides (Broth.) Par., from Lord Howe, is similar if not conspecific.

D. billardieri (Brid.) Par. (including *robustum* of the Handbook)
Sainsbury confessed himself defeated by the polymorphism of this species or group of species (Handbook, p. 129) and we have fared no better, although Australian plants seem to be rather less variable than those of New Zealand. Commonly the stems are quite short, 2 cm or so on average, with leaves 12 mm long, crowded and usually falcate, yellow-green and glossy, often with a red tinge, tapering rather gradually to a long, fine fragile subula, the leaves almost tubular throughout by inrolling of the margins. The subula is denticulate above, as in *menziesii* but with the toothing getting stronger towards the tip where there are a few large teeth. The nerve is characteristically thin and weak especially at the base, c 30–70 μm (*billardieri* type) rarely almost totally lacking, but can be quite strong and wide (*robustum* type) and there seems to be no satisfactory boundary between the two conditions. The cells of the upper lamina and subula are elongated and porose, c 45–60 × 9 μm; those in the leaf-base 70–150 μm long, the walls very thick and porose except sometimes for thin cross-walls. There is a border of narrow hyaline cells, especially some distance up from the leaf insertion, and a conspicuous alar group.

The sheathing perichaetial bracts may or may not have a short hair-point; there is a moderately long seta, c 2 cm, with a slightly curved capsule c 2·5 mm long.

CHROMOSOME NUMBER: $n=12$ (NSW), 13 (TAS); D. *robustum* var. *setosum*, $n=7$ (NSW).

DISTRIBUTION: TAS, SA, VIC, NSW, ACT, QLD; also in Africa, S. America and New Zealand.

ILLUSTRATIONS: Studies, Plate 4; Wilson and Hooker (1845), Plate 58 (as *D. setosum*), Plate 59 (as *D. pungens*).

It is rather surprising that Sainsbury did not merge *billardieri* and *robustum* for the distinctions between them are not at all consistent. The weak nerve of the former can be found in conjunction with the hair-pointed perichaetial bracts of the latter and there seems no point in keeping separate two species that are so connected. Most Australian material seems to have the weak nerve

PLATE 25. *Dicranoloma menziesii* VIC—Fruiting plant × 7, leaf × 50, cells of leaf-base (bottom right) and subula × 1000

of typical *billardieri* but it is a variable feature. There may be several species involved in the complex group but the task of sorting out the mixture is a formidable one and not to be undertaken in the herbarium alone, nor in Australia in isolation from the rest of its distribution. The name *billardieri* precedes *robustum.*

D. perichaetiale Sainsb. (TAS). "Vegetatively this is inseparable from *D. billardieri,* but in fruit is distinguished at once by the extremely long perichaetium." (Sainsbury, 1955b, p. 9).

D. platycaulon (C.Muell.) Dix. could easily be passed over for *billardieri* in the field, but it differs in having the rather broad (*c* 1·6 mm) leaves *undulate* about mid-leaf, although this feature is sometimes not very evident. The cells in the base of the subula are short and very irregular; *c* 12–30 × 12 µm, with irregularly but not strongly thickened walls, not or scarcely porose. There may be whitish tomentum at the leaf bases. This species, supposedly endemic to New Zealand, has recently been found in the Cradle Mountain region of Tasmania, and the identification confirmed by K. W. Allison. DISTRIBUTION: TAS; New Zealand. ILLUSTRATIONS: Studies, Plate 3.

D. dicarpum (Nees.) Par.
Unlike the golden colouration of most Dicranolomas this tends to be a clear green with a whitish or greyish tint that is quite easily picked out in the field. The stems, which are commonly branched, extend to 5 cm or more, with rather falcate-secund leaves which are slightly contorted and spirally twisted when dry, and *deeply plicate* on both sides of the strong dark nerve, which is *c* 110 µm wide at the base, narrower but with two abaxial ridges above. The subula is long and fine and very *strongly toothed* on both margins and on the back of the nerve, the teeth evident even with a hand lens. The uppermost subula cells are quadrate or oblong with rounded corners, 15–24 × 12–15 µm. Diagnostically in mid-leaf or above there is a region with short square cells next to the nerve (*juxtacostal cells* in the Handbook) contrasted with long cells nearer the margins. The cells in the leaf-base are very long and narrow and porose with several marginal rows forming a hyaline border. The alar cells are enlarged, square and orange forming a conspicuous group.

The perichaetials are long (*c* 7 mm) sheathing and conspicuous, with a short terminal hair-point. The capsule is *c* 3 mm long, erect and narrow, on a seta projecting 0·5–1·0 cm above the perichaetia. Commonly there are 1–4 capsules in one perichaetium.

CHROMOSOME NUMBER: n=7 (NSW).

DISTRIBUTION: TAS, VIC, NSW, ACT, QLD, Lord Howe; also in New Zealand.

ILLUSTRATIONS: Studies, Plate 3; Brotherus (1924–5), Fig. 166.

The strongly toothed plicate leaves are sufficient to distinguish this species but when the plication is weak the patch of short cells next to the nerve, usually near the base of the subula, is a useful confirmatory character, although not always as obvious as the Handbook seems to suggest.

The remaining species in the Index, none of which we know, include:

D. angustifolium (Hook.f. & Wils.) Watts & Whitelegge (TAS)

D. angustiflorum Mitt. ex Dix. (TAS)

D. austrinum (Mitt.) Watts & Whitelegge (TAS, WA, QLD)

D. burchardtii (Par.) Par. (TAS)

D. diaphanoneuron (Hampe) Par. (WA, VIC)

D. nelsonii (C.Muell.) Par. (TAS)

D. punctulatum (Hampe) Ren. in Par. (VIC, NSW)

D. sullivanii (C.Muell.) Par. (VIC)

and the tropical:

D. argutum (Hampe) Par. (NSW, QLD)

D. austro-scoparium (Broth.) Watts & Whitelegge (QLD)

D. bartramioides (Broth.) Par. (Lord Howe)

D. elimbatum Dix. (QLD)

D. serratum (Broth.) Par. (NSW, QLD)

D. spiniforme (Broth.) Par. (QLD)

D. wattsii (Broth.) Par. (QLD)

We have preferred to treat *D. trichopodum* (Mitt.) Broth. as a *Dicranum* [20, q.v.].

Dicranoweisia Lindb. ex Mild.

Vegetatively there is a superficial resemblance between this genus and *Weissia* and its relatives because of the curled leaves with small upper cells and elongated lower cells, but the peristome is typically dicranoid with 16 teeth split almost to half way into papillose segments (except *antarctica*). The dense thick cushions usually of a sombre green are easy enough to recognize for all that, even when not in fruit. They are plants of cold regions, in Australia mainly restricted to montane regions and above.

D. microcarpa (Hook.f. & Wils.) Par.

Except perhaps in alpine habitats this is a rare species recognizable by the dull blackish olive-green cushions of tightly packed shoots c 0·5 cm tall and 1 cm across, rather like a small *Grimmia* at first sight, but with leaves very tightly curled when dry, as in *Tortella cirrhata*. The leaves are 1·5–3·0 mm long when straightened out, in our plants, very narrow and channelled, widest in the basal ¼ but even there only c 0.3 mm wide. The leaf margin is variably recurved in the mid-region and the lamina bistratose in parts. The uppermost cells are squarish to shortly rectangular, c 9–16 × 9 µm, but rather variable in shape and wall thickness; those in the leaf base much longer, to 60 µm, commonly thin-walled and hyaline in a large patch at the margins forming a not very distinct group, and very thick-walled and porose near the nerve. Cells in mid-leaf are c 30 µm long, but also thick-walled and porose or sinuose.

The perichaetial leaves are inconspicuous. The capsule is erect and cylindrical when empty, c 0·8 × 0·3 mm, more ovoid when full, on a seta c 7 mm long. The peristome is of 16 brown teeth inserted well below the rim, strongly and closely barred in the lower two thirds, pale and widely barred and sparsely papillose above, where each tooth is split into two. There is a persistent annulus of two rows of cells.

DISTRIBUTION: TAS, VIC; to be expected also in NSW; endemic.
ILLUSTRATIONS: Wilson (1859), Plate 171 (as *Weissia*).

This is much the same size as the New Zealand *spenceri* but that plant is yellowish, the leaves less tightly coiled, and the peristome teeth are obliquely striolate.

D. cirrata (Hedw.) Lindb., a northern hemisphere species, is recorded by the Index presumably in error. It has the peristome teeth undivided as in *antarctica*.

D. riparia (Hampe) Par. from SA we do not know.

D. antarctica (C.Muell.) Par. is recorded from TAS by Sainsbury (1955b) and has recently been found in VIC and NSW (I.G.S.). It is often bigger than *microcarpa*, has strong alar cells, conspicuous perichaetial leaves and entire, coarsely papillose peristome teeth.

ILLUSTRATIONS: Handbook, Plate 20.

Holomitrium Brid.

H. perichaetiale (Hook.) Brid.

Once this species is known the tightly curled narrow leaves, when dry, shining and yellow on the abaxial surface of the nerve, are almost sufficient to identify it at once. The stems are *c* 1–3 cm long of which the upper half is green and the lower portion brown and usually tomentose with characteristically curled smooth rhizoids. The leaf base is large, *c* 1·0–1·5 × 0·5–0·8 mm, out of a total leaf length of 2–3 (–6) mm. The leaf base narrows quickly to a long channelled subula (2–3 cells thick at the margins) through which the broad strong nerve (70–80 µm wide or more) usually runs to the tip, often excurrent as a tiny hyaline apiculus. Adaxially the nerve is partly covered by short lamina cells. The rectangular cells in the leaf base (60–85 × 9–10 µm) are very thick-walled and porose on the longitudinal walls, rather thin on the transverse walls; the marginal cells there are much narrower and shorter, forming a large patch of rather thin-walled, more or less hyaline cells sometimes extending in one or two rows some distance up the margin of the leaf. Alar cells may be quite well developed, conspicuously enlarged and coloured, but may be little differentiated. Cells of the subula have very thick walls and rounded cavities, *c* 6 µm long × 6–10 µm wide, with an area of irregularly shaped, often triangular, cells at the junction of the leaf base and subula. The cells are very finely papillose in our material from NSW.

The sporophyte is highly characteristic, with very long, sheathing perichaetial bracts round the base of the seta, an erect, cylindrical capsule, long fine beak and very short peristome teeth.

DISTRIBUTION: TAS, VIC, NSW, QLD, Lord Howe; also in New Zealand, and New Caledonia.

ILLUSTRATIONS: Handbook, Plate 20.

This species is perhaps most commonly an epiphyte but also grows on rock outcrops in wet forest.

From the type description *H. dietrichiae* C.Muell., QLD, sounds much the same.

Leucobryum Hampe

The extraordinary leaf structure is diagnostic of this genus. Long and narrow green cells are joined at the tips to form an open network layer which is sandwiched between two sheets of much bigger, dead, empty, hyaline cells. In transverse sections through the leaf (Plate 26) the long cells show up as tiny, dark squares at the points of intersection of the hyaline cells; the photo-synthetic tissue is thus very sparse. As the hyaline cells are interconnected by numerous pores and have some openings to the outside, the leaf is capable of absorbing a great deal of water, acting like a sponge, much as in the quite unrelated genus *Sphagnum* which has a somewhat comparable leaf structure. In *Leucobryum* the margins of the leaves are only one cell thick and probably represent the true leaf lamina; the greater part of the leaf (with its sandwich-like structure) therefore probably represents the nerve, an analogous de-velopment to the leaf of *Polytrichum*.

Despite the unusual leaf anatomy, the capsule in *Leucobryum* is typical of the Dicranaceae; curved, deeply grooved, and with 16 papillose peristome teeth, split in two nearly to the base. The capsule neck has a pronounced *struma*.

Leucobryum is dioicous and is one of the few moss genera in which the male and female plants are very different. The males are tiny plants consisting of a very short stem anchored by rhizoids and a few leaves surrounding a group of 2–3 antheridia (Plate 26). Being minute and identical in colour to the female plants, males are hard to find but where there are capsules male plants cannot be far away.

L. candidum (P.Beauv.) Wils.

This is the commonest species in temperate Australia, frequent in rain-forest, wet sclerophyll forest and fern gullies, mainly on the ground or rotten logs but sometimes epiphytic.

Shoots are very variable in length, from 1 to 5 or more cm long (of which

PLATE 26. *Leucobryum candidum* VIC—Fruiting shoot showing 2 dwarf male plants perching on the leaves below and to the left of the base of the seta ×7. Enlarge-ment of male plant, cut away to show the antheridia, perching on part of a full-sized leaf ×50, sections through parts of leaves ×100

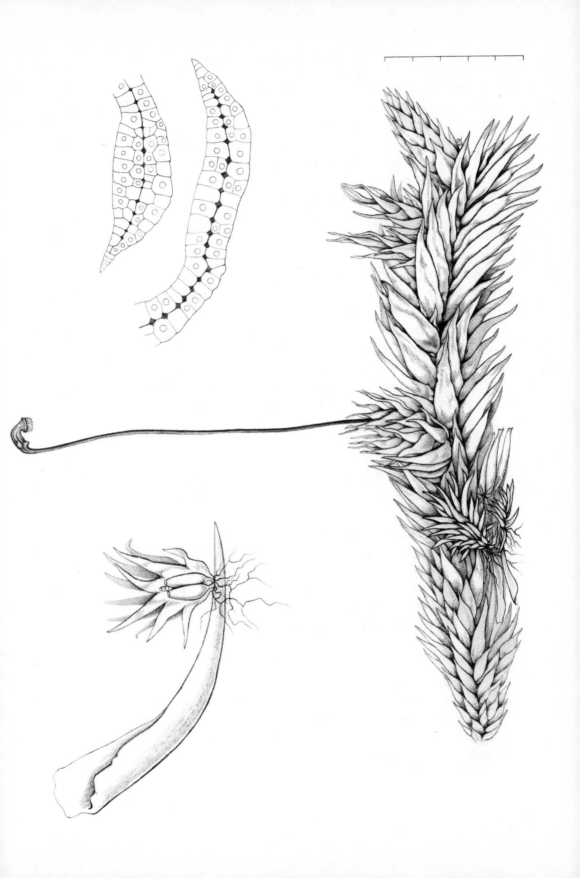

the terminal 0·5–1·0 cm is fresh growth), forming loosely packed mats or sometimes quite large cushions. The leaves are a dirty whitish green, with a tinge of blue when dry, arranged either in a spiral or in 5 ranks (var. *pentastichum*) usually with a spiral twist. In suitably moist habitats the tips of the leaves can bear tassels of rhizoids. The two layers of hyaline cells are each basically one cell thick but the cells are commonly divided to give 2 or 3 layers of cells, especially towards the base of the leaf. In the upper half of the leaf the cells of the abaxial layer are considerably deeper than those of the adaxial layer and the end walls protrude with adjacent cells slightly overlapping. As the cells tend to be in transverse rows this part of the leaf is strikingly corrugated when dry, although some forms exist where it is not at all conspicuous. Towards the apex the protrusion and overlapping of cells becomes increasingly pronounced and at the apex the whole back of the leaf becomes quite strongly toothed. The margins are entire but not easily seen as the upper half of the leaf is curved into the form of a dentate, conical tube.

CHROMOSOME NUMBER: n=11 (NSW).

DISTRIBUTION: TAS, WA, VIC, NSW, extending into QLD and Lord Howe; also in Oceania, East Indies and New Zealand; ?New Guinea.

ILLUSTRATIONS: Plate 26; Handbook, Plate 16.

L. sanctum (Brid.) Hampe is a subtropical species from QLD, distinguished from *L. candidum* by a smooth abaxial surface and hyaline layers consistently 1 cell thick.

L. ballinense Broth. is restricted to NSW and QLD.

L. wattsii Broth., from NSW and QLD is a distinct-looking little species which has small stems (0·5–1·0 cm long) in tight cushions, leaves curved inwards towards the stem (especially noticeable at the apex) and smooth leaves ending in short sharp points. There are two layers of hyaline cells abaxially and one layer of smaller cells adaxially in the leaf.

L. turgidulum C.Muell. from NSW, we have not seen.

L. confusum Thér., from TAS and New Zealand seems, on New Zealand material, just to be *L. candidum*.

NSW material of *L. teysmannianum* Doz. et Molk. in the National Herbarium of Victoria, is apparently *L. candidum*.

Trematodon Michx.

This genus of small earth-mosses is distinguished by the great length of the capsule neck, usually at least half the total length of the capsule. At times the capsule looks as if it ought to belong to *Tayloria*, but the subulate leaves and narrow cells at once separate them.

The leaves contract from a concave sheathing base to a fine subula, nerved to the tip and often rough with projecting cell ends near the apex; the perichaetial leaves are much larger than the vegetative. The cells are narrowly rectangular, clear and lax below, with no distinct alar cells; shortly rectangular and more obscure above. The marginal cells are sometimes narrower but do not form a distinct border. Both the vegetative shoots and the capsules are very variable in size. The peristomes and spores seem to be the crucial features in recognizing species; the calyptra is long and usually side-split at the base, but may be 2–4-lobed.

Key to species

1.	Capsule cleistocarpous or gymnostomous . . .	2
	Capsule peristomate	3
2.	Capsule gymnostomous; spores 50-60 µm; seta long (*c* 1 cm)	*mackayii*
	Capsule often cleistocarpous; spores 25-32 µm; seta short (*c* 3 mm)	*amoenum*
3.	Peristome teeth 150-230 µm, irregularly split or entire, sometimes very imperfect; spores 25-32 µm. Dioicous	*flexipes*
	Peristome teeth *c* 300-400 µm, split into segments which join to form narrow tips; spores 20-25 µm. Autoicous	*suberectus*

T. amoenum (C.Muell.) Stone & Scott (*Bruchia* in the Index)†

This little moss of silty soils, rotting logs, tree roots etc. in damp upland areas can form quite large patches, conspicuous by the massed orange capsules. The capsules are gymnostomous, or cleistocarpous with an operculum which sometimes falls off and sometimes does not. There may be a conspicuous annulus of very thick-walled cells, or it may be reduced or even absent. The spores are very coarsely warted, 25–30 × 30–32 µm.

DISTRIBUTION: VIC, NSW, ACT; endemic.

ILLUSTRATIONS: Roth (1911), Plate 13 (as *Bruchia*).

With the discovery that the operculum is sometimes dehiscent, the species has had to be transferred from *Bruchia*. To some extent it bridges the gap between the two genera.

T. mackayii (R.Br.ter.) Broth., a larger species with stems 6–9 mm high, usually has a long seta (up to 1 cm), and a fully dehiscent operculum but no peristome. The spores are large, 50–60 µm, and heavily warted. The spore size readily distinguishes the species from others in the genus.

DISTRIBUTION: TAS, VIC, ACT; also in New Zealand.

ILLUSTRATIONS: Handbook, Plate 14.

T. flexipes Mitt. in Hook.f. is probably the commonest of our species, and perhaps the smallest, the stems only 2–3 mm high. It is dioicous. There is a distinct short peristome, 150–230 µm, usually slightly *incurved*. The teeth may be entire, or irregularly perforated, or split, faintly striate on the outer surface and more or less papillose on the inner. There is a broad annulus of 3–4 cell rows and a characteristically flexuose seta of 1·5–4·0 mm, sometimes quite sharply bent. There is a long beak on the operculum, about as long as the theca. The calyptra is variably split and sometimes inflated below. The spores are 25–32 µm in diameter.

DISTRIBUTION: TAS, VIC, NSW, ACT; also New Zealand.

ILLUSTRATIONS: Handbook, Plate 14; Roth (1911), Plate 25; Wilson (1859), Plate 172.

T. alpinus Willis seems to be a minute alpine form of *flexipes*, a smaller edition of that species in all respects but spore size. We consider it conspecific.

T. suberectus Mitt. in Hook.f.

This species tends to be larger than *flexipes*, with stems 5–10 mm high and longer, more acute leaves. It is autoicous. The peristome is long and erect, *c* 400 µm tall, the outer surface of the teeth strongly ornamented with vertical striations and the inner surface papillose. The 16 teeth are margined with hyaline papillae, and are split into 2 unequal portions which are joined at the tips. The spores are rather small, 20–25 µm, and there is a complete annulus.

DISTRIBUTION: NSW, QLD; also in New Zealand.
ILLUSTRATIONS: Handbook, Plate 14; Roth (1911), Plate 25; Roth (1913), Plate 2 (as *adaequans*).

T. adaequans Geh. ex Roth. is probably the same thing, although New Zealand material of *suberectus* has a slightly shorter peristome (*c* 300 µm).
T. longescens C.Muell. (NSW, QLD) is a beautiful species with a very long neck.
T. baileyi Broth. (QLD) and *T. brachyphyllus* C.Muell. (QLD) we do not yet know.

DICNEMONACEAE

Includes: *Mesotus acutus* Mitt. (?TAS)

Eucamptodon

E. muelleri Hampe & C.Muell.
The prostrate shoots of this species extend for 5 cm or more, with numerous erect branches *c* 0·5–1·0 cm tall, densely encased in concave, mostly oblong–lanceolate, broadly acute nerveless leaves. The leaves are yellowish brown, chaffy in texture, and widely spreading except at the shoot apex. The margins are inflexed, especially near the apex, to give a conical point to the leaf. The cells measure 40–90 (–120) × 8–10 µm in mid-leaf and are spindle-shaped and thick-walled; those at the leaf base with thick sinuose and porose walls forming a *yellow band* across the base; those at the angles larger, quadrate and thickened forming conspicuous and slightly decurrent alar groups; those at the apex shorter. The capsule is shortly emergent above the very tall sheathing tubular perichaetial bracts which stand up *c* 5–10 mm above the rest of the plant; erect and cylindrical with undivided peristome teeth which characterize the genus, and a long-beaked operculum.
DISTRIBUTION: VIC, NSW, QLD; endemic.
ILLUSTRATIONS: Brotherus (1924–5), Fig. 173.
    This is a northern species which is said to be a not uncommon epiphyte in NSW just extending into eastern VIC. Vegetatively it is very similar to *Dicnemon semicryptum* of New Zealand, but differs in not having the divided peristome teeth of that genus. *E. squarrosus* Besch. is unknown to us.

Dicnemon calycinum (Hook.) Schwaegr., a very similar plant which has leaves single-nerved almost to the apex, and *D. rugosum* (Hook.) Schwaegr. are both given by the Index, presumably following Watts and Whitelegge's Census (1902, p. 54) but both these records seem to be doubtful.

PLEUROPHASCACEAE
Pleurophascum Lindb.

P. grandiglobum Lindb.

In Tasmania this famous bryological rarity grows in subalpine button-grass plains either on wet shaded peat between tussocks or, much more abundantly, in boggy pools or flushes, often in a carpet of *Campylopus introflexus*.

The shoots are often almost solitary or, when very abundant, forming loose open turfs, characteristically yellow-green above and dark brown below, about 1 cm high. The nerveless leaves, closely overlapping but not tightly pressed together, are very concave, either blunt or shortly pointed at the apex or sometimes with a short corkscrewed hair-like tip and measure *c* 2–3 × 1 mm; the margins are characteristically slightly recurved all round. Cells are large, very thick-walled and porose, rhomboid at the leaf apex, increasing in length farther down the leaf and reaching 90–150 × 15–20 μm (sometimes even longer) at the leaf base; a few in the angles are short and square. The cells contain abundant inclusions resembling oil-globules and the whole structure of the living cell is very reminiscent of *Dicranum*.

The capsule is the plant's chief claim to fame. This is pale green (orange when ripe), large and spherical, 3–4 mm in diameter, on top of a long (1–2 cm) seta, but is cleistocarpous. There is a small point on top but no separate operculum; the capsule splits open at the side. The calyptra is small, smooth and black, falling off early.

DISTRIBUTION: TAS; also in New Zealand.

ILLUSTRATIONS: Plate 27; Handbook, Plate 24; Brotherus (1924–5), Fig. 174.

PLATE 27. *Pleurophascum grandiglobum* TAS—Shoot with inset to show calyptra × 7. Basal cells (lower right) and upper cells (upper left) × 1000

In the field, when not fruiting, the plants are rather reminiscent of *Bryum laevigatum* [66]. Under the microscope, or with capsules, it can be mistaken for nothing else.

CALYMPERACEAE

Includes: *Arthrocormus schimperi* (Doz. & Molk.) Doz. & Molk. (QLD)
Calymperes armatum Broth. (= *Syrrhopodon incompletus*)
C. kennedianum Hampe (QLD)
C. latifolium Hampe (WA, QLD)
C. moluccense Schwaegr. (WA, QLD)
C. motleyi Mitt. in Doz. & Molk. (QLD)
C. nigrescens Broth. & Geh. (QLD)
C. repens (Harv.) C.Muell. (QLD)
C. tenerum C.Muell. (QLD)
Calymperopsis wattsii (Broth.) Fleisch. (NSW)
Exodictyon subscabrum (Broth.) Card. (QLD)
Leucophanes australe Broth. (QLD)
L. octoblepharidioides Brid. (QLD)
Octoblepharum albidum Hedw. (WA, QLD, NT)
O. exiguum C.Muell. (Torres Str. Is.)
Syrrhopodon amoenus Broth. (QLD)
S. cairnensis Broth. & Watts. (QLD)
S. croceus Mitt. (QLD)
S. fimbriatulus C.Muell. (NSW, QLD)
S. incompletus Schwaegr. (NSW)
S. mammillosus C.Muell. (QLD)
S. muelleri (Doz. & Molk.) Lac. (QLD)
S. novae-valesiae C.Muell. (NSW, QLD)
S. parvicaulis C.Muell. ex Broth. (QLD)
S. platycerii Mitt. (Lord Howe)
S. revolutus Doz. & Molk.
Thyridium fasciculatum (Hook. & Grev.) Mitt. (NSW, QLD)
T. repens (Harv.) Mitt. (QLD)
T. subfasciculatum (Hampe) Jaeg. (QLD)
T. undulatum (Broth. & Geh.) Fleisch. (QLD)

POTTIACEAE

Includes: *Didymodon dubius* (Schwaegr.) Par.
D. rubiginosus (C.Muell.) Broth.
D. wildii (Broth.) Broth. (NSW)
Hymenostomum olivaceum C.Muell. ex Geh. (NSW)
H. perpusillum (C.Muell.) Par. (QLD)
H. sullivanii C.Muell. ex Geh. (NSW)
Hymenostylium recurvirostre (Hedw.) Dix.
Hyophila involuta (Hook.) Jaeg. (QLD)
Trichostomum aristatulum Broth. (NSW)
T. leptotheca C.Muell. (VIC)

POTTIACEAE—POTTIOIDEAE)

Pottia (Reichenb.) Ehrh. ex Fuernr.

These tiny earth mosses present some difficulty because of the polymorphism of most of the species. It is fortunate that they fruit freely as some of the species are most satisfactorily separated on spore characters. The species are very heterogeneous and some have been the subject of a revision by Chamberlain (not yet published). The Victorian species were revised by Willis (1954a, 1957a), but we still know far too little about the Australian plants.

The individuals are scattered or in small groups, the leaves erect or spreading, usually broad above the middle, obovate or oblong, and with the nerve variable in length and strength; the margin plane, incurved or recurved. The cells of the lower part of the lamina are larger, clear, smooth, rectangular and usually with thin walls; the upper smaller, clear or obscure, variably papillose or sometimes smooth, shortly rectangular, hexagonal, quadrate or rhomboidal with thin to firm walls.

The capsule is exserted and erect, either cleistocarpous or, more commonly, opening by an operculum (stegocarpous), the peristome absent or rudimentary or of 16 more or less divided straight teeth. The operculum is shortly conical, obtuse or with a long beak which is often oblique; the calyptra side-split and beaked, either smooth or rough; the spores varying from smooth, to variously papillose or with large vesicles.

Key to species

1. Capsule cleistocarpous. Leaves very broad, obtuse,
    often cucullate . . . . *drummondii*

Capsule stegocarpous. Leaves oblong or obovate, usually more or less acute (exc. *scabrifolia, brevicaulis*)   2

2.   Upper cells clear, smooth, 17–20 μm wide. Leaf margin usually plane. Operculum suddenly narrowed to an oblique beak; peristome absent. Stem usually over 5 mm . . . .   *truncata*

    Upper cells more or less opaque, very papillose, less than 18 μm wide (usually much less). Margins of leaves recurved. Peristome present or absent. Stem usually less than 2 mm . . .   3

3.   Operculum with a well defined beak; peristome absent. Leaves short, more or less obtuse; nerve not usually excurrent. Spores papillose but not spiny   4

    Operculum conic, obtuse, sometimes umbonate; peristome present or absent. Leaves longer, more or less acute, nerve often excurrent, not broadened above. Spores with spines or large tubercles   5

4.   Nerve broadened distally; cells bulging from adaxial surface of nerve . . . .   *brevicaulis*

    Nerve not broadened; cells not protuberant . .   *scabrifolia*

5.   Spores opaque, with spiny papillae; peristome absent . . . .   *davalliana*

    Spores clear with large vesicles on the surface; peristome with short pale papillose teeth (rarely absent) . . . .   *starckeana*

P. drummondii (Wils.) J. H. Willis

These little plants of bare earth in low rainfall areas are bulb–like, the stems 1–2 mm high with rather few erect or spreading leaves less than 1 mm long, incurved and slightly twisted when dry. The leaves are very broad, obtuse and concave to convolute and sometimes cucullate. The cells in the upper half are 17–25 μm wide, smooth or with a few papillae. The nerve usually fails below the apex.

The seta, 1·5–2·0 mm long, is pale reddish brown; the capsule ovate, elliptic or pear-shaped with a long oblique beak, red-brown when ripe and with no sign of a dehiscence line. The oblique calyptra usually covers all except the neck of the capsule. The spores are pale and filled with oil globules, almost smooth or with irregular patches of very small granules on the surface.

The variety *obscura* Willis is described as having more acute leaves with the nerve reaching the apex, rather obscure cells, and slightly warty larger spores (30–40 μm). The situation is not at all clear, for these features are subject to much variation and are combined differently in different specimens. We would prefer to include the variants provisionally within one polymorphic species until a careful study of more specimens from different areas is made (the New Zealand *P. maritima* would also fall within the same range of variation).

DISTRIBUTION: TAS, WA, SA, VIC, NSW, ?QLD; ?endemic.
ILLUSTRATIONS: Plate 49.

It is a not uncommon plant of clay-pans and salt-marshes, both inland and coastal.

P. truncata (Hedw.) Br. & Schimp.

By far the largest of our Pottias, this is the commonest and the most easily recognized species, growing on bare soil on waste ground or in suburban gardens. The stems are 3–10 mm (sometimes to 2 cm) high, often with a few branches; the leaves 1–3 mm long, soft, spreading, oblong, spathulate or oblanceolate, twisted when dry; the margins more or less entire, plane or slightly recurved; the nerve shortly excurrent in a point. The lower lamina cells are clear and rectangular, those above the middle smooth (rarely minutely papillose) clear, hexagonal or shortly rectangular, about 17–20 μm across.

The seta is *c* 2–6 mm long with a gymnostomous, ovate to obovate capsule which is wide at the mouth after the operculum is shed. The very characteristic operculum is low conical with a sudden long narrow oblique beak (like *brevicaulis*). There is no peristome and the calyptra is smooth, covering half the capsule. The spores are 22–34 μm, very finely and closely papillose so that they appear slightly rough.

DISTRIBUTION: TAS, SA, VIC, ACT, (?elsewhere); a cosmopolitan species.
ILLUSTRATIONS: Plate 28; Watson (1955), Fig. 46.

P. davalliana (Sm.) C. Jens.

These tiny plants with stems about 2 mm high grow on calcareous soil. The leaves are 1–2 mm long, ovate, acute with the margins revolute usually right to the apex; the nerve, often with numerous papillae on the abaxial surface, is excurrent in a short, sharp reddish point of several cells. The upper lamina cells are more or less hexagonal, *c* 12–18 μm wide, and very papillose.

The seta is about 2–4 mm long, the capsule shortly oval and gymnostomous, with a conical operculum slightly apiculate but not beaked; the calyptra is rough. The spores, 28–30 (–40) μm are opaque with sharp papillae.

DISTRIBUTION: TAS, WA, SA, VIC; ?cosmopolitan.

ILLUSTRATIONS: Watson (1955), Fig. 47.

P. brevicaulis (Tayl.) C.Muell. has often been included with *davalliana* (Willis, 1954a) possibly because of mixed specimens, but the type material, according to Dr D. Chamberlain (pers. comm.), differs from *davalliana* in the short stumpy leaves with occasionally acute but mostly obtuse apices; the margin usually recurved but not right to the apex, and the nerve (expanded towards the apex as in *Desmatodon convolutus* and with protuberant very papillose cells on the adaxial surface) ending just short of the apex or percurrent, rarely excurrent. The abaxial surface of the nerve is much less conspicuously papillose than in *davalliana* and *scabrifolia*. The cells in mid-leaf are hexagonal, 11 μm or a little more, densely papillose; the margins entire except for the fine crenulations of the papillae.

The seta is often comparatively long, the capsule narrower than in *davalliana* and the operculum has a well marked beak as in *truncata*. The spores are 25–28 μm with papillae rounded not spinose.

Plants attributable to this species have been found in northern VIC and part of the description has been derived from them.

DISTRIBUTION: WA, VIC, NSW; endemic.

ILLUSTRATIONS: ?

PLATE 28. *Pottia truncata* VIC—Moist plant × 22. Basal cells (below) and upper cells × 1000

P. scabrifolia Bartr.

This species is similar in habit to *davalliana* and *brevicaulis* and the description has been compiled from an examination of part of the type material from UWA herbarium.

The stems are 1–2 mm tall with basal leaves very tiny, increasing to little more than 1 mm long at the apex. The leaves are mostly obovate with an obtuse apex, occasionally rounded or indented but usually ending in a short apiculus, topped by a single cell. The margin, entire except for the crenulation of papillae, is recurved practically to the apex; the nerve, strongly papillose on the abaxial surface and covered by very papillose lamina cells on the adaxial side, usually finishes in a very short single-celled point, or sometimes fails below the apex. The upper leaf cells are densely papillose, often thick-walled, 10–12 µm wide and less than 2 × 1; those below are smooth, rectangular and 3–7 × 1.

There is usually one perichaetial leaf which has a plane margin and very rounded apex. The seta is 3–4 mm long; the capsule erect, with theca 0·5–1 mm × 0·25–0·4 mm, cylindrical and tapering to the seta (mainly narrower than *brevicaulis*), peristome absent; the low conic operculum with a short rather narrow oblique beak about 0·2 mm long. The calyptra is narrow, side-split at the base, slightly rough with sparse rounded papillae; the spores very finely papillose to almost smooth, not spiny as in *davalliana*, 20–22 µm.

It is autoicous with the male bud-like in a leaf axil below the apex.

DISTRIBUTION: WA, SA; endemic.

ILLUSTRATIONS: ?

P. scabrifolia may be distinguished from *davalliana* by the obtuse apex and short abrupt point, not a reddish sharp mucronate point of several cells; from *brevicaulis* by the lamina cells on adaxial surface of the nerve not protuberant.

A specimen from SA (Monash 677) appears to be closest to this species except for a shorter capsule and more papillose spores.

P. starckeana (Hedw.) C.Muell.

These minute plants, like *davalliana*, grow on calcareous soil. The leaves usually have strongly recurved margins and the nerve excurrent in a brief mucro; the upper lamina cells are obscure and papillose, *c* 12–15 µm. It differs from *davalliana* in the presence of a peristome with pale, short, blunt papillose teeth. The operculum is obtuse and conical and the calyptra scabrid as in *davalliana* but the spores (25–34 µm) are smooth, translucent and with large vesicles ("resembling in miniature bags filled with apples": Venturi quoted by Dixon, 1924a, p. 187).

This is also a very variable species, in southern Australia in particular. The peristome shows various degrees of reduction and specimens having the other typical characters of the species may have no peristome—in that case the clearest difference from *davalliana* is the character of the spores but the spore ornamentation is also variable. Clearly further study is necessary in the group of *davalliana*, *brevicaulis*, *scabrifolia* and *starckeana*.

DISTRIBUTION: TAS, WA, SA, VIC, ?NSW, ACT; ?cosmopolitan.

ILLUSTRATIONS: Dixon (1924a), Plate 23.

It is common in mallee country.

P. tasmanica Broth. sounds suspiciously similar from Brotherus' description, and material in Herb. BM appears identical.

Pottia sp.

What appears to be a new species has been found on saline clay around alkaline lakes at Underbool and Dimboola, VIC, under *Salicornia* and *Arthrocnemon*, sometimes with the liverwort *Carrpos*. Vegetatively the plants have a superficial resemblance to Funariales.

The stems are about 2 mm long, the upper leaves larger than the lower, *c* 2 mm long, concave, erect or slightly spreading when wet and more or less crumpled when dry, spathulate, oblanceolate or obovate with the apex obtuse (occasionally acute) and apiculate. The margin is entire and plane or irregularly and weakly recurved; the nerve weak, usually finishing short of the apex. The lower cells are very large, rectangular, lax and thin-walled, the upper cells clear, smaller (20 µm wide) shortly rectangular to quadrate or rhomboidal, smooth to variably papillose, with 5–6 small papillae per cell; the marginal rows of cells mostly smooth.

The seta is about 3 mm long; the capsule elliptical, oblong or obovate, *c* 1 mm long, the operculum, only just covered by the calyptra, is conical tapering to a long oblique beak. There is no peristome. The spores are 36–48 µm with irregularly granular ornamentation.

We prefer not to describe this as a new species until we know much more about this challenging genus.

P. willisiana Sainsbury (1956c) has been described for central Australia, but overlooked by the Index.

P. heimii (Hedw.) Fuernr., a widely distributed gymnostomous species, has been recorded for Tasmania but has not been seen by us. It is usually distinguished by a bordered leaf, slightly denticulate in the upper margin, and a columella which adheres to the operculum at dehiscence.

Acaulon C. Muell.

The tiny bulb-like plants are only 1–2 mm high, including the leaves, and usually grow in patches on bare ground. The extremely short stem has a few erect, overlapping leaves which are broad, concave and nerved, the upper much larger than the lower, convolute and completely enveloping the cleistocarpous capsule.

The seta is very short, the foot short and thick and the vaginula rounded and obovate; the capsule is globose with an extremely small obtuse apiculus and no evidence of a line of dehiscence. The calyptra is very small, conical, entire or irregularly torn at the base, and rather like a tiny night-cap.

Key to species

1. Leaf margin plane or incurved; leaf as wide as long,
   or wider. Cells not papillose . . . . . .　　2
   Leaf margin partly recurved; leaf longer than wide.
   Some cells usually papillose . . . . . .　　4

2. Leaves with 2–3 irregular (inconspicuous) longi-
   tudinal lamellae on adaxial surface of the nerve.
   Plants apparently confined to low rainfall areas　　*Acaulon* sp.
   Leaves without lamellae . . . . . .　　3

3. Leaves rounded on abaxial surface, pale green,
   (brownish white when old). Common, widely
   distributed . . . .　　*integrifolium*
   Leaves strongly keeled, usually reddish gold. Plant
   markedly triquetrous seen from above. Rare .　　*triquetrum*

4. Plant usually reddish in colour. Nerve red-brown
    or yellow, excurrent with recurved arista. Cells
    mostly papillose. Spores 30–40 μm .   .   .   .   *crassinervium*
    Plant green. Nerve excurrent, arista straight some-
    times long. Cells variably papillose. Spores
    25–30 μm .   .   .   .   *robustum*

A. integrifolium C.Muell. (=*apiculatum* of the Handbook)

Several species have been reduced to *integrifolium* by Willis (1954a, 1957a). The species is very variable but in the typical plant the pale green leaves (brownish white when old) are rounded, not keeled, on the abaxial surface, very broadly ovate, apiculate, and with a practically entire margin. The nerve either finishes before the apex or is percurrent or very slightly excurrent. The cells of the upper lamina are quadrate to shortly rhomboidal, thick- or thin-walled, 12–15 μm wide; those in the lower lamina shortly rectangular and thin-walled.

The spores are smooth to granular and measure 28–40 μm.

DISTRIBUTION: TAS, WA, SA, VIC, NSW, ACT, ?QLD; also in New Zealand.

ILLUSTRATIONS: Plate 48; Handbook, Plate 30; Roth (1911), Plate 22; Wilson (1854), Plate 83 (as *Phascum apiculatum*).

This species is common and may even be found on undisturbed soil in Melbourne suburbs.

A. triquetrum (Spruce) C.Muell., recently found for the first time in Australia, on bare red soil, is even smaller, mostly less than 1 mm high, and has a marked triquetrous appearance from above because the strongly nerved upper leaves are sharply keeled. The plant has a reddish colour, the leaf margin near the apex is more or less serrated, and the golden-brown nerve is shortly excurrent in a reflexed arista.

The spores with small spiny papillae are smaller, 25–35 μm.

DISTRIBUTION: WA, SA, northern VIC, NSW; also Europe, N. Africa, N. America.

ILLUSTRATIONS: Dixon (1924a), Plate 22.

Another *Acaulon* species, as yet without a specific name, occurs in the dry north-west of Victoria (also WA, SA, NSW). It is very variable with regard to the length of the excurrent nerve and presence or absence of a hair-point but is consistently characterized by the irregular, green lamellate outgrowths from the upper adaxial surface of the nerve, a condition not previously found in *Acaulon* and resembling *Pterygoneurum* [38]. Like *Pterygoneurum* these plants also have an extremely long and strong rhizoidal system. Further study is required.

A. robustum Broth. ex Roth.

Several early specimens of *robustum*, one of which is labelled "co-type", are in the NSW Herbarium and specimens attributable to this species have recently been found in NSW and northern Victoria. Again the species is rather variable but different from *A. integrifolium*. The short stem often branches so that the plants are in clusters of 2 or 3; the leaves, of a deeper green, are longer and more pointed and, when dry, appear keeled and sometimes slightly twisted. The leaf margin, which is sometimes slightly denticulate above, is usually recurved from just below the apex, and the strong nerve is excurrent in a short or long arista which may be hyaline. The cells are smooth to variably papillose.

The seta is slightly longer and may be a little bent and the small calyptra is side-split. The spores are *c* 25 μm and the capsule is often exposed when ripe, by spreading of the leaves.

DISTRIBUTION: VIC, NSW; endemic.

ILLUSTRATIONS: Roth (1913), Plate 2.

A. crassinervium C.Muell.

The co-type of this species also is in the NSW herbarium. Specimens which we have found in dry areas of VIC and SA agree with the type material and must be ascribed to this species.

The ovate–lanceolate leaves, longer and more pointed than *integrifolium*, are reddish and have a more or less entire margin which is recurved, often to more than halfway down the leaf. The strong red-brown nerve is excurrent in an arista which is often recurved and sometimes hyaline at the tip. The cells of the upper lamina are variably papillose on either or both surfaces and on the margin. Papillae may also occur on the adaxial surface of the nerve.

The calyptra is tiny but split on one side; the spores measure *c* 35–40 μm. The leaves spread and expose the capsule at maturity.

The type of *Phascum tasmanicum* Dix. & Rodw. was examined at the British Museum, and appears to be referable to this species.

DISTRIBUTION: TAS, WA, SA, VIC, NSW, ?QLD; endemic.

ILLUSTRATIONS: Dixon and Rodway (1923).

Acaulon appears to grade into *Phascum* and plants with some features of both genera, like *A. crassinervium* and *A. robustum*, are difficult to place. Typically *Phascum* (e.g. *P. cuspidatum*) is larger than *Acaulon* with leaves gradually increasing in length upwards, concave, ovate to lanceolate, definitely longer than wide, with narrowly recurved margins; the nerve excurrent in a cuspidate point. The seta is longer and the foot narrower; the capsule with a more defined apiculus and a small side-split (asymmetric) calyptra rather than the tiny conical type of *Acaulon* which just perches on the very top of the capsule. Papillose cells may be present in both genera but are more common in *Phascum*.

In the northern hemisphere *A. floerkeanum* is considered to be an intermediate between the two genera and has been credited to Australia but this is possibly *A. crassinervium*. Roth (1911) put both these species in a subgenus of *Phascum* called *Microbryum*, with intermediate characters.

Similarly the very variable *A. robustum* seems to approach the northern hemisphere *Phascum cuspidatum*, but without more study it appears preferable to leave both *robustum* and *crassinervium* with their present status.

Aloina Kindb.†

A. sullivaniana (C.Muell.) Broth.

At first sight similar to *Crossidium* [36] and often mixed with it, this can be distinguished by the inflexed leaf margin with no recurvature.

The stems are *c* 2 mm with leaves more or less concave, very thick, very dark green or with a reddish tinge, broadly oblong, 0·75–3·0 mm long, rounded and cucullate at the apex, about 10 in a rosette at the stem apex with a few smaller below. The leaves are incurved when dry, spreading when moist, and have a very long hyaline point and a short clear sheathing base; the adaxial surface of the very broad nerve in the upper lamina is covered by branching photosynthetic uniseriate filaments of cells. These are overarched by the widely inflexed colourless margin of the leaf (similar to *Polytrichum juniperinum*). The margin is entire or minutely crenulate and the apex cucullate. The nerve is very broad and flat, not showing on the abaxial surface but excurrent in a long flexuose, hyaline, more or less smooth arista about 1 mm long in the upper leaves; in the lower leaves only cuspidate. The cells of the incurved lamina are mostly without chlorophyll, thick-walled, elliptic and elongated transversely. Cells are smooth; those on the abaxial surface of the leaf mostly small, isodiametric to transversely elongated and thick-walled (a few cells in the lower part of the leaf shortly rectangular).

The seta is long (to 1·5 cm) and flexuose; the capsules more or less cylindrical but slightly curved and tapered slightly to the mouth, 3·0–3·5 mm long, erect. The operculum is long-beaked, usually ½ or more the length of the theca. There is a peristome of 32 filiform papillose teeth from a basal membrane, and a long narrow calyptra with a short split at the base; it covers only the top of the theca. The spores are green, more or less smooth, 10–13 μm.

DISTRIBUTION: WA, SA, VIC, NSW; endemic.

ILLUSTRATIONS: ?

In the field it is distinguished most readily from *Crossidium* by the lack of any recurving of the leaf margin and by there being no conspicuous nerve showing on the abaxial surface, a feature easily seen with a lens in dry specimens.

From *Pterygoneurum* [38] with a hair-point it can be distinguished by the much longer, peristomate, capsules. *Pterygoneurum* is a paler green, the nerve is apparent on the abaxial surface and is sometimes golden and it has lamellae instead of just filaments on the adaxial upper part of the nerve.

A. ambigua (B.S.G.) Limpr.

This second species which is mainly northern hemisphere in distribution has been found in the area and is most easily distinguished from *sullivaniana* by the absence of a hair-point and by the presence of elongated cells on the abaxial surface of the nerve in the mid-region of the lamina. The upper leaves are markedly hooded.

DISTRIBUTION: SA, VIC; also in Europe, Asia, N. America, S. Africa; mainly northern hemisphere.

ILLUSTRATIONS: Dixon (1924a), Plate 24 (as *Tortula*).

PLATE 29. *Calyptopogon mnioides* VIC—Vegetative shoot moist (left) and dry ×7. Display of cells extending inwards from the leaf margin (right) including intramarginal border × 1000

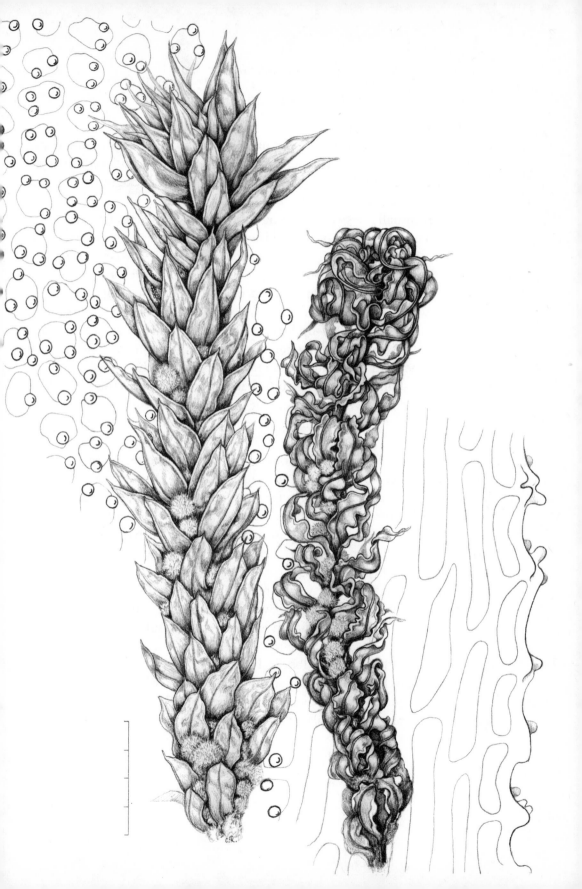

Calyptopogon (Mitt.) Broth.

C. mnioides (Schwaegr.) Broth.

This is a fairly common moss on trunks, branches and twigs of trees and shrubs in montane to subalpine districts, growing in small rather open cushions like a big *Orthotrichum* [53] or *Ulota* [57] but distinguished by the peculiarly crisped leaves when dry.

Stems are usually 1·5–2·0 cm long; the leaves with undulate margins, spreading when moist and tightly twisted and contorted when dry in a way which, once known, can easily be recognized in the field. The best confirmatory characteristic is provided by the clusters of gemmae on the adaxial surface of the nerve at the leaf tips and below, visible especially when the leaves are wet and opened out but usually detectable even when dry and partly covered by the rolled-up leaves. The leaf at the apex is folded into a keel or sometimes a tube which helps to contain the gemmae. The tufts of rhizoids on the stem (see Plate 29) are not always detectable and serve more to anchor the stem to the substratum than to bind the stems together. The nerve is strong and brown, projecting on the abaxial surface of the leaf, especially when dry, and extending beyond the leaf apex as a short, blunt point.

Cells in mid-leaf are roughly isodiametric, c 12 μm across, with rounded corners, elongated to 50–90 × 10 μm in the leaf base; those at the margins thick-walled, narrow and hyaline, forming a very wide conspicuous border which is usually intramarginal above, having one or two rows of green quadrate lamina cells outside it. The isodiametric cells are papillose on both surfaces with several papillae per cell, more pronounced on the adaxial surface. Elongated cells in the leaf base and margins are smooth.

Capsules, which seem to be very rare in Australia, are erect and cylindrical, c 2 mm long, almost equalled or overtopped by the long, straight perichaetial leaves. The seta is similar in length to the capsule.

DISTRIBUTION: TAS, VIC, NSW; also in New Zealand and South America.

ILLUSTRATIONS: Plate 29; Handbook, Plate 30.

Both in growth form and leaf structure this species is very distinct and cannot be mistaken for any other once it is known. Gemmae almost always seem to be present and the plants are quickly identified once they are seen.

Crossidium Jur.

C. geheebii (Broth.) Broth.

These minute calcicolous plants grow singly or in groups on bare earth in dry areas. The stems, *c* 2 mm high, are usually buried in soil so that only the top tuft of leaves is exposed. The leaves are concave, oblong or ovate, 1·0–1·5 mm long, with an apex which is obtuse, acute or almost cucullate, with entire, narrowly revolute or recurved margins; the lower leaves smaller and often awnless, the upper larger and usually with a long smooth hyaline arista. When moist they are erecto-patent, when dry incurved, slightly twisted and overlapping. The nerve, which is conspicuous on the abaxial surface, is wider in the upper lamina where it is covered on the adaxial surface with short branching filaments of cells which are deep green and variably papillose. The cells in the lower lamina are clear, more or less rectangular, 23–40 μm wide, and thin-walled; those above quadrate to shortly oblong, 12–15 μm wide, firm-walled and ranging from smooth or indistinctly mamillose to papillose.

The seta is *c* 8 mm long; the brown capsule *c* 1·5 mm, erect to slightly curved, oblong–elliptic, with a narrow mouth and a peristome of 32 papillose filiform teeth borne on a low basal membrane. The operculum is conical or beaked, less than half the length of the capsule, which is ¾ covered by the calyptra which, in turn, has a long subula at the top and is split at the base. The spores are brown, 10–15 μm, rough with fine papillae.

DISTRIBUTION: WA, SA, VIC, NSW, ?NT; also in New Zealand.

ILLUSTRATIONS: Handbook, Plate 30.

There is a good deal of variation in leaf shape, length of arista, papillosity of cells, filaments, shape of operculum and perfection of the peristome; more study is necessary to determine if there is just one polymorphic species, but the indications are that *C. geheebii* is dioicous and there is, in addition, a monoicous species and also a gymnostomous one.

Desmatodon Brid.

D. convolutus (Brid.) Grout

On rather dry forest soils or clay banks, mallee soils etc., this is one of the commonest species, but is seldom conspicuous. The very short stem,

1–2 mm tall, is often almost entirely buried in silt or sand, only the tips of the leaves, tightly coiled down in a flattish disc, projecting above soil level. The leaves are wholly distinctive, *c* 1·3–2·0 × 0·3–0·5 mm, very thick and spathulate at the tips with the nerve projecting as a short reddish spine-like mucro. The leaves in cross section are U-shaped, but with the margins tightly recurved along the upper ⅔ of the leaf. The nerve is yellowish and *widened* below the apex forming a pad, densely covered by isodiametric separate papillose cells. The upper lamina cells are irregularly quadrate or hexagonal, roughly isodiametric, 10–15 μm, papillose but often with the papillae low enough to be inconspicuous; at other times very dense and obvious. The basal cells are hyaline, rectangular, 35 × 20 μm, thickened on the cross walls and at the corners, narrower at the margins but not forming a distinct border.

The yellowish seta, *c* 3–10 mm long, carries a narrowly ovoid, erect capsule, *c* 1·4 × 0·3–0·4 mm, dark purplish brown with age; the peristome teeth slightly oblique, 0·2 mm long (sometimes to 0·4 mm) including a basal membrane of *c* 40 μm, orange fading to yellowish, finely papillose. The 16 teeth are split to the membrane into 32 segments which break off early to leave the capsule open. The operculum has a thick oblique beak, *c* 0·5 mm long. The perichaetial leaves are not distinct.

DISTRIBUTION: TAS, WA, SA, VIC, NSW, ACT, NT, QLD; apparently absent from most of Asia but otherwise world-wide.

ILLUSTRATIONS: Dixon (1924a), Plate 24 (as *Tortula atrovirens*).

The leaf structure is very characteristic. For other species with "granular" nerves, see the Key to Genera.

D. adustus Mitt. differs from *convolutus* in the light brownish-green club-shaped shoots, *c* 5 mm tall, with tiny *incurved* lanceolate leaves (margins recurved). The concave perichaetial leaves are conspicuous, twice as long as, and much broader than, the vegetative leaves.

DISTRIBUTION: WA, VIC; endemic.

ILLUSTRATIONS: ?

D. reflexidens (Hampe) Jaeg. (VIC) is the other Australian species. We do not yet know it.

Pterygoneurum Jur.

These tiny bud-like plants grow scattered or in groups on bare earth. The leaves are usually obovate, very concave, and diagnostically the nerve in the upper adaxial part is furnished with 2–4 longitudinal green lamellae. The upper cells of the lamina are roughly square, those below shortly rectangular. The emergent capsule is ovoid to shortly cylindrical, *c* 1 mm long, the peristome absent or rudimentary.

P. ovatum (Hedw.) Dixon

The stems are thick but very short, less than 1 mm long, often with a very strong and extensive rhizoid system which is much longer than the whole plant including the sporophyte. The leaves are less than 1 mm long, concave, broadly obovate with the apex obtuse or sometimes cucullate. The margin is more or less entire, widely and strongly inflexed at the summit and partly covering the irregular longitudinal lamellae on the upper part of the nerve. The nerve is narrow, often golden, and may be excurrent in a strong, more or less smooth, hyaline spike or long curved hair-point. Cells of the upper portion of the lamina are clear, squarish, *c* 10–20 μm (usually smallest towards apex) with firm but not thick walls; the lower sheath-like part of the leaf consists of larger, very shortly rectangular, sometimes lax cells. The lamellae are up to 8 cells high, chlorophyllose, and sometimes with very short filamentous outgrowths.

The seta is 2·0–2·5 mm, sometimes flexuose; the capsule erect, shortly cylindrical with a small neck and constricted below the mouth, irregularly wrinkled when dry. There is a rather flat or low-conic operculum with a long, often oblique beak (like *Pottia truncata*, 32), the calyptra side-split and subulate covering most of the capsule. There is no peristome and the spores are 40–50 μm (28–32 μm in some specimens), dark brown, densely granular.

DISTRIBUTION: WA, SA, VIC, NSW; widespread in the northern hemisphere.

ILLUSTRATIONS: Dixon (1924a), Plate 23 (as *Tortula pusilla*); Grout (1928–40), 1, Plate 98.

Some of our material matches European *P. ovatum* very closely, but it is very variable and there may be more than one species involved.

P. kemsleyi J. H. Willis, described from WA, has no hair-point to the leaves, and lamellae with a thick outgrowth of filaments. The spores are 25–40 μm, very finely papillose.

DISTRIBUTION: WA, SA, VIC, NSW; endemic.
ILLUSTRATIONS: Willis (1954b) p. 9.

Tortula Hedw.

This genus of common, cushion-forming mosses is usually recognizable, even vegetatively, by the wide obtuse leaves mostly widest well above the middle, excurrent nerve, densely papillose upper cells (not all species) and empty hyaline cells in the leaf base. Most of the species fruit freely; the long narrow capsule has a pale orange-pink or reddish peristome of 32 long thread-like papillose teeth spirally twisted above and joined at the base to form a tube. None of the Australian species, that we know of, has the leaves clearly bordered. Tortulas are common mosses of a wide variety of well drained habitats—stones, soil and trees. *Pterygoneurum* [38], *Desmatodon* [37], *Aloina* [34], *Crossidium* [36] and perhaps also *Calyptopogon* [35], which are treated in this book as separate genera, used to be classified as Tortulas; they have very similar peristomes but the first four differ in the thick fleshy leaves in which the nerve has outgrowths on the adaxial surface: filaments, lamellae, granules etc. Other genera which might be confused with *Tortula* are *Barbula* [46], *Tortella* [43], *Pottia* [32] and *Encalypta* [51]. They are distinct in fruit but can give trouble vegetatively.

Key to species

1. Gemmae present either on leaves or at the shoot
   apex   2
   Gemmae absent   3

2. Spherical gemmae on adaxial surface of leaves.
   Abaxial surface of nerve usually papillose . .   *papillosa*
   Gemmae like miniature leaves in a cluster at the
   stem apex. Nerve smooth abaxially . . . .   *pagorum*

3. Leaf apex sharply toothed; nerve scarcely excurrent   *rubra*
   Leaf apex entire; nerve excurrent in a hair . .   4

4. Nerve reddish brown; hair-point denticulate.
    Stems usually more than 0·5 cm long; forming
    yellowish brown turfs often in sandy soil . . *princeps*
    Nerve usually green or yellowish; hair-point
    smooth. Stem usually less than 0·5 cm long;
    forming hoary green cushions often on brick-
    work . . . . *muralis*

T. princeps De Not.

Probably the commonest species of *Tortula* in Australia, this moss mainly
inhabits dry soils but is also epiphytic and is often very abundant in the
mallee, dry sclerophyll forest and sandy areas. It is an attractive plant with
stems 0·5–2·0 cm high and leaves *c* 2 mm long or more, partly wound round
the stem when dry, showing the shiny reddish brown nerve which contrasts
strikingly with the green lamina. The margins are usually revolute below.
The cells above are regularly arranged in rows and files, and are densely
pluri-papillose; those below, in the basal *c* ⅓ of the leaf, are much enlarged,
hyaline and empty, except for the marginal few rows of rather narrower
cells which tend to stay green.

The capsule is *c* 2–3 mm long or more, on a seta 0·5–1·0 cm long, twisted
when dry with a right-hand thread. The long peristome (to 0·8 mm or more)
of slender reddish teeth is tightly twisted when dry, in the same direction,
and with a conspicuous tube, for about half its length, at the base. The operculum is long and pointed.

DISTRIBUTION: TAS, WA, SA, VIC, NSW, ACT; almost world-wide in distribution,
except for parts of the tropics.

ILLUSTRATIONS: Handbook, Plate 31; Wilson (1859), Plate 172 (as *T. antarctica, T. cuspidata, T. rubella*).

This is a very variable moss indeed. On the whole, Australian material
seems to be considerably smaller than New Zealand plants and more
variable. The leaf arrangement, characteristic of European plants, where the
leaves tend to occur in dense clusters at intervals up the stem (interruptedly
comose) is not often seen in either Australian or New Zealand specimens.
The variability, which extends to all the vegetative characteristics, is much in
need of thorough investigation. Some specimens may be referable to the
New Zealand *T. tenella* Broth.

This species is most likely to be confused with *Barbula pseudopilifera* [46], which can be distinguished from it by the yellow nerve, and with *Tortula muralis* (see key).

Ramsay (1967b) has recorded *T. bealeyensis* R.Br.ter. from Tasmania but we have seen no material. It has unbordered leaves and a smooth, red excurrent nerve.

T. rubra Mitt.

This moss is about as robust as *T. princeps* although up to 5 cm tall, but quite distinct in the leaf apex which is clearly and irregularly toothed with a shortly excurrent nerve which is often expanded just as it leaves the lamina. The leaves are further distinct in the basal hyaline cells of which the uppermost tend to be seriately papillose with one or two lines of papillae down each cell. The nerve is papillose on both surfaces.

The narrowly cylindrical capsule is up to 5 mm long, slightly curved and with a conspicuous tubular peristome base.

DISTRIBUTION: VIC, NSW; also in southernmost South America and New Zealand.

ILLUSTRATIONS: Studies, Plate 8.

Not uncommon at high altitudes, over 5000 ft, on damp grassy ledges in open grassland or scrub where it forms rather loose large cushions, this plant is very like *Leptodontium* [48] which may also have seriately papillose basal cells. However, that species has very undulate leaves, a nerve which does not reach the apex, and thick-walled, not hyaline basal cells.

T. muralis Hedw.

A very common moss of old brick walls, even in the centre of the largest cities, this species forms greyish-green cushions, hoary with the long hyaline hair-points. It is usually much smaller than either of the two preceding species, and distinguishable by the smooth hair-point and green or yellow (sometimes rusty) nerve which does not contrast in colour with the lamina cells. The margins are revolute almost to the apex.

The peristome is very short (*c* 150 μm) and tightly twisted, with a tiny basal membrane.

PLATE 30. *Tortula muralis* VIC—Two shoots, one dry (left) and one moist ×7, cells from leaf-base (lower) and upper cells × 1000

CHROMOSOME NUMBER: n=48 (TAS, NSW).

DISTRIBUTION: TAS, WA, SA, VIC, NSW, ACT, ?QLD; throughout the world, although it is now virtually impossible to tell where it is native and where introduced. Together with *Bryum argenteum* this is one of the very few successful urban mosses. It is superficially not unlike *Grimmia pulvinata* [11].

ILLUSTRATIONS: Plate 30; Watson (1955), Fig. 44.

T. papillosa Wils.

This very distinct species forms dull dark greenish or brown patches on the trunks of trees, but also sometimes on walls. Globular, pale green gemmae on the upper leaf surface are diagnostic. The leaves are concave with the margins plane or incurved. The nerve is excurrent and usually papillose on both surfaces and the cells, usually with single high papillae on the abaxial surface, are large (20 μm) and sometimes smooth.

Capsules are not uncommon in Australia, although rare elsewhere. The seta is sheathed by large perichaetial leaves at the base and is twisted (left hand) above (Stone, 1971).

CHROMOSOME NUMBER: $n=7(6+ m)$ (NSW).

DISTRIBUTION: TAS, WA, SA, VIC, NSW, ACT, ?QLD; widespread in the temperate regions of both hemispheres.

ILLUSTRATIONS: Plate 31; Stone (1971).

T. pagorum (Milde) De Not.

This species is very similar in size and colour and habitat to *T. papillosa* with which it is usually mixed, but from which it differs in the gemmae which occur as a dense cluster of minute leaf-like bodies at the stem apex, nesting in the apical leaves; the leaf margins are narrowly recurved about the middle. It is also different in the smaller cells (*c* 12 μm), the nerve which is smooth abaxially, and in the papillae which are single and conical in *papillosa* but 3–4 per cell and crescentic or circular in *pagorum*.

Sporophytes are rare or overlooked and indeed have only recently been described but differ from those of *papillosa* in having no enlarged perichaetial leaves and in the seta which is twisted (right-hand) the opposite way. Full details of the sporophyte, with illustrations, can be found in Stone (1971).

PLATE 31. *Tortula papillosa* SA—Fruiting shoot (dry) and vegetative shoot (moist) × 22, gemma (bottom right) and basal and upper cells (left) all × 1000

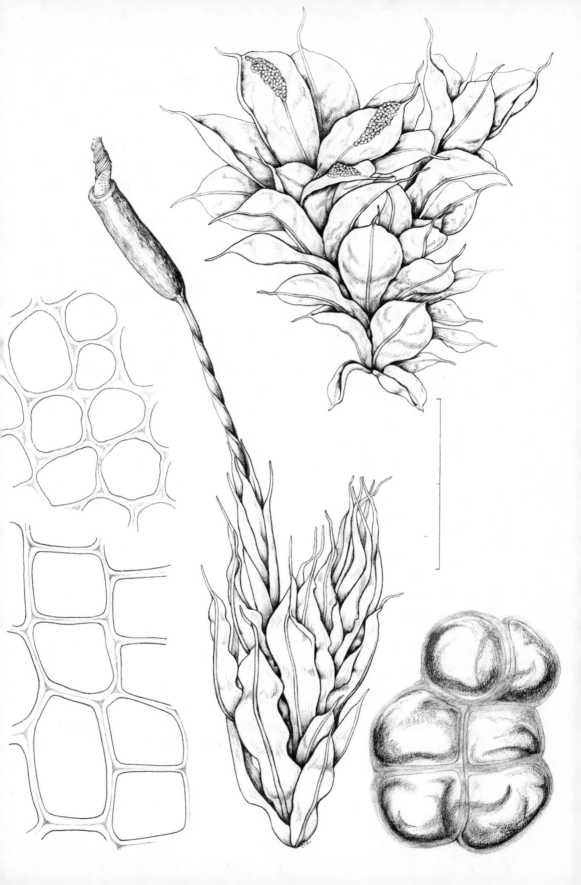

DISTRIBUTION: WA, SA, VIC, NSW, (see details in Catcheside, 1967); Europe, N. America.

ILLUSTRATIONS: Plate 32; Stone (1971).

This seems to be a plant more restricted to dry habitats than *papillosa*, and mixtures of the two, so common in most of Australia, tend to be almost all *papillosa* in wetter and colder climates. Both species occur as epiphytes on the "trunks" of *Atriplex* on the Nullarbor Plain.

Broadly speaking these five species are the only ones likely to be collected. Any *Tortula* on dry sandy soils and grassland, especially in lowland regions, is likely to be *princeps*; in alpine regions, on damper sites it may be *rubra*; any on brickwork in town or country will probably be *muralis*; epiphytes will usually be either *papillosa* or *pagorum* or, in most cases, a mixture of both.

Of the remaining species recorded for Australia, *T. baileyi* Broth., found only once (in Adelaide) has gemmae like *pagorum* but is a distinct species according to Catcheside (1967), differing in having a nerve which is papillose abaxially, much larger cells in the upper part of the leaf (17–20 instead of 12 µm) and in the distribution and shape of the leaf papillae. It ought to be looked for in the Adelaide district, on trees. All other records of the species are *pagorum* (Catcheside, 1967).

T. adusta (Mitt.) Mitt. we have treated under *Desmatodon* [37, q.v.].

We have not seen material of any of the remaining species recorded for Australia in the Index:

T. brevisetacea (F.Muell.) Thér. (SA, VIC)

T. crawfordii (Par.) Watts (NSW)

T. evanescens Broth. (NSW)

T. parramattana Mitt. (NSW)

T. robusta Hook. & Grev. (VIC)

T. ruralis (Hedw.) Gaertn. (SA)

T. serrulata Hook. & Grev. (?VIC)

T. subbrunnea Broth. & Watts (NSW)

PLATE 32. *Tortula pagorum* SA—Fruiting shoot ×15, gemmiferous shoot both moist (above) and dry ×37. Gemma (bottom right) ×500, cells from leaf-base (lower left) and upper cells ×1000

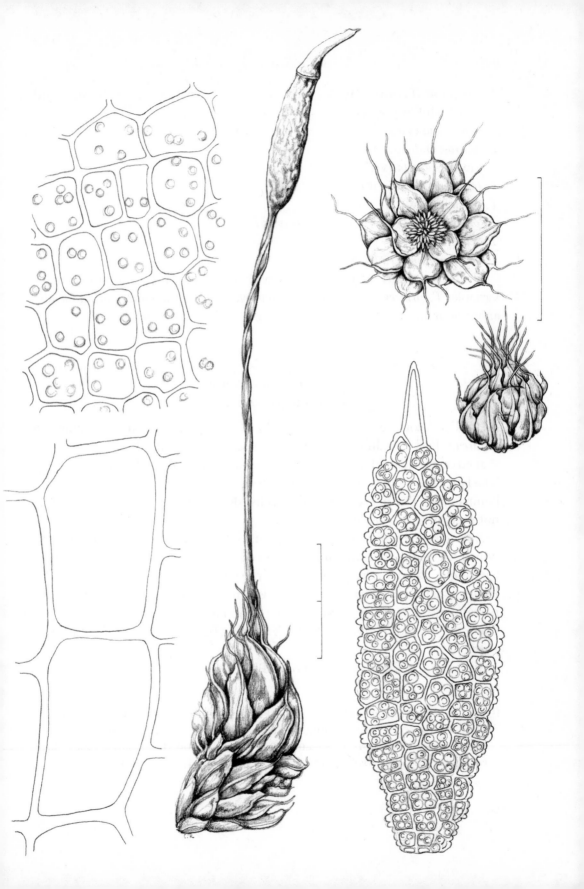

T. subspirals (Hampe) Broth. (VIC), which appears to be a *Desmatodon*, and the Carl Müller species:

T. androgyna (VIC)

T. austro-ruralis (QLD)

T. flexomarginata (VIC)

T. latrobeana (TAS, VIC)

T. murina (NSW)

T. panduraefolia (=*princeps, fide* Rodway, 1914)

T. readeri (NSW)

T. streptopogoniacea (NSW)

No doubt many of these names will refer to forms of the immensely variable *T. princeps*. The species: *brevisetacea, flexomarginata, panduraefolia* and *subspiralis* are illustrated in Mueller (1864) in the genus *Barbula*.

(POTTIACEAE—EUCLADIOIDEAE)
Anoectangium Schwaegr.

A. bellii Broth. ex Dix. is rather hard to separate from *Gymnostomum calcareum* [41] but differs in the transverse section of the leaf which is W-shaped on the adaxial surface in *Gymnostomum* but V-shaped in *Anoectangium*. The latter has convex wings and the nerve smooth adaxially, not covered by lamina cells. The nerve conducting cells (deuter) are on the adaxial surface, not median as in *Amphidium* [54, q.v.].

Anoectangium bellii is also liable to confusion with *Zygodon intermedius* [58] which has leaves of a similar appearance in cross-section, but *Anoectangium* has longer, narrower leaves with the nerve papillose abaxially for most of the length whereas in *Zygodon* the nerve is smooth abaxially, except at the tip. DISTRIBUTION: VIC; also in New Zealand. So far it is known in Australia only from the Byaduk Caves in S.W. Victoria (Beauglehole and Learmonth, 1957).

ILLUSTRATIONS: Plate 33; Handbook, Plate 25.

PLATE 33. *Gymnostomum calcareum* SA—Fruiting plants, extracted from cushion, ×22, leaf ×50; sections through leaf-base (lowermost section) and upper part of leaf (middle section) upper and basal leaf cells (top right), all ×1000; *Anoectangium bellii* VIC—Section through leaf (upper section) ×1000

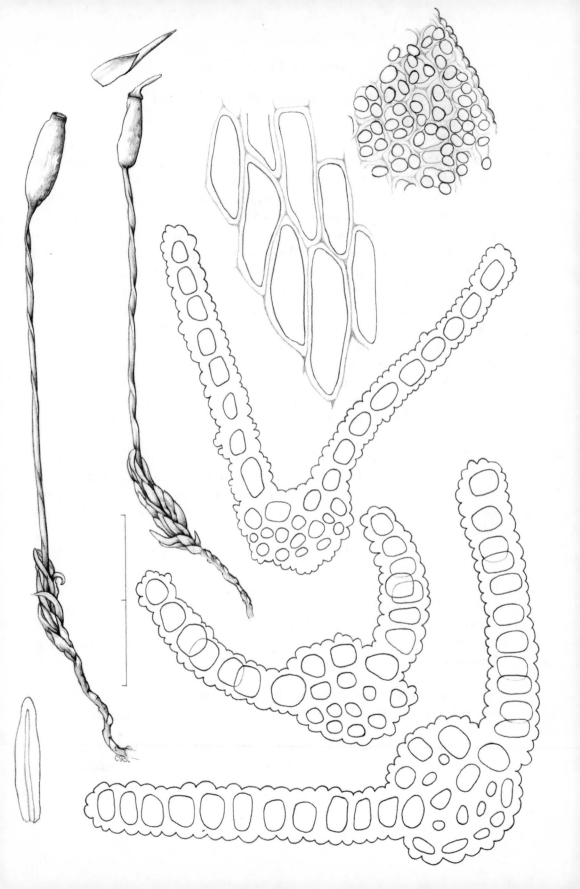

Gymnostomum Nees & Hornsch.

G. calcareum Nees Hornsch.

Not uncommon on the roofs of caves and overhanging cliffs, especially on calcareous rock, this species forms bright, pale green patches, often large and often matted together with tomentum into a hard corky pad.

The stems are 2–25 or more mm high, parallel and closely packed, often tomentose, the upper 1 mm or less a vivid yellow-green, the rest brown. Leaves are narrow, almost parallel-sided, 0·5–1·0 mm long × 0·1–0·25 mm wide, arched back away from the stem when moist, lightly twisted and incurved when dry, with plane entire margins. The leaf apex is particularly variable, commonly rounded or even semicircular, but sometimes tapering to a sharp point. The nerve is strong and papillose abaxially, failing in or below the apex or just excurrent, covered adaxially by a layer of papillose cells, hard to distinguish except in transverse section. The adaxial surface of the upper half of the leaf is deeply grooved, W-shaped in transverse section with slightly concave wings, giving a somewhat tubular appearance to the leaf. Cells above are squarish hexagonal, *c* 7–10 μm across, densely papillose with approximately 4 papillae per cell; those below elongated, thick-walled, smooth, 17–20 × 7 μm.

The yellowish capsule is elliptical, usually *c* 1 mm long but sometimes shorter, on a straw-coloured seta 3–5 mm long; the mouth red, peristome absent (gymnostomous); operculum with a long, usually oblique beak almost half the length of the rest of the capsule; calyptra side-split, soon falling off.

DISTRIBUTION: TAS, WA, SA, VIC, NSW, ACT, QLD, NT; almost world-wide in distribution except in the tropics.

ILLUSTRATIONS: Plate 33; Handbook, Plate 26.

It is a strongly calcicolous moss, closely resembling *Amphidium* [54] and *Anoectangium* [40]. Any large deep pad of vivid pale green moss on a rock wall or cave is likely to be one of these three; by far the commonest is *Gymnostomum*.

Amphidium cyathicarpum differs in the duller colour, longer, more slender, curlier leaves and having the nerve smooth, instead of papillose adaxially.

Anoectangium bellii (*q.v.*) is much more troublesome to separate and *Zygodon intermedius* [58] has been known to occur in similar habitats also.

(Pottiaceae—Trichostomoideae)
Tetrapterum Hampe ex Jaeg.

T. cylindricum (Tayl.) Jaeg. (= *Astomum cylindricum* (Tayl.) Mitt.)
Typically this species forms a low dense turf on bare earth and usually fruits freely, the cylindrical, shining, red–gold capsules, splendidly coloured, showing just above the leaves. The stems are usually less than 1 mm high but sometimes sparsely branched; the leaves are shortly tongue-shaped, the upper to 1·5 mm long, *c* 2·5–3·0 × 1, erect or erecto-patent, the lower smaller and often somewhat reflexed. The leaves are wider in the basal hyaline part, narrowing just below the middle where the papillose region begins and so slightly waisted; when moist they are more or less concave, sometimes rather undulate at the margin and at times slightly revolute or involute. When dry the leaf lamina is folded along the mid-rib so that the leaf is channelled and the leaves become incurved and twisted in a part spiral around the stem. The margin is plane and entire except for the crenulations of the papillae. There is a strong nerve, papillose on both surfaces except at the base, just excurrent and ending in a small apiculus of usually one pointed, thick-walled, golden-coloured cell. The cells below are clear, shortly rectangular, 8–10 μm wide and mostly 2–3 (–5) × 1, firm-walled; above very opaque, densely chlorophyllose and with walls covered with horseshoe-shaped papillae, more yellow than those below, small, more or less quadrate, *c* 10 μm wide.

The seta is about 1 mm (sometimes 2 mm) and the capsule *c* 1·0–1·3 mm long, cleistocarpus, usually slightly emergent, oblong, elliptical or cylindrical with a short thick apex. Often there are 4 slight ridges extending from the seta on to the capsule base, and 4 also at the apex; or, when young, the capsule may be irregularly 8-ridged. The calyptra is side-split, covering ⅓–½ the capsule and with a long subulate apex. The spores are large, 30–35 μm, brown, very coarsely papillose, the papillae flat or more often indented at the apices and sometimes merging.

The bud-like male shoot, near the base of the female shoot, has similar but shorter leaves.

DISTRIBUTION: TAS, WA, SA, VIC, NSW, ACT, QLD, ?NT ?Lord Howe; probably endemic.

ILLUSTRATIONS: Plate 34.

This is a very distinct species and, when fruiting, is unlikely to be confused with other mosses.

The following Müllerian species are also recorded:

T. brachypelma C.Muell. (SA)

T. sullivanii (C.Muell.) Broth. (VIC, NSW)

T. tetrapteroides (C.Muell.) Broth. (VIC)

T. weymouthii (C.Muell.) Broth. (TAS)

    T. cylindricum was formerly in the genus *Astomum* of which *A. brisbanicum* (C.Muell.) Broth., *novae-valesiae* Broth. ex Roth and *wattsii* Broth. ex Roth are in NSW. *A. brisbanicum* should be in the genus *Trachycarpidium* (Stone, 1975)

Tortella (Lindb.) Limpr.

From the rather similar genus, *Barbula* [46], this is distinguished by the *plane or incurved*, never recurved leaf margins and usually by the strong contrast between the hyaline cells of the leaf base and the densely papillose cells above; the junction between these two cell types is frequently sharp, passing obliquely upwards and outwards to the leaf margin. The leaves are typically undulate when moist, and tightly curled, or sometimes wound round the stem, when dry. They are usually yellowish-green in colour. The nerve is strong and stands out abaxially as a distinct rib. The capsules are similar to those of *Barbula*, with densely papillose peristome teeth split down the middle to give 32 filiform segments.

Key to species

  1. Leaves, or at least leaf tips, helically wound round
      the stem. Perichaetial leaves very long (*c* 3 mm)
      conspicuous and sheathing . . . . .   *calycina*
      Leaves strongly curled when dry but not wound
      round the stem. Perichaetial leaves not distinct.   2

  2. Peristome teeth almost straight. . . . .   *cirrhata* (+*dakinii*)
      Peristome teeth spirally twisted . . . .   *knightii*

PLATE 34. (upper) *Tetrapterum cylindricum* VIC—Fruiting plant × 30, upper and lower leaf cells × 1000; (lower) *Bryobartramia novae-valesiae* VIC—Fruiting plant × 30, cells × 1000

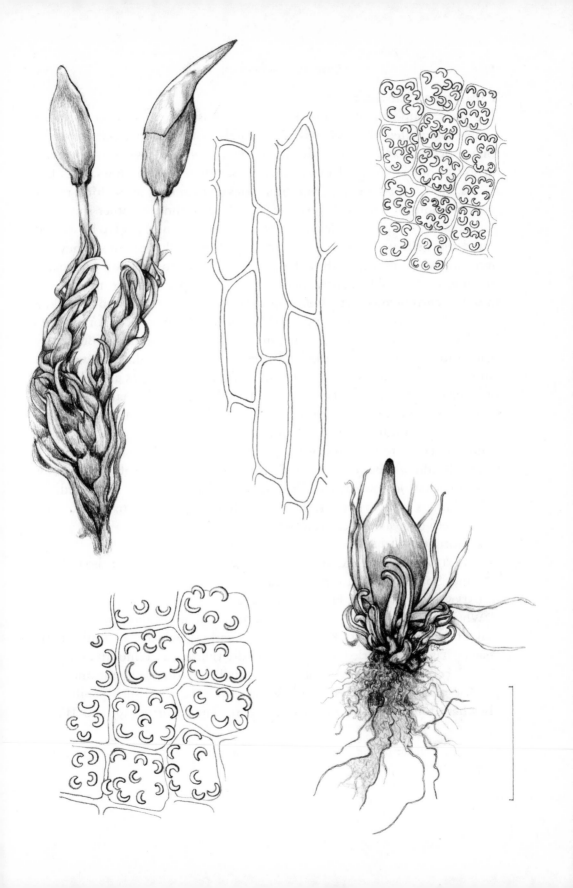

T. calycina (Schwaegr.) Dix.

The most abundant of all the Tortellas in Australia, as well as the most widely distributed, this species is immediately distinct in fruit because of the long perichaetia and is usually distinct vegetatively also. It is a common plant on soil in dry sclerophyll forest, mallee, sea dune communities, etc. The shoots, mostly 1·0–1·5 cm tall but sometimes up to 2 cm or more, form rather loose cushions of yellowish green colour. Unlike those of other species in Australia, the long narrow leaves (c 2·5–4·0 × 0·6–1·0 mm) tend to be wound round the stem with a corkscrew twist, at least at the leaf tips. They are undulate when moist, with a shallow V-shaped cross section and plane margins, occasionally slightly incurved near the apex. The strong nerve is mostly excurrent in a short, blunt cusp and is papillose, or covered by papillose cells, on both surfaces in the upper part of the leaf. In the upper ⅔ or so of the leaf the cells are isodiametric, c 8–11 μm across, and their outlines obscured by dense multiple papillae. In the leaf bases the cells are hyaline, smooth and empty, greatly elongated and wider, measuring c 100 × 18–25 μm. The junction between the two cell types is not sharp, but is usually marked near the nerve by very characteristic elongated, often golden cells which have single or double longitudinal rows of papillae.

Equally characteristic, in fruiting plants, are the perichaetial leaves which form a sheathing tube, 3–5 mm long, round the reddish base of the very long (2–5 cm) straw-coloured seta. The capsule is slim, c 2 mm long, with an operculum as long or longer. The long, densely papillose, hair-like, peristome teeth have a corkscrewed twist.

CHROMOSOME NUMBER: n = 13 (NSW), 30 (TAS), 52 (TAS).

DISTRIBUTION: TAS, WA, SA, VIC, NSW, ACT, QLD, NT, Lord Howe; also in New Zealand, South America and S.E. Asia.

ILLUSTRATIONS: Plate 35; Handbook, Plate 29.

When sporophytes are present, the perichaetial leaves and very long setas make this almost unmistakable. *Holomitrium* [27], which has similar perichaetia, has tightly curled leaves, not spirally twisted, and smooth or almost smooth cells. Vegetatively *T. calycina* is usually identifiable by the uniseriate papillae near the upper leaf base. Within our experience such papillae are elsewhere found only in *Leptodontium* [48], *Tortula rubra* [39] and *Wijkia*

PLATE 35. *Tortella calycina* VIC—Fruiting plant × 15, leaf × 50, basal and upper cells × 1000

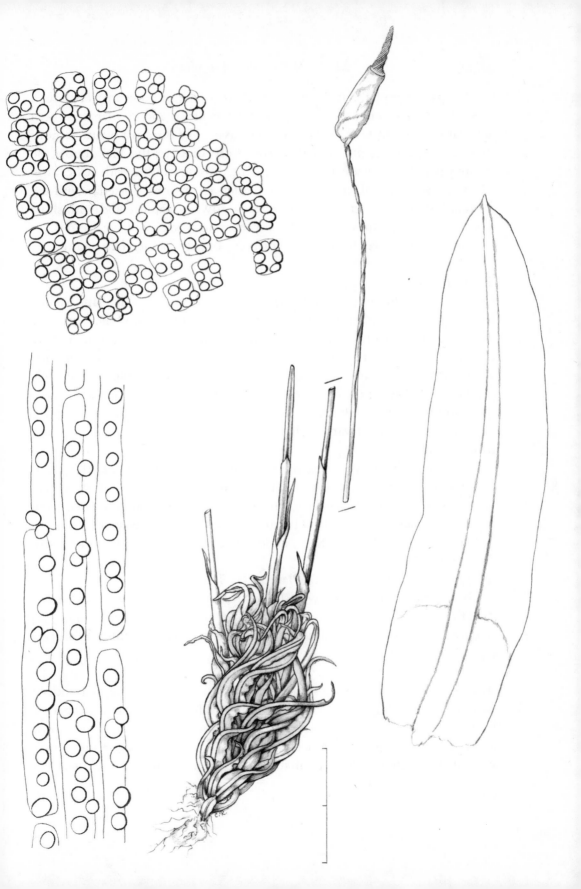

[132]. Apart from habitat differences, the first two of these are very much rarer; *Leptodontium* has the margins widely recurved and *T. rubra* has them noticeably denticulate at the apex of the leaves. The third is a pleurocarpous moss of quite different growth form. According to the Handbook the seriately papillose cells are sometimes absent, causing trouble in identification. There are certainly plants in Victoria which seem to be Tortellas with the *calycina* leaf arrangement but lacking these rows of papillae and with short setas (1·0–1·5 cm); they may well be a separate species, but we are not yet convinced. They match material (not type) of *Barbula subcalycina* [46].

T. cirrhata Broth.

Quite different in appearance from the previous species, this plant has the leaves tightly and separately curled, not spirally twisted round the stem. The leaves tend to be slightly smaller than in *calycina*, 1–3 × 0·2–0·3 mm, and are undulate and channelled, U-shaped in transverse section in the upper part, with the margins often inrolled almost to form a complete tube. The papillose upper cells may be a trifle smaller, typically 7·5–10·0 μm across, the transition to the basal hyaline cells is more abrupt and the line of junction passes upwards and outwards obliquely to the margin. There are no seriately papillose cells.

The perichaetial leaves are similar to the vegetative leaves and the seta is rather short, *c* 1·5 cm. The capsule, 1·5–2·0 mm long × 0·4–0·5 mm, is erect with a very long slender beak, 1 mm or more; the reddish peristome teeth are almost straight, or slightly oblique, but not helically twisted.

DISTRIBUTION: WA, SA, VIC, NSW; endemic to Australia. It is commonest in the south and west.

ILLUSTRATIONS: Hampe (1844), Plate 28 (as *Trichostomum*).

The strongly curled yellowish leaves resemble *Holomitrium perichaetiale* [27], the much smaller *Weissia controversa* [45] and *Ptychomitrium australe* [52] from all of which the oblique boundary of the basal hyaline cells gives unambiguous separation.

T. dakinii Willis, with a diagnostic pale pink peristome, is very similar indeed although Willis (1955c, p. 7) believes that it differs "in the far less tapering leaves, narrower capsules (about 3 × 0·3 mm) and rather larger spores which average about 13 μm in diameter". We have not found these differences helpful. Vegetatively the type material is extremely similar to *cirrhata* but perhaps differs mainly in the cross section through the upper region of the leaf which tends to be V-shaped with the extreme margins just incurved, unlike the U-shaped tubular cross section in *cirrhata*.

The salmon-coloured (not red) peristome seems also to be a rather weak character and we have found the spore size too variable in both species to be useful. But there does seem to be a major difference in the outermost (exothecial) cells of the capsule wall. These measure *c* 45–90×20 μm in *dakinii* (type specimen) compared with 100–170 (average about 120) ×25 μm in *cirrhata*. Moreover the small thick-walled cells at the capsule mouth are rather regularly arranged in vertical files and about 4–5 horizontal rows all round, forming an even, regular band round the mouth in *cirrhata*. In *dakinii* these cells are very irregularly arranged forming a rather scalloped band, uneven on both sides, and not at all distinct. This gives the capsule mouth a ragged appearance under low magnification which seems quite distinctive. One cannot help feeling that a difference of a single allele in the genetic complement of *cirrhata* would be capable of turning it into *dakinii* but such an accusation could be levelled at many another, generally accepted, species of bryophyte.

DISTRIBUTION: VIC; endemic.

ILLUSTRATIONS: Willis (1955c).

T. knightii (Mitt.) Broth. is similar to *cirrhata* but the leaves are narrower and finer, and perhaps less tightly twisted. The main difference lies in the peristome which is strongly twisted helically. There are no evident perichaetial leaves.

DISTRIBUTION: TAS, ?VIC, NSW; also in New Zealand.

ILLUSTRATIONS: Wilson (1859), Plate 172 (as *Tortula*); Allison and Child (1971), p. 64.

T. novae-valesiae Broth. from NSW is close to *knightii* but said to differ in the leaf shape. They may well be conspecific.

T. subflavovirens Broth. & Watts, from Lord Howe, is very similar to and perhaps conspecific with *cirrhata*.

T. inclinata (Hedw.f.) Limpr. is listed for Australia in the Index but the record is said by Watts and Whitelegge (1902, p. 70) to need confirmation.

Tridontium Hook.f.

T. tasmanicum Hook.f.

This is quite a big aquatic or semi-aquatic plant, seldom found but probably not uncommon on stream banks, on stones in rivers and in the spray of waterfalls where it can form quite large cushions.

The shoots are 2–7 cm high, with dark, blackish olive or sometimes yellowish olive green, shiny leaves and black stems but often bare of leaves below. The leaves are lingulate (tongue-shaped), individually incurved or rolled up at the tips when dry, sometimes corkscrewed; 2–3 mm long × 0·5–0·7 mm broad in mid-leaf; when moist, rather secund and slightly falcate. The apex is round in profile and cucullate, and the single nerve fails in the apex. Cells, almost throughout, are roughly isodiametric, hexagonal, thick-walled with the cavities rounded or slightly wider than long, rather irregular in size, *c* 6–10 μm across. Those in the leaf base are rectangular, up to 4×1 or longer. The border of elongated thick-walled cells in the basal quarter or more of the leaf is *intramarginal* with a marginal row of quadrate cells outside it. It is usually, but not always, conspicuous and is sometimes continued well up the leaf as a band of slightly enlarged, often coloured, but not greatly elongated cells. The margin is narrowly recurved.

The capsule, on a long reddish seta, is top-shaped, with a wide thickened mouth.

DISTRIBUTION: TAS, VIC, NSW; also in New Zealand where it appears to be much commoner than in Australia.

ILLUSTRATIONS: Plate 36; Handbook, Plate 28; Brotherus (1924–5), Fig. 225.

The dark cushions of this moss are likely to be mistaken, in its aquatic habitat, only for *Rhacomitrium crispulum* [12] or *Cryphaea tasmanica* [88] but it is quite distinct from these in the lingulate leaves with round, cucullate apices, in the corkscrewing of individual leaves when dry and, microscopically, in the intramarginal border. The appearance of the plant, when dry, is fully distinctive.

PLATE 36. *Tridontium tasmanicum* VIC—Shoot ×7, leaf ×50, basal and upper cells ×1000

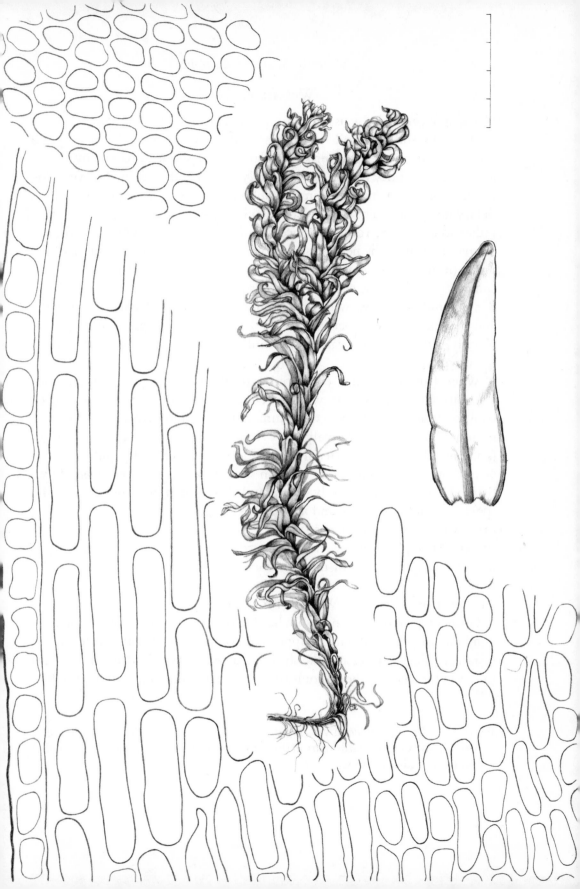

Weissia Hedw.

W. controversa Hedw. is a common and confusing little moss of rather dry earth banks and silty soils in grassland, open woodland, dry sclerophyll forests etc. It forms open turfs or isolated short shoots, 2·5–4·0 mm long, identified when dry by the very strongly curled and contorted leaves, 2·0–2·5 mm long × 0·1 mm wide in mid-subula, but can be very puzzling in the field when moist. The leaves have a wide base gradually contracted to a rather long subula, roughly twice as long as the base, with very tightly *incurved* margins. The nerve is strong and yellow, excurrent in a sharp, short point (mucro), and covered over adaxially by papillose lamina cells but smooth abaxially. The cells of the subula are squarish and very small, *c* 6 μm, densely multi-papillose with low, rounded papillae; the basal cells lax, wide and hyaline. At the upper part of the leaf base the margins are characteristically, although not invariably, crenulate by the projecting cross-walls of the marginal cells.

When fruiting, as they commonly are, the plants are much more conspicuous, with masses of straw-coloured setas, 6–8 mm long, with oblong or oval erect capsules, *c* 0·8 × 0·3–0·4 mm, narrowed just below the *red-rimmed* mouth and with a relatively very long, obliquely beaked operculum (*c* 1 mm). The peristome teeth are slender, short, red and papillose, up to 90 μm long but breaking off easily. In some forms (var. *gymnostoma* (Dix.) Sainsb.) the teeth are entirely absent and the plant is gymnostomous. The spores are finely papillose and measure 15–18 μm.

CHROMOSOME NUMBER: n=13 (NSW).

DISTRIBUTION: TAS, WA, SA, VIC, NSW, ACT, QLD, Lord Howe; cosmopolitan, in all the major regions of the world except Madagascar, according to the Index.

ILLUSTRATIONS: Plate 37; Watson (1955), Fig. 58.

Once this plant is known it is not hard to recognize, even when sterile, by a combination of the habitat, the appearance of the contorted and curled leaves of rather thick texture, and the short excurrent nerve. *Tortella cirrhata* [43] differs in the sharp, oblique boundary between hyaline and papillose cells in

PLATE 37. *Weissia controversa* VIC—Fruiting shoots × 15, leaf × 50, cells from leaf-base half-way out towards margin (lower right), leaf-base near the nerve (middle) and from mid-leaf (upper) × 1000

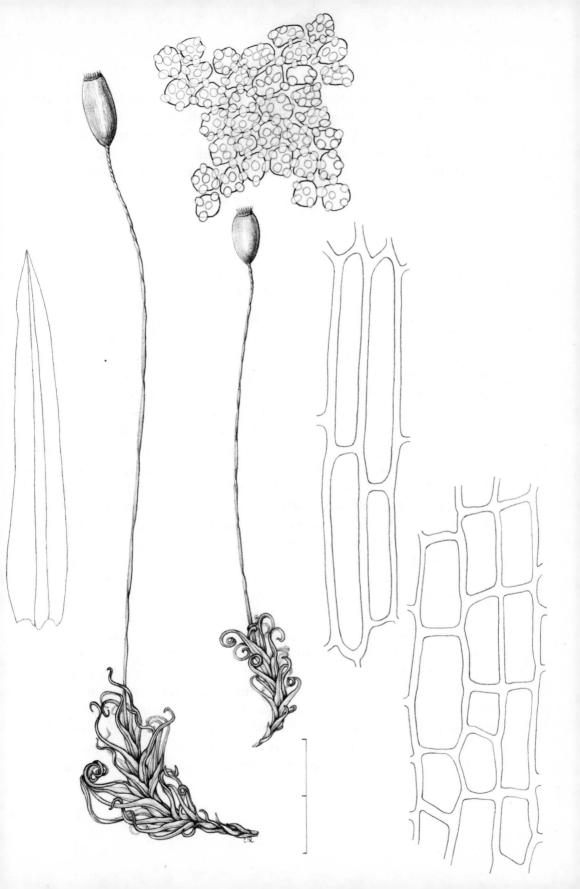

the leaf base, and both *Ptychomitrium* [52] and *Dicranoweisia* [26] differ in the habitat, which is usually rock rather than soil, and in the smooth cells.

A gymnostomous specimen (VIC) with broad, plane-margined leaves has been tentatively identified for us by A. C. Crundwell as *W. microstoma* var. *brachycarpa*. It may be an introduction and requires further study.

(POTTIACEAE—BARBULOIDEAE)
Barbula Hedw.

This difficult genus is characterized by rather narrow leaves, which are widest below the middle and mostly curled when dry, with recurved margins, and papillose cells. The peristome teeth are divided nearly to the base into two hair-like papillose segments and the peristome is usually helically twisted. *Tortula* [39], which is similar, differs in the broader leaves, widest in the middle or above, and in the usually longer peristome teeth set on a more distinct basal membrane. The hyaline, wide basal leaf cells of *Tortula* are shown to some extent by some of the species of *Barbula* but are normally much less marked. The genus *Trichostomopsis* has been erected to include species such as *australasiae* with lax basal cells, bistratose upper leaf margins, straight peristome teeth and a particular nerve structure (Robinson, 1970). It does not seem to us to be a strong genus since its characteristics are neither consistent nor exclusive. It is worthy of comment that Hooker and Greville's illustration of the type of *australasiae* shows a clearly twisted peristome.

It is a particularly difficult genus taxonomically because of the variability of some species or species groups, notably *torquata* where modifications, apparently attributable to habitat factors, are common and are likely to obscure more fundamental species distinctions. One of the major difficulties is that there are numerous distinctive growth forms, based on the shape and degree of twisting of the leaves when dry, which do not seem to be consistently correlated with anatomical features. We have spent much time and effort in trying to recognize *subtorquata* and *luehmannii* as species distinct from *torquata* and have failed to find consistent discriminating features, while feeling all the time that success was just around the corner. Dixon, in the Studies, admitted a wide range of variability in *torquata* and that is the view we have been forced to adopt here. We are unlikely to make much more progress without further very extensive collecting and have to concede, like Manton: "in the unequal contest between *taxonomist* and plant, the plant has in this case so far won handsomely".

There are a number of evidently distinct species, on the other hand, of which we have found only single specimens, wrongly identified in herbaria, but they have to be left until more and better material comes to light.

Ecologically this is a diverse group, mostly terrestrial, with a preference for sandy soils and gritty, rather messy habitats such as waste ground, gravelly roadsides, streets etc., as well as sandy forest soils.

Key to species

1. Basal cells thin-walled, hyaline and smooth (except at margins), quite different from the rest of the lamina 2
   Basal cells usually neither thin-walled nor hyaline, except at the very base of the leaf or in perichaetial and adjacent leaves . . . . . . *torquata*

2. Leaf contracted abruptly at the very tip; nerve usually excurrent 3
   Leaf tapering continuously to the narrow apex; nerve excurrent or not 4

3. Hair-point usually very long. Margins strongly recurved almost to the apex . . . . . . *pseudopilifera*
   Hair-point merely a cusp. Margins plane for some distance below the apex. Introduced. . . . *unguiculata*

4. Leaves, at least in some shoots, helically wound round the stem when dry, not individually crisped. Margins not bistratose . . . . . *torquata* (*luehmannii* form)

   Leaves curled and twisted, but not round the stem. Margins bistratose, at least above . . . . *australasiae*

B. australasiae (Hook. & Grev.) Brid. (= *Trichostomopsis australasiae* (Hook. & Grev.) Robinson)

Either this is a much less common species in Australia than in New Zealand or it is generally overlooked. The green or brownish green turfs have the leaves quite strongly curled but not wound round the stem. The leaves are very narrowly triangular, *c* 1·0–2·5 × 0·3 mm, widest just above the base, with a strong nerve failing in the apex or sometimes shortly excurrent, covered on both surfaces by quadrate lamina cells, and sunk in a broad shallow channel. The margins are quite entire and are variably revolute above mid-leaf, bistratose above (in some cases along the whole of the upper half of the leaf) and merge with the nerve imperceptibly, at a low angle, at the apex. The upper cells are more or less square, 7–9 μm across, multipapillose with low, rounded papillae, while the basal cells, especially away from the margins, are much wider and longer, lax, thin-walled, hyaline and smooth. The outer stem cells, in transverse section, are thick-walled but with the lumen not sufficiently small to class them as stereids, and the outermost walls (on the stem surface) thin.

Capsules are not uncommon and tend to be dull reddish or crimson-brown. They are cylindrical with slightly convex walls in profile, *c* 2·0 × 0·5 mm, on a seta 10 mm long; the operculum with a long conical beak, *c* 1 mm long. The peristome is rather short and slightly twisted.

DISTRIBUTION: TAS, WA, SA, VIC, NSW; also in New Zealand and S. America. ILLUSTRATIONS: Handbook, Plate 29; Hooker and Greville (1824b), Plate 12; Wilson (1854), Plate 85 (as *Trichostomum fuscescens*).

We are far from happy with the extensive synonymy proposed for this species.

From *B. torquata*, which is very similar, it differs in the curled but not helically twisted leaves, the nerve less excurrent and not forming such a wide point, in the area of hyaline cells in the leaf base, bistratose upper margins, and in the different stem structure. *Weissia controversa* [45] is similar in appearance when dry, but has incurved margins.

B. torquata Tayl.

The helical twisting of the leaves round the stem and the scarcely altered cells in the leaf base, are usually sufficient to distinguish this species from the previous one, but there are forms which are hard to place. It is typically very dark brown in colour with the leaves, at least in some shoots, strongly wound round the stem. The nerve is thick, excurrent in a rather broad flattish short

PLATE 38. *Barbula torquata* VIC—Shoot extracted from cushion × 11, leaf × 50, basal and upper cells × 1000

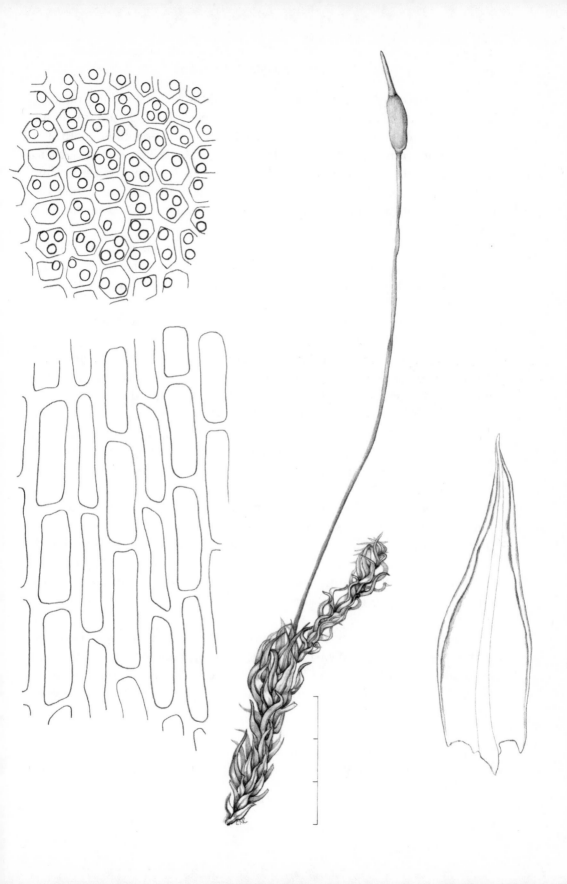

point. The strongly revolute margins, one cell thick throughout, gradually merge with the nerve at a low angle, at the apex. The cells are quadrate–hexagonal above, *c* 6–12 μm, rather obscure, multi-papillose with 1–3 rather blunt domed papillae per cell; the cells in the leaf base are longer, to *c* 24×9 μm, shortly rectangular and thick-walled but usually not hyaline and not appearing distinct from the rest of the leaf. In cross section the mature stem shows an outer ring of narrow-lumened, thick-walled stereid cells.

The seta is *c* 1–2 cm tall, the capsule cylindrical, *c* 1·5 × 0·3–0·5 mm, with a long conical-beaked operculum of *c* 1·2 mm. The perichaetial leaves are sheathing but short, not much longer than the adjacent leaves, often not conspicuous.

DISTRIBUTION: TAS, WA, SA, VIC, NSW, ACT, NT; also in New Zealand.

ILLUSTRATIONS: Plate 38.

This is the most variable of the Australian Barbulas. It is very common on sandy soils in grassland, dry sclerophyll forest, mallee, sand hills etc. It is closest to *B. australasiae* and *B. pseudopilifera* (q.v.), but also can resemble *Ceratodon* [15] vegetatively, although clearly distinct from it in the papillose cells. Almost always some degree of twisting of the leaves is evident, on at least some stems, but the degree and nature of the twisting is very variable and the appearance of the plants is correspondingly diverse.

B. luehmannii Broth. & Geh. is a plant which has been recorded not infrequently but we have seen no material which exactly matches the type collected from Victoria. It has the areolation of *australasiae* combined with the twisted leaves of *torquata*, although often rather more weakly twisted, and usually coiled at the tip, not wound round the stem, giving a rather different appearance to the plants.

B. subtorquata C.Muell. & Hampe has been the subject of some confusion with *Leptodontium* [48]. We have not yet found specimens to match the type material from Mt Gambier and at present have to class it as a robust form of *torquata* with the leaves less twisted than usual. It is illustrated in Mueller (1864), Plate 3.

PLATE 39. *Barbula pseudopilifera* VIC—Vegetative shoot dry (below) and moist (above) × 7, basal cells (below) and upper cells × 1000

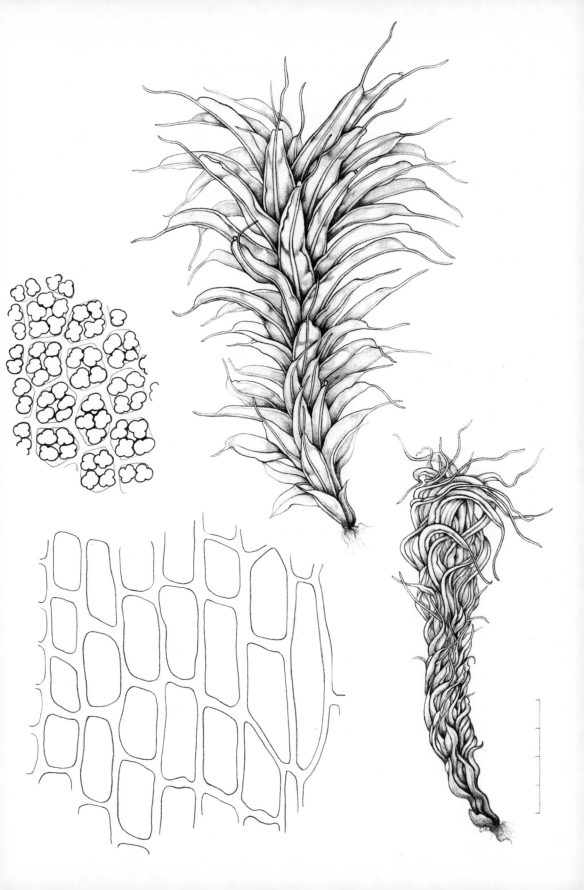

B. pseudopilifera C.Muell. & Hampe†

In mallee, desert scrub and dry forest this moss commonly forms quite extensive patches, recognizable usually by the long, golden hair-points of the leaves which are tightly wound round the stem with the points spreading spirally from the stem apex. The leaves are 1·7–2·5 × 0·3–0·6 mm, widest *c* ¼–⅓ way up the leaf, with a hair-point measuring another 0·8 up to 1·5 mm, although sometimes this is much shorter, as it is in the type material. The margins are widely and tightly recurved and form a very obtuse apex by converging on the nerve almost at right angles. This is usually seen in profile because the tip of the leaf is tightly folded along the nerve so that it can scarcely be opened out flat. The upper cells are very papillose, commonly quite obscured by high multiple papillae (which are themselves warty). They measure *c* 9–10 μm across (the Handbook gives 15 μm). The cells in the lower half of the leaf are more elongate, and those of the basal quarter are hyaline and smooth, especially towards the nerve, forming a very distinct area margined by *c* 6 rows of quadrate, non-hyaline cells.

The seta is yellow, 1–2 cm tall, bearing a narrowly cylindrical capsule 2·0–2·5 mm long, with a very long-beaked operculum of up to 2 mm. The perichaetial bracts are long, sheathing, exserted well above the adjacent leaves but not twisted.

DISTRIBUTION: TAS, WA, SA, VIC, NSW, ACT; also in New Zealand.

ILLUSTRATIONS: Plate 39.

When the hair-point is fully developed this plant is unmistakable. When short, it has some resemblance to *torquata* but differs most obviously in the obtuse, not pointed lamina which meets the nerve at a high instead of a low angle; it can also bear some resemblance to *Leptodontium* [48].

B. unguiculata Hedw., an introduction from the northern hemisphere, is a small plant, *c* 0·5–2·0 cm high, with wide blunt leaves widest near the middle, with the margins rather undulate and slightly recurved but flat for some distance below the apex where the strong nerve is just excurrent in a small cusp. The leaves are tightly curled up, rather as in *Weissia controversa* [45]. The upper cells are very obscure, densely papillose with low papillae, *c* 7–10 μm. The basal cells are more elongate and smooth but still thick-walled.

DISTRIBUTION: TAS, VIC; widely distributed throughout the northern hemisphere.

ILLUSTRATIONS: Watson (1955) Fig. 51; Dixon (1924a), Plate 28.

The other Australian species in the genus are:

B. acrophylla C.Muell. (VIC, NSW)

B. calcicola (Hampe) Broth. (WA) (illustrated in Hampe, 1844, Plate 29)

B. chlorotricha (Broth. & Geh.) Par. (VIC, NSW). From the type description this sounds very like *pseudopilifera*.

B. chrysochaete C.Muell. (VIC)

B. chrysopus C.Muell. (TAS)

B. cylindrangia C.Muell. (VIC)

B. ehrenbergii (Lor.) Fleisch. (=*Hydrogonium ehrenbergii* (Lor.) Jaeg.) (WA)

B. geminata C.Muell. (VIC)

B. hampeana Par. (TAS, VIC)

B. speirostega C.Muell. (NSW)

B. subcalycina C.Muell. (VIC, NSW, QLD). See *Tortella calycina* [43].

Bryoerythrophyllum Chen
(=*Erythrobarbula* of the Handbook)

The generic name *Erythrobarbula* of Steere is illegitimate, since Chen's *Bryoerythrophyllum* is an earlier name for the same thing.

B. binnsii (R.Br.ter.) Wijk & Marg.

The shoots, about 1 cm tall, form dull reddish brown turfs on dry earth banks. The leaves, 1·4–1·6 × 0·3–0·4 mm at most, are widest near mid-leaf, tongue-shaped and obtuse with a tiny sharp, sometimes reflexed, hyaline tip. There is a prominent nerve, papillose or overlain by papillose cells, especially above, and running in rather a wide channel down the leaf from just below the apex. The margins are *plane*, only occasionally slightly recurved. The cells are square above, c 9 × 9 µm, but completely obscured by dense multiple papillae, commonly 4, or in 4 groups, per cell. In the leaf base, next to the nerve, there are longer and wider smooth hyaline cells forming a conspicuous median group; the marginal cells there are narrower and green but scarcely papillose.

We have not seen sporophytes; New Zealand material is dioicous with a small erect cylindrical capsule on a slender seta, and a single peristome of 16 short teeth split into 2 papillose filaments.

DISTRIBUTION: TAS, VIC; also in New Zealand.

ILLUSTRATIONS: Handbook, Plate 27.

In the absence of capsules there is no certainty that Victorian material is correctly identified, but the plane margins and constant presence of a hyaline apiculus at the leaf apex should be sufficient to distinguish it from the next species. It is an unobtrusive plant recognizable by the reddish tint of the turfs and the tongue-shaped, intensely papillose leaves. Probably it is much commoner than it seems.

B. recurvirostre (Hedw.) Chen is very similar indeed but differs in having the leaf margins recurved almost to the apex, the hyaline tip less constantly present and the peristome teeth entire. It is also synoicous. We have never found it in Australia.

DISTRIBUTION: TAS?, VIC, NSW?; recorded by Watts and Whitelegge (1902, p. 68) but treated with some scepticism by Sainsbury (1953b, p. 88). Willis (1955a, p. 160) has provided a firm record for Victoria. It is quite widespread in the Northern Hemisphere and also in New Zealand and Oceania.

ILLUSTRATIONS: Watson (1955), Fig. 53; Dixon (1924a), Plate 26(H).

(POTTIACEAE—LEPTODONTIOIDEAE)
Leptodontium (C.Muell.) Hampe ex Lindb.

Leptodontium sp.

In Victoria this species is not uncommon, quite distinct, easily recognizable and much misunderstood, frequently masquerading as *Barbula subtorquata*. It is a big plant, 1·5–4·0 cm tall, yellowish green or glaucous above and brown below and with the leaves, when moist, slightly undulate, usually *clearly in 3 ranks* which are themselves lightly twisted helically round the stem. The leaves at the base of the stem are virtually straight, those near the apex have a slight curl round the stem which tends to obscure the 3-ranked arrangement.

The leaves are 2·0–2·5 × 0·8–1·0 mm, *ovate–lanceolate*, widest at c ⅕ or less up from the leaf base. The margins are rather widely recurved in the lower half and quite entire, even at the apex. The leaf shape, like the leaf arrangement, is very much that of *Triquetrella*, flattish or weakly folded in the broad leaf base, tightly folded about the nerve above, so that the leaf base spreads out under a coverslip but the two sides of the lamina in the upper half stay pressed together. The bottom corner of the leaf, on each side, is decurrent down the stem in a long narrow strip, several cells wide. The nerve is very strong, channelled, finely multi-papillose on the abaxial surface at the base, reaching the apex or shortly excurrent. The cells are small, *c* 6–7 µm in diameter, hexagonal, with rounded cavities, densely multi-papillose with low papillae, but not quite obscure, sometimes almost smooth. In the base they are long or shortly rectangular but still papillose, rather thick-walled and slightly porose, usually papillose right to within 1 or 2 cells of the leaf insertion. Some of those next to the nerve may have the papillae in longitudinal rows. The cells in mid-leaf are very variable in shape.

We have not seen capsules.

DISTRIBUTION: TAS, SA, VIC, NSW, ACT; ?endemic.

ILLUSTRATIONS: ?

On dry soils this is not uncommon, often with *Barbula pseudopilifera* [46]. It is rather like a very big *Anomodon tasmanicus* [49] but that has leaves only up to 1·2 mm long, a nerve failing clearly below the apex, and a hyaline apiculus. This seems a very consistent species morphologically and is quite easy to recognize under the dissecting microscope once it is known. It can be confused with forms of *Barbula pseudopilifera* which lack hair-points, but is distinct from that species in the smaller cells, in the leaf shape and in the gradually tapering leaf apex.

It differs from the New Zealand *L. interruptum* (the name by which it has hitherto generally been known) in the decurrent leaf bases, entire leaf apex, much less undulate leaves, and in the outer cells of the stem which are small and thick-walled in our species, but enlarged, later collapsing, in *interruptum*. The seriate papillae in the leaf bases are also usually much more conspicuous in *interruptum*. The correct genus for this species cannot be finally confirmed until capsules are found.

Triquetrella C.Muell.

T. papillata (Hook.f. & Wils.) Broth.

The wiry, irregularly branched, yellowish shoots of this common moss, usually 2–5 cm long but only 0·5–1·0 mm wide when dry, are abundant in sandy and rocky soil throughout the drier forests, grassland and coasts, forming soft open cushions. Paradoxically it is also found in semi-aquatic habitats bordering rocky streams but probably always where there are intermittent droughts.

The ovate leaves, *c* 1·0–1·5 mm long × 0·7 mm wide, are tightly appressed to the stem when dry, producing very slender shoots, but spread very widely when wet and take on a much more vivid colouration. The appearance of the plants, wet and dry, is astonishingly different. The arrangement of the leaves is in 3 ranks, usually with a slight spiral twist but sometimes so twisted that the 3-ranked arrangement is scarcely maintained; the margins are entire, tightly revolute below; the nerve strong, failing in or below the apex. Cells throughout are small (*c* 7 μm) roughly hexagonal and usually conspicuously papillose with 1–3 spiculose and sometimes forked papillae per cell.

Sporophytes are rather rare but often abundant where they occur; the capsule is narrowly cylindrical on a long, flexuous seta.

DISTRIBUTION: TAS, WA, SA, VIC, NSW, ACT; also in New Zealand.

ILLUSTRATIONS: Plate 40; Handbook, Plate 28; Brotherus (1924–5), Fig. 217.

When dry the very slender wiry stems are likely to be mistaken only for the much rarer *Anomodon* which it resembles superficially in size and shape, and which grows in similar habitats.

Anomodon tasmanicus Broth. (?Thuidiaceae), at first glance very similar to *Triquetrella* and formerly known as *T. curvifolia* Dix. & Sainsb., is quite distinct on closer inspection in the hyaline tips to the leaves, in the anchoring rhizoids along the stem, and more obviously in the leaf arrangement; the leaves are curled when dry and partly twisted helically round the stem, giving a quite different appearance under the lens, not unlike a *Thuidium* to which it is probably closely related. It is sometimes rather glaucous. Sporophytes have never been found so the genus to which this species belongs is not fully established.

DISTRIBUTION: TAS, WA, SA, VIC, ACT; also in New Zealand.

ILLUSTRATIONS: Handbook, Plate 28 (as *Triquetrella curvifolia*).

PLATE 40. *Triquetrella papillata* SA—Fruiting shoot dry (below) and moist × 15, leaf × 50, cells × 1000

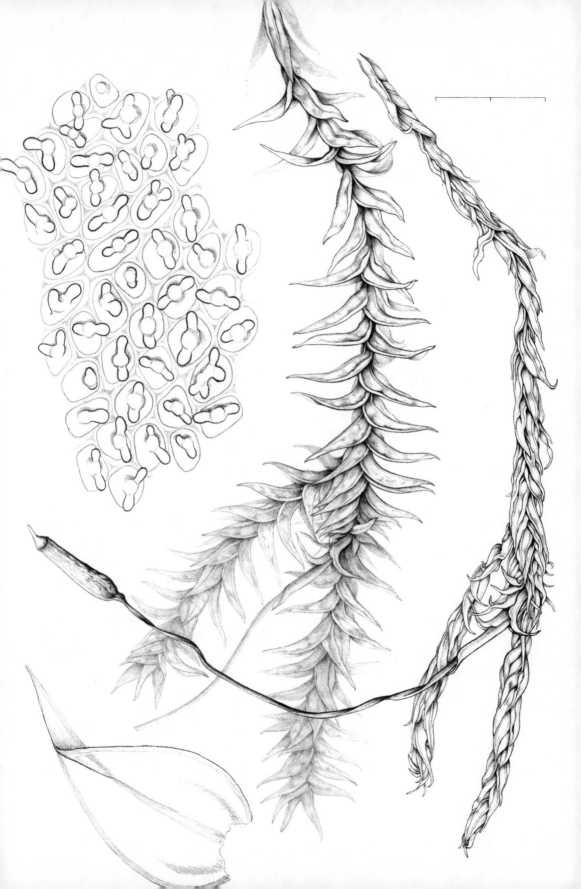

T. fragilis C.Muell. (VIC) and *T. richardsiae* C.Muell. (SA, NSW) we have not seen.

T. preissiana (Hampe) C.Muell. (WA) is just *T. papillata* according to Watts and Whitelegge (1902).

BRYOBARTRAMIACEAE

Bryobartramia Sainsb.

The genus is characterized by a greatly inflated calyptra which does not break away from the vaginula in the usual manner but encloses the cleistocarpous capsule and the short seta even at maturity.

B. novae-valesiae (Broth.) Stone & Scott, 1973

This plant has been known in Victoria as *B. robbinsii* Sainsb. and because of its unique character Sainsbury created the new family Bryobartramiaceae. The same moss, however, had been described earlier from NSW as *Trachycarpidium novae-valesiae* Broth. ex Roth. It does not belong to the genus *Trachycarpidium*, from which it differs fundamentally, so the new combination, *B. novae-valesiae*, is necessary.

The short rather thick stem, 1–2 mm long, develops several very short branches, has no central strand and is often partly buried in the soil where there is a colourless and very extensive underground rhizoidal system. The vegetative leaves are smallest near the base of the stem, 0·3–0·5 mm long, and increase in size to about 1·0 mm near the top. They are usually flat to ± concave, curved inwards when dry, variable in shape, more or less tongue-shaped, obtuse or broadly acute, the margin occasionally irregularly lobed but usually plane and entire except for crenulations of the papillae on the cells. The single nerve ends just below the apex and, except at the base of the leaf, is very well developed and conspicuous on the abaxial surface.

The cells of the upper part of the leaf are quadrate to rounded, 15–18 μm, with firm walls and variably papillose, the papillae high and circular or crenate; cells at the leaf base hyaline and smooth with thinner walls. The perichaetial leaves are much longer, and almost straight when dry; to 2 mm long, variable in shape, sometimes strap-shaped throughout but usually with an oblong base and strap-like subula above; often the subula with a single twist and curved inwards. The margins are entire below, usually distantly toothed from the mid region upwards; the cells mostly smooth but with a few papillae on upper cells.

The seta is very short and the cleistocarpous capsule is globose to oval with a very small apiculus; both are completely enclosed even at maturity by a *greatly inflated calyptra which does not break away at the base in the usual manner.*

At first glance there is a superficial resemblance to *Goniomitrium* because of the inflated calyptra but in *Bryobartramia* the calyptra lacks longitudinal plicae and has a long straight beak and the leaves have a strong nerve and papillose cells, features which are absent in *Goniomitrium*. This species is generally believed to be rare, but is probably overlooked because of its tiny size.

DISTRIBUTION: WA, SA, VIC, NSW; also in S. Africa.

ILLUSTRATIONS: Plate 34; Sainsbury (1948), Plate 11; Lee (1952); Roth (1913), Plate 2 (as *Trachycarpidium*).

It seems to be not uncommon in red-gum country along the Murray River, and also in north-central and eastern (C. Beauglehole pers. comm.) Victoria, in NSW, and has been recently found for the first time in WA; also in SA (L. Williams, pers. comm.). It has also very recently been identified in S. Africa, the first extra-Australian record (Stone and Schelpe, 1973).

ENCALYPTACEAE

Encalypta Schreb. ex Hedw.

E. vulgaris Hedw.

This delightful little moss is unmistakable, when fruiting. The very long straw-coloured calyptras, often said to resemble the extinguishers formerly used to put out candle flames, are found in no other genus in our flora. It tends to grow in small dense turfs, up to *c* 5 mm tall, in fragmentary soil on rock ledges and crevices etc., usually on calcareous rock.

When sterile it is not easily—or often—found, but may be identified by the very wide tongue-shaped, vivid green opaque leaves, *c* 1 × 3·0–3·5 mm and *c* 30–40 cells wide on each side of the nerve, either very obtuse and rounded at the tip with the strong brown nerve failing well below the apex, or pointed with the nerve reaching almost to the very tip. The leaves are tightly folded (keeled) about the nerve, and are curled when dry, often slightly wound round the stem, with the nerve showing prominent and glossy. The cells are very densely multi-papillose with "hobnail" papillae which project all round the leaf profile giving a characteristically rough margin; the upper cells are hexagonal or squarish, c 12 μm in diameter, each with 1–4 clusters of dense, often circular, papillae; the basal cells are much larger, less densely papillose, and more or less hyaline. Those at the margins of the hyaline area are much longer, narrower and rather thicker-walled, forming an evident border. The nerve is covered adaxially, in the upper half, by papillose lamina cells but is usually smooth (sometimes papillose) abaxially. The rhizoids are smooth.

On young sporophytes, the relatively huge calyptra, 2 mm wide at the base × 4–5 mm long including a narrow tip of c 2 mm derived from the neck of the former archegonium, completely encloses both the capsule and most of the short, 3–5 mm, seta; it is irregular but not fringed at the base. The capsule is *gymnostomous*, narrowly cylindrical, slightly wider at the *base*, c 2–5 mm long. The spores are large, 30–40 μm, irregularly angled and with extraordinary projecting papillose ridges and bosses.

DISTRIBUTION: TAS, SA, VIC, NSW, ACT; almost world-wide.

ILLUSTRATIONS: Plate 41; Handbook, Plate 32; Watson (1955), Fig. 37; Allison and Child (1971), Plate 14; Grout (1928–40), I, Plate 69.

E. ciliata Hedw. differs in having a calyptra fringed at the base with narrow filaments and in having a peristome. It is given by the Index from Australia but we do not know the source of the record. *E. novae-valesiae* Hampe (VIC, NSW) has been recorded from Gippsland and the Blue Mountains.

ERPODIACEAE

Includes: *Aulacopilum glaucum* Wils. (NSW)
         A. hodgkinsoniae (Hampe & C.Muell.) Broth. (NSW, QLD)
         Wildia solmsiellacea C.Muell. & Broth. (QLD)

PLATE 41. *Encalypta vulgaris* VIC—Dry plant ×15 with two immature capsules and one old capsule showing fungal attack. Basal and upper cells (left) and cells of upper leaf margin (right) ×1000

PTYCHOMITRIACEAE

Ptychomitrium Fuernr.

These mosses of rocks and sometimes gritty soils are not particularly distinctive, yet, for all that, áre surprisingly easy to recognize when dry. The symmetrical capsule, calyptra evenly split all round at the base instead of one-sided, and bistratose leaf tips, are all distinctive. The leaf margins are entire in *australe* and toothed in *mittenii*.

P. australe (Hampe) Jaeg. is a common plant forming small compact cushions on rocks, usually in forests. The short stems (0·5–1·0 cm) when dry have very tightly curled leaves completely circinate at the tips but not wound round the stem, olive or blackish green and with a strong nerve failing just below the apex and covered for much of its length on both surfaces by squarish lamina cells. The leaves are mainly lanceolate, 2–5 × 0·4–0·7 mm, the margins entire but sometimes rather sinuous, not quite toothed. The cells are fairly thick-walled, squarish–hexagonal with rounded cavities, in rather regular longitudinal rows, c 7–8 µm above, becoming more elongated and thicker-walled below; the cells in the basal ¼ or less of the leaf are rectangular, hyaline, much enlarged (c 20 × 10 µm) and rather thin-walled. The lamina is bistratose at the margins from about mid-leaf upwards, and across the whole of the tip of the leaf.

The erect capsule is narrowly oval, c 1·0–1·5 mm long, rather sharply contracted to the 3–5 mm long seta, *pale brown* except for the red mouth and with a single peristome of 16 teeth, orange at first (becoming pale) papillose and irregularly split or grooved along the mid-line, rather widely spreading when dry. The spores are 12–20 µm, densely and coarsely ornamented with irregular short ridges. The operculum has an erect beak, c 0·5 mm long, under the markedly plicate calyptra.

CHROMOSOME NUMBER: n = 13 (NSW, QLD).

DISTRIBUTION: TAS, WA, SA, VIC, NSW, ACT, QLD; also in New Zealand.

ILLUSTRATIONS: Handbook, Plate 32.

P. acutifolium Hook.f. & Wils. (TAS, SA, VIC) may be conspecific according to Sainsbury (1955c). It is illustrated in Wilson (1859), Plate 173.

P. mittenii Jaeg., which sometimes grows mixed with *australe*, is usually appreciably taller, to 3 cm, and has the leaves much less tightly curled, scarcely circinate at the tips, and much more like a *Macromitrium* [55]. The ovate-lanceolate leaves, 2·5–3·5 × 0·7–1·0 mm, are strongly plicate with usually a single fold on each side of the midrib; the margins loosely recurved below and slightly undulate, sinuose below becoming more strongly sinuose and coarsely toothed above. The nerve fails just below the apex, and the lamina is bistratose near the apex and farther down along the margins. The upper cells are much as in *australe* but those below are narrowly rectangular and thick-walled in the middle of the leaf-base, becoming quadrate towards the margins.

The capsule is like that of *australe* but the teeth are longer, and erect not spreading when dry; the operculum has a very long slender beak *c* 1 mm long; and the spores are 7–10 μm, much less strongly ornamented. Polysety is common in *mittenii* but we have not observed it in *australe*.

DISTRIBUTION: TAS, VIC, NSW; confined to Australia.

ILLUSTRATIONS: Wilson (1859), Plate 173 (as *P. serratum*).

It is far less common than the previous species, but quite probably over-looked. It seems to have a preference for rocks and gritty soils by rivers in eastern regions, but more observations are needed.

P. muelleri (Mitt.) Jaeg., recorded from VIC (Watts and Whitelegge, 1902) is mainly found in NSW and QLD. It seems very similar to *australe*, but larger and with a longer calyptra.

CHROMOSOME NUMBER: n=24 (NSW).

P. laxifolium (C.Muell.) Par. and *P. microblastum* (C.Muell.) Par., both confined to NSW, are unknown to us.

All these species, except *australe*, are endemic to Australia.

ORTHOTRICHACEAE
Orthotrichum Hedw.

The species of this genus form small open cushions on bark or rock, very similar to *Ulota* [57] and differing most obviously in having the basal cells scarcely distinct from the adjacent marginal cells of the leaf, but this difference is one of degree and is not always hard and fast. The Handbook claims that the stomata in the capsule wall are confined to the neck and base of the capsule in New Zealand species of *Ulota*, but extend to the middle and upper parts of the capsule in *Orthotrichum*. Sometimes, however, they can be found at the base in *O. tasmanicum* and in the upper part in *U. cochleata* so this does not provide a rigid separation. Malta (1933) comments that this makes the delimitation of the two genera still more difficult.

O. tasmanicum Hook.f. & Wils.

Sainsbury (Handbook, p. 210) says that "this is a most perplexing plant, owing to great variability coupled with a lack of correlation of its fluctuating characters" and anyone who has hunted for other species will agree. No matter how different in appearance, the plants collected almost always seem to key out to *O. tasmanicum*. Generally it forms open, rather small, freely fruiting cushions, of shoots *c* 5–15 mm long of a dull greenish or yellowish brown, in humid forest and fern gullies. The leaves are typically 2–4 × 1 mm, flexuose or slightly crisped when dry and either oblong or tongue-shaped or ovate, with a strong nerve failing well below the apex and deeply sunk in a rather wide channel which is prominent on the abaxial leaf surface. The margins are recurved and rather undulate or even contorted and have a sinuose (almost toothed) outline above, denticulate in parts and contracted to a narrow triangular point which is sometimes sharp.

　　If the leaf shape is variable, the cells are even more so. They are extremely irregular in size and shape, mostly isodiametric throughout the greater part of the leaf with heavy thickenings, especially at the corners, to give irregularly rounded cell cavities. The upper cells typically are narrow, *c* 9 μm across, and may be densely papillose with several domed papillae per cell, but these are often much reduced and the leaves may be nearly smooth. In the leaf base the cells are less papillose, very long and narrow, and thick-walled, especially the longitudinal walls; those at the margins of the leaf base in 2–6 rows, quadrate and often hyaline forming an inconspicuous border. The nerve may be papillose or smooth on the abaxial surface.

　　The capsule, 2·0 × 0·3 mm on a seta *c* 3 mm long but very variable in length, is normally deeply 8-grooved, occasionally almost smooth, the

PLATE 42. *Orthotrichum tasmanicum* VIC—Fruiting plant × 15, upper and lower leaf cells × 1000

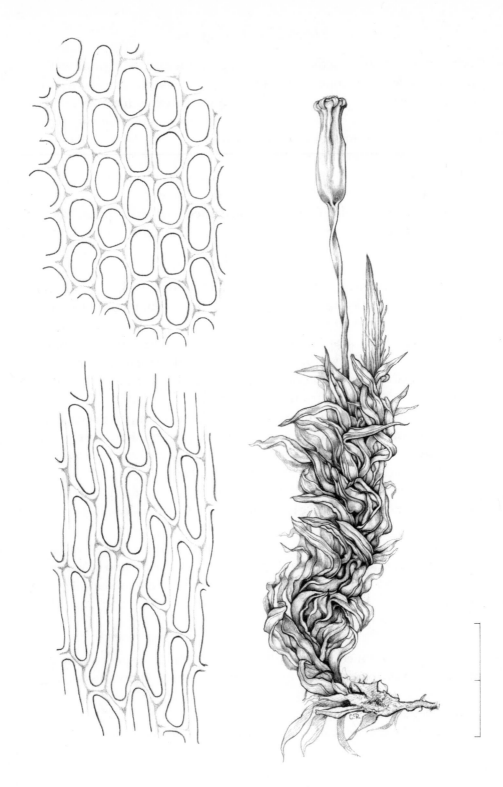

peristome teeth short, rather orange, reflexed back on the capsule wall when dry, as in *Ulota*. The inner teeth are yellowish, *similar in size*, curled in over the spores when dry. The operculum is low-convex with a short erect beak, and the large plicate calyptra is conical and either smooth or slightly hairy. The spores, 12–15 μm, are coarsely warted.

CHROMOSOME NUMBER: n=11 (NSW).

DISTRIBUTION: TAS, SA, VIC, NSW, ACT; also in New Zealand.

ILLUSTRATIONS: Plate 42; Allison & Child (1971), p. 71.

The basal marginal cells and almost hairless calyptra separate this plant from *Ulota lutea*, the only common species with which it is likely to be confused.

O. rupestre Schleich. ex Schwaegr., a plant of rocks in dry and wet sclerophyll forest, not uncommon in upland country, is appreciably larger, *c* 2–3 cm, dark greenish, and with stiffly erect leaves closely pressed to the stem; and the cells have *high spike-like* papillae. The dark brown capsules are usually *immersed* (i.e. overtopped by the leaves) short and wide, *c* 2 × 1 mm, with the *inner peristome absent or vestigial*, and the calyptra densely and coarsely hairy. It differs from *O. tasmanicum* in the habitat, usually rocks instead of bark, in the straight erect leaves and wide, oval, immersed capsules, in the reduced inner peristome and the shaggy calyptra. Fortunately capsules are common on both species so these are useful characters. Vegetatively the plant looks rather like a dark green *Rhacomitrium*.

DISTRIBUTION: TAS, VIC, ACT; elsewhere it is almost world-wide. There seems to be no substantial difference between this and *O. pulvinatum* R.Br.ter. from New Zealand.

ILLUSTRATIONS: Dixon (1924a), Plate 32; Grout (1928–40), II, Plate 45.

What appears to be *O. alpestre* Hornsch. has also been found in Victoria and ACT; it is similar to a small *rupestre*, but grows on trees and has the stomata *immersed* i.e. sunk below and almost covered over by the outermost cell layer of the capsule wall.

The genus is much under-collected in Australia, especially in upland regions, and the other species given in the Index are known from very few specimens:

O. acroblepharis C.Muell. (VIC, NSW)

O. albidum Hedw. appears to be an error for *Octoblepharum* in the Index.

O. angustifolium Hook.f. & Wils. is illustrated in Wilson and Hooker (1845), Plate 57. It is customarily put in the genus *Muelleriella*. The Austr. 1 reference in the Index is from Kerguelen.

O. calvum Hook.f. & Wils. (?VIC)

O. encalyptaceum C.Muell. (VIC, NSW)

O. lawrencei Mitt. in Hook. (TAS) differs from *tasmanicum* in the short, not elongated, basal cells. It is illustrated in Wilson (1859), Plate 172.

O. sullivanii C.Muell. (VIC, NSW)

O. whiteleggei C.Muell. (NSW)

O. campbelliana Watts & Whitel. was published without description and is therefore a *nomen nudum*, not a valid name.

Amphidium Schimp.

A. cyathicarpum (Mont.) Jaeg. forms dense pads on cliffs and rock ledges in upland regions, the shoots 2–3 cm tall, rusty brown below and deep bright green above.

The leaves are long, *c* 3 mm, very narrow, keeled and very curly when dry. The strong narrow nerve is exposed and either papillose on both surfaces or smooth adaxially, usually not quite reaching the apex. The margins in our specimens are often somewhat undulate and generally have distant irregular step-like teeth, each corresponding to a reduction in width of the lamina (which is usually no more than 12 cells wide on each side of the nerve at the base) by one cell (Sainsbury describes the margins as mostly entire in New Zealand). The cells are small, more or less quadrate (sometimes wider than long near the margins) thick-walled with rounded cavities, granular with rounded low papillae which are evenly distributed over the whole leaf surface; the extreme basal cells are rectangular, clearer, usually thinner-walled.

On fruiting plants the seta is short, 2–3 mm, and the gymnostomous capsule, almost overtopped by the side branches of the fruiting stem, is characteristic, small *c* 1 mm wide and long, straight-sided, widest at the mouth and deeply 8-grooved. The plants are monoicous with short, bud-like male branches in the leaf axils.

It appears to be not uncommon in SA and VIC.

CHROMOSOME NUMBER: $n = 16$ (TAS).

DISTRIBUTION: TAS, WA, SA, VIC, NSW; also in S. Africa, S. America, New Zealand and Oceania.

ILLUSTRATIONS: Handbook, Plate 36.

It is liable to confusion with *Gymnostomum* [41, q.v.].

Macromitrium Brid.

At present this is perhaps the most difficult genus, taxonomically, in the Australian moss flora. The characters of gametophyte and sporophyte show respectively too much and too little variation to be satisfactory and it is a puzzle where to turn for better; possibly biochemical criteria might hold the answer. It is not even known with certainty whether we are dealing with a few very variable species or a very large number of invariate microspecies, or both, although the first of these seems most likely on present evidence. Only in the field, not the herbarium, lies the hope of discovering the full range of variability of the species. We cannot pretend to know more than a handful of species and take comfort that Sainsbury found himself in the same position: "The following treatment is purely tentative and aims at doing little more than clearing the ground for future investigators" (Sainsbury, 1955c, p. 17). The genus is mainly tropical and subtropical in distribution, predominantly growing on tree trunks although occasionally found on rocks. It is usually very easy to recognize, mainly because of the growth form of radiating prostrate stems, bearing dense, usually short, erect branches like the tufts of a pile carpet. The fruit too is usually an oval capsule often contracted at the mouth, with or without a peristome, and with a symmetrical, plicate, often hairy calyptra. The cells are very like those of *Ulota* [57] and *Orthotrichum* [53] and its other relatives. *Schlotheimia* [56] is very similar in growth form but has a big, bell-shaped, smooth calyptra and usually a characteristically deep green and rusty red colouration which is distinctive in the field. Many of the species have dwarf male plants (as in *Leucobryum*, 28), the two sizes of plant, male and female, arising from differently sized spores, but the extent of this phenomenon in the Australian species remains to be investigated (see, however,

Ramsay, 1966b). In most species the spores seem to be finely and densely papillose.

The species given in the key seem to us, at least provisionally, to be distinct either as individual species or as representatives of groups of closely related species.

Key to species

1. Leaves straight and appressed when dry, neither
   wound helically round the stem, nor curled at
   the tips . . . .   *tenue*
   Leaves curled when dry, either inrolled at the
   tips or wound round the stem   2

2. Leaves individually curled but not wound round
   the stem   3
   Leaves wound round the stem, at least on some
   branches   7

3. Cells smooth throughout. Shoots often rope-like.   *weymouthii*
   Cells papillose or at least very convex. Shoots not
   rope-like   4

4. Cells smooth but very convex. Seta very short,
   usually shorter than the capsule . . . . .   *malacoblastum*
   Cells multi-papillose above, with spiculose papillae
   lower down. Seta longer than the capsule . .   5

5. Upper cells densely papillose, obscure   6
   Upper cells slightly papillose, cell walls perfectly
   clear . . . .   *archeri*

6. Leaves conspicuously dimorphic; the erect branches
   have the leaves tightly inrolled at the tips whereas
   the leaves of prostrate shoots are mostly straight
   when dry. Perichaetial bracts long and narrow,
   overtopping the branch leaves . . . . .   *scottiae*
   Leaves not obviously of two sorts. Perichaetial
   leaves clustered round the base of the seta, but
   not projecting much above the branch leaves
   and not long and finely pointed . . . . .   *involutifolium*

7.   Erect shoots spear-pointed with tightly packed
     leaves   8
     Erect shoots rounded, with loose or irregularly
     arranged leaves   10

8.   Leaves bright green. Stems tomentose with reddish
     tomentum . . . .   *Schlotheimia*
                                           spp. [56]

     Leaves yellowish. Stems not tomentose or with
     brown tomentum   9

9.   Cells smooth . . . .   *rodwayi*
     Cells papillose . . . .   *whiteleggei*

10. Leaves very wide and obtuse at the apex (c 0·4 mm
     across) and with the nerve excurrent in a short
     cusp. . . . .   *wattsii*
     Leaves narrow at the apex (c 0·2 mm across) and
     usually acute . . . .   *archeri*

M. tenue (Hook. & Grev.) Brid. (=*eucalyptorum* of the Handbook)
Of all the species, or groups of species, in this troublesome genus, this is
the easiest to recognize and one of the commonest. The stems are very
slender, with the leaves erect and densely appressed when dry, not at all
twisted or curled; a form with abundant axillary bulbils among the upper
leaves has recently been found on King Island, in Bass Strait. The leaves are
small, 0·7–1·0 × 0·3 mm wide, almost triangular, with a strong yellow
nerve failing below the apex and sunk in a deep channel which is narrow
in the upper half of the leaf but expands greatly in the lower half to form
a spoon-shaped depression (well shown in Hooker and Greville's drawing,
1824, of "*Orthotrichum microphyllum*"). When moistened, the leaves, as in
Rhacomitrium, flex outwards over-far, returning slightly towards the stem
in taking up their final spread position. The cells above are usually strongly
convex, rounded-hexagonal and rather thick-walled, c 7–10 μm diameter
but scarcely enlarged even in the leaf base.

PLATE 43. *Macromitrium tenue* VIC—Fruiting shoot ×15, leaf ×50, cells from
mid-leaf ×1000

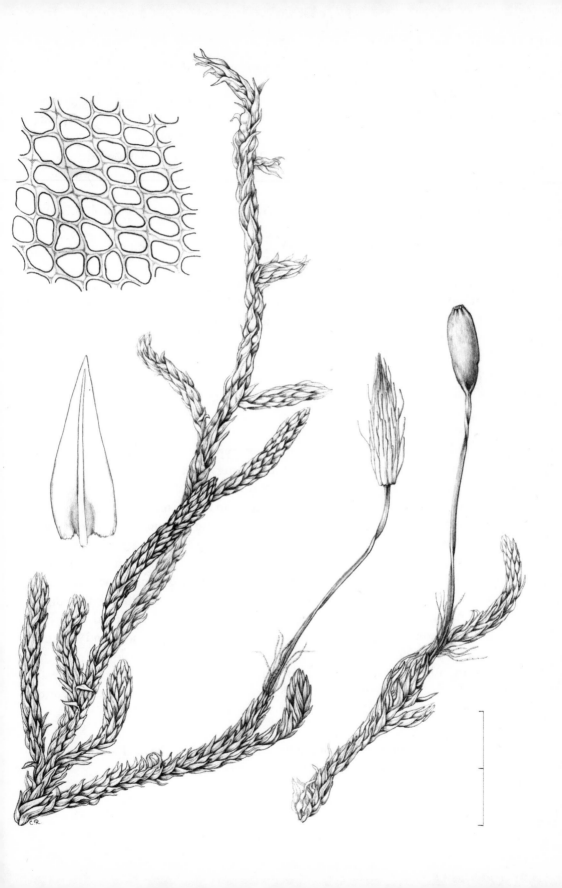

The capsule, c 1·2 × 0·5 mm, on a seta 2–5 mm long, is deeply grooved and pleated at the mouth. The outer peristome is missing and the inner present only as a hyaline, papillose membrane which shows as a whitish lining to the capsule mouth. The plicate calyptra is hairy and of a pale golden straw colour; the Handbook says that it is sometimes almost glabrous.

CHROMOSOME NUMBER: $n = 11$ (NSW).

DISTRIBUTION: TAS, SA, VIC, NSW, ACT, QLD; also in New Zealand, S.E. Asia, S. Africa.

ILLUSTRATIONS: Plate 43; Handbook, Plate 37; Hooker and Greville (1824a), Plate 6; Vitt (1973), Figs 8–10.

The species *eucalyptorum*, *microphyllum* and *tenue*, long suspected to be conspecific, have recently been equated by Vitt (1973) but we have not yet felt able to support his use of the sub-genus *Macrocoma* as a full genus, to contain the straight-leaved species, without a fuller revision of *Macromitrium* as a whole.

M. geheebii C.Muell. in Hampe may also be synonymous.

M. weymouthii Broth.

A dull olive brown colouration seems typical of this species, at least in one of its common forms. The rather long erect stems, c 0·5–1·0 cm tall, form moderately dense cushions like a long-pile carpet. Characteristically the leaves are erect, not at all wound round the stem, but tightly rolled up at the tips and very constant in length and rolling so that the helical files round the stem, of leaf bases and rolled-up leaf tips, give a very rope-like appearance. The leaves are c 1·0–1·4 × 0·2–0·4 mm, often contracted to a short, abrupt asymmetric point. The nerve is strong, reaching nearly to the apex, and there is usually a single longitudinal fold running down one side of the leaf; the margins are plane or slightly recurved. The cells above are isodiametric, thick-walled, with rounded ends, and measure c 5–7 μm diameter; those at the base of the leaf are much longer, to c 30–50 μm, almost straight with very thick walls, except the short cross-walls which stay rather thin; in mid-leaf the cells are intermediate in length. All cells are smooth.

PLATE 44. *Macromitrium weymouthii* VIC—Fruiting shoot × 15, leaf × 50, lower and upper cells × 1000

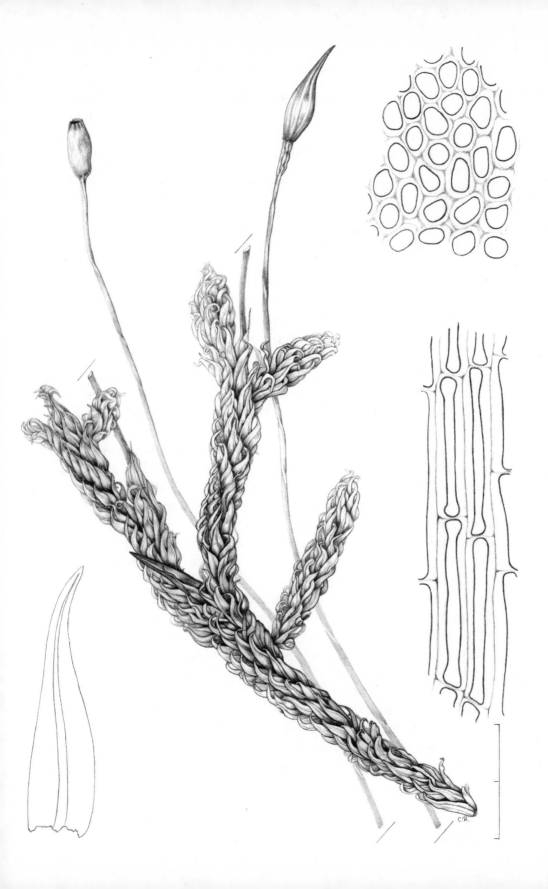

The capsule is of the common *Macromitrium* size, c 1·2 × 0·6 mm, on a seta c 1 cm long, constricted and 8-grooved in the upper quarter or less. The peristome, of 16 hyaline papillose teeth, is inserted well below the rim so as to be partly covered over by the contracted capsule mouth. The calyptra is glabrous and becomes split down one side when mature.

DISTRIBUTION: TAS, VIC, NSW; also in New Zealand.

ILLUSTRATIONS: Plate 44.

The smooth cells, rope-like stems and, to some extent, the contracted capsule mouth are the distinguishing features of this species. There are specimens, very similar in other respects, which have papillose cells but whether these belong to a related species or must be included in the range of variation of this one, is still conjectural. A form also exists in which the peristome teeth are inserted on the rim, and reflex outwards when dry, quite possibly a different species.

M. microstomum (Hook. & Grev.) Schwaegr. and perhaps *reinwardtii* Schwaegr., are supposedly conspecific with *weymouthii* in which case the world distribution extends over S. America, S.E. Asia and Oceania.

M. archeri Mitt. in Hook. represents perhaps the biggest and most puzzling group of all, characterized by the leaves being wound helically round the stem and by the papillose cells. The erect branches tend to be short, c 3 mm, although sometimes nearer 1 cm, and the leaves rather strap-shaped, often with almost parallel sides, c 1·2–2·0 × 0·3–0·4 mm with the nerve deeply channelled and usually shortly excurrent in an abrupt apical cusp. The upper cells are rounded–hexagonal, c 10 µm across, and are highly convex or convex with a small nipple on top (mamillate) or with flat tops or even forked papillae, showing especially on the sides of the channel in which the nerve lies. The basal cells are elongated but commonly not as much as in *weymouthii*, reaching c 20–30 µm long, and have single, high, even spike-like, papillae mainly overlying the cell junctions and conspicuously displayed by the slight recurvature of the leaf margins. Occasionally the papillae are very long and bent forwards towards the leaf apex.

The capsules have a plicate and contracted mouth and either have a hyaline peristome, inserted below the capsule rim (*M. ligulare?*) or else are gymnostomous, but whether the two conditions are found within a single species, as in *Weissia*, we do not know. The calyptra is glabrous.

A striking feature, whose taxonomic weight we have not yet felt confident to assess, is that the capsule, in some forms at least, is deciduous, breaking off at the neck and leaving 4–5 mm of seta persisting, which gives the plants a very distinctive appearance. It seems likely to be a constant and valuable characteristic.

CHROMOSOME NUMBER: n=8, 9 (NSW).

DISTRIBUTION: TAS, WA, VIC, NSW, ACT, QLD; also perhaps in New Zealand. The only species of the genus so far recorded from WA (Willis, 1955d).

ILLUSTRATIONS: Wilson W (1859), Plate 173.

M. asperulum Mitt. in Hook. (NSW, ACT), *pusillum* Mitt. in Hook. (TAS, NSW, QLD), *weisioides* C.Muell. (NSW, QLD), belong to the same group and may be conspecific; *ligulare* Mitt. (NSW, QLD) is fairly similar vegetatively but has a peristome and is perhaps the peristomate form mentioned above. How many species there are in this group, all told, is anyone's guess.

M. rodwayi Dix. in Weymouth and Rodway (1922) has been recorded from southern Tasmania only. It has rather smooth erect branches, tapering to a spear point. The leaves have quite long fine points, sometimes hyaline, and tend to be straight at the tips, not inrolled but wound helically round the stem. The cells are smooth, except for some irregular lumpy surfaces, at the margins especially, but are often quite convex, giving a slightly crenulate margin. The lower cells are long and very thick-walled, except sometimes for a marginal row of squarish cells. We have not seen sporophytes.

DISTRIBUTION: TAS; endemic.

ILLUSTRATIONS: ?

The helically wound leaves and smooth cells are the main features of this species.

M. longirostrum (Hook.) Schwaegr., from TAS, New Zealand and S. America, is similar.

M. wattsii Broth. seems to be a very distinct species with very wide and obtuse, rather flat, tongue-shaped leaves, tightly coiled down round the short stem. The cells above are low-papillose or mamillose, c 10 μm; those below elongate, thick-walled, but *not* very narrow, usually oval or rectangular with rounded corners, c 11–16 × 8 μm, sometimes narrow and longer.

The capsule, 1·4–1·6 × 0·5 mm, on a rather short seta c 3–4 mm long, is smooth or slightly grooved throughout, sometimes contracted at the mouth but usually not. The peristome teeth are *erect*, hyaline to whitish, but falling off easily. The calyptra is hairless.

DISTRIBUTION: NSW, QLD; endemic to Australia.

ILLUSTRATIONS: ?

"Common on sea coast rocks" (Burges, 1952). The very broad and obtuse helically wound leaves and the short basal cells distinguish this species from all except *M. aurescens* Hampe (NSW, QLD) from which it differs, according to the type description, in the smooth capsule and short, glabrous calyptra. But the capsule varies even on the same plant, and the hairiness of the calyptra is a notoriously unreliable character in some *Macromitrium* species so the known differences between these two species, at present, are scant.

M. scottiae C.Muell.

This species is rather robust, with erect stems, 1·5–2·0 mm in diameter and 5–10 mm tall; it has the branch leaves tightly rolled up at the tips, sometimes slightly twisted round the stem but usually not. The leaves are strongly dimorphic, with the stem leaves usually not at all rolled but straight, conspicuous and persistent making a striking contrast with the curled up branch leaves. The upper cells are quadrate and densely papillose with multiple papillae, those lower down shortly rectangular, c 2 × 1, with single high spiculose papillae. The nerve is excurrent in a short cusp.

The perichaetial bracts are very long, slender, and conspicuous, projecting well above the adjacent leaves, although as the sporophytes develop after fertilization they become overtopped by side branches and are then less conspicuous.

CHROMOSOME NUMBER: n = 11 (NSW).

DISTRIBUTION: NSW, QLD; endemic.

ILLUSTRATIONS: ?

This is apparently very distinct in the dimorphic leaves, short basal cells and in the striking perichaetia.

M. involutifolium (Hook. & Grev.) Schwaegr. has leaves rather similar in size and arrangement to those of *scottiae*, rolled up individually at the tips but scarcely wound round the stem. The upper cells are hexagonal, dark and densely multi-papillose, and the basal cells narrowly rectangular, long and thick-walled, mostly with tall spiculose papillae. There is no obvious difference between stem and branch leaves.

The perichaetial bracts are clustered round the base of the seta but are rather blunt-tipped, and do not project above the branch leaves. The peristome teeth are hyaline and papillose, attached below the capsule mouth. The calyptra is slightly hairy.

DISTRIBUTION: NSW, QLD; also in Oceania, ?New Zealand.

ILLUSTRATIONS: ?

This seems to be a common species, especially in the north, characterized by the short but distinct perichaetial leaves.

M. exsertum Broth. (NSW, QLD) is very similar. The leaves have similar tightly rolled-up tips with a slight spiral twist round the stem. The upper cells are isodiametric and almost smooth, the basal cells very long and narrow and thick-walled, with high spiculose papillae. The perichaetial bracts are like those of *involutifolium* but slightly longer and emerging distinctly above the adjacent stem leaves.

This may be just a form of *involutifolium*.

M. whiteleggei Broth. & Geh. has sharply pointed, very small shoots, *c* 2 mm tall × 1 mm across, the leaves being wound tightly round the erect stem, as in *rodwayi*. The leaves are very small, 0·8–1·0 × 0·2–0·5 mm, and the cells, too, are very small and smooth, *c* 4–9 μm in the upper half or less of the leaf. The basal area of elongated cells is extensive; they are very long and thick-walled with rather sparse spiculose papillae. The calyptra is glabrous and the perichaetial leaves not exserted.

DISTRIBUTION: NSW, QLD.

ILLUSTRATIONS: ?

This species differs from *rodwayi* in the small size of leaves and shoots and in the upper cells which are smooth, small and relatively few in number.

M. malacoblastum C.Muell. seems to be very distinctive when fruiting. The capsule is relatively large, *c* 2·0 × 1·0 mm, the seta very short, *c* 1 mm or less, and the calyptra hairy. The stems are quite tall, to 1 cm, with the large leaves, c 2·5 × 0·5 mm, tightly coiled at the tips but scarcely wound round the stem. The nerve fails below the sharply pointed apex. The cells, which are highly convex, giving a crenulate leaf margin, measure c 12 μm above, and are not much enlarged in the leaf base where they reach c 15 × 6 μm.

DISTRIBUTION: NSW; endemic.

ILLUSTRATIONS: ?

This seems very distinct in the cells and in the short seta and large leaves and capsule.

The remaining Australian species allowed by the Index are:

M. baileyi Mitt. (QLD)

M. brachypodium C.Muell. (?QLD)

M. brevisetaceum Hampe (Lord Howe, ?NSW)

M. caloblastoides C.Muell. (NSW, QLD)

M. carinatum Mitt. (?NSW, QLD)

M. circinicladum C.Muell. (NSW)

M. cylindromitrium C.Muell. (QLD)

M. daemelii C.Muell. (NSW, QLD). Illustrated in Brotherus (1924–5), Fig. 455.

M. diaphanum C.Muell. (NSW, QLD). Illustrated in Brotherus (1924–5), Fig. 453.

M. dimorphum C.Muell. (NSW, QLD)

M. funiforme Dix. (QLD)

M. incurvifolium (Hook & Grev.) Schwaegr. (QLD)

M. intermedium Mitt. (QLD)

M. ligulaefolium Broth. (NSW)

M. longipes (Hook.) Schwaegr. which is distinguished by the usually long seta and by the curved cavities of the basal leaf cells. It is doubtfully recorded from TAS and is illustrated in the Handbook (Plate 37).

M. luehmannianum C.Muell. (VIC)

M. novae-valesiae C.Muell. (NSW)

M. owahiense C.Muell.

M. pallido-virens C.Muell. (QLD)

M. peraristatum Broth. (Lord Howe). Illustrated in Brotherus (1924–5), Fig. 450.

M. platyphyllaceum C.Muell. (NSW, QLD)

M. prolixum Bosw. (NSW)

M. prorepens (Hook.) Schwaegr. (?VIC, ?NSW)
M. pugionifolium C.Muell. (NSW, QLD)
M. repandum C.Muell. (QLD)
M. sordide-virens C.Muell. (QLD)
M. subbrevicaule Broth. & Watts (Lord Howe)
M. subhemitrichodes C.Muell. (NSW)
M. subulatum Mitt. (TAS, VIC)
M. tongense Sull. (Lord Howe)
M. torquatulum (C.Muell.) C.Muell. & Broth. (TAS)
M. viridissimum Mitt. (NSW, QLD)
M. woollsianum C.Muell. (NSW)

Schlotheimia Brid.

The characters which separate this genus from *Macromitrium* [55] may seem rather trivial—the smooth, hairless and rather large calyptra is the main one—but it is a surprisingly easy genus to recognize, at least in Australasia. The shoots are a peculiarly vivid dark green above and red beneath, with almost no trace of the orange or yellow which is so typical of *Macromitrium*. Possibly this reflects a fundamental generic difference in pigmentation but, as far as we know, this has never been investigated biochemically. The cells too have a distinct appearance which is hard to define; probably it lies in a difference in composition between two layers of the walls, both of which remain distinct so that the cell outlines are clear despite the heavy superimposed thickenings. The growth form is very much that of *Macromitrium* although the prostrate stems are less in evidence than in many species and the erect stems predominate.

S. funiformis Tayl. ex Dix.

The most striking feature of this plant, when moist, is the strong undulation of the leaves which, together with the bright green colour, immediately separates it from any species of *Macromitrium*, at least in Australasia. The stems, 1·0–1·5 cm tall, are matted together with tomentum of a peculiar reddish, almost pink, colour. The leaves, *c* 1·3 × 0·3–0·4 mm, are tongue-shaped, almost parallel-sided, and rather quickly contracted to give an obtuse leaf with a short slender point, but sometimes more gradually tapering. When dry they are tightly wound helically round the stem. The margins are plane or slightly recurved and the leaf is not plicate but grooved by a central channel in which the nerve runs out almost to the apex where it fades out. The upper cells are oblong or rhomboidal, often obliquely elongate, and typically measuring 15×9 μm. In mid-leaf the cells are longer, narrowly hexagonal with very thick and sometimes sinuose walls, c 30×10 μm, and in the base of the leaf the cells extend to 45×7 μm with such thick walls that the cell cavities are reduced to mere longitudinal slits. The primary walls, and the pattern they form, are still distinct despite the extra thickenings. The cells are all completely smooth.

The perichaetial bracts are slightly larger than adjacent leaves, projecting clearly above them. The seta, c 100 μm in diameter, bears a more or less erect capsule capped with a long fine erect beak and a large smooth hairless calyptra which is not plicate but is split at the base into 5–6 strips and later split right down one side.

DISTRIBUTIONS: NSW; endemic.

ILLUSTRATIONS: ?

We have found this on trees in the Sydney region but have little idea of how common it is. According to Dixon (1950, p. 94) it "differs from *S. brownii* and *S. baileyi* Broth. (which I take to be only a small form of *S. brownii*) in the funiform leaf arrangement, longly exserted bracts, thicker seta, and smooth capsule". We have not seen mature capsules but the seta is scarcely thicker and the bracts scarcely longer than in *brownii* so that only the rope-like (funiform) twisting of the dry leaves and the "gently corrugated" leaves recorded in the type description, separate it from that species. However, although *brownii* can show both these features to a slight extent, they are so strikingly developed in *funiformis* that it seems worth while retaining it as a separate species in the meantime.

S. brownii Schwaegr. is probably not common; it differs most conspicuously in the smooth leaves, scarcely twisted.

DISTRIBUTION: VIC, NSW, QLD; also in New Zealand.

ILLUSTRATIONS: Handbook, Plate 35.

We have not seen this in the field, except in New Zealand, and would prefer to see much more material before being sure that there are two distinct species in Australia.

Ulota Mohr.

In the Australian region this genus is almost confined to Tasmania, where several species have been described. As with *Orthotrichum* it is much under-collected. The Australian species have been reviewed by Malta (1933).

U. lutea (Hook.f. & Wils.) Mitt.
The small cushions of this moss are characteristically found on small branches and twigs in humid forest, in rather similar situations to *Orthotrichum tasmanicum* [53] but much less common. The shoots, bright green or yellowish brown, are 1–2 cm long and are very densely foliate with leaves *c* 2·0 × 0·4 mm, variably twisted when dry, sometimes almost erect and only slightly flexuose, more commonly quite strongly twisted and contorted. The leaf is long and narrowly triangular with a slightly or much widened concave ovate base. The strong single nerve is sunken in a deep channel, prominent abaxially, and fails just below the apex. The upper cells are very thick-walled, roughly isodiametric but very irregular in size and shape, roughly 10 μm across, becoming longer towards the base of the leaf except at the margins. In the leaf base the cells are long and narrow and very thick-walled, often with a slight sigmoid twist, *c* 30–60 × 6 μm, those next to the stem tinged with orange-yellow. The margins are entire and plane or very slightly recurved and, at the base, the marginal cells (in 1–3 or more rows) are quadrate and hyaline with thick cross walls, those of adjacent cells tending to lie in alignment to form continuous transverse bars of thickening. The cells are sometimes papillose, or smooth but often with the surfaces lumpy and irregularly thickened.

The capsule is deep straw-coloured, *c* 1 mm long, narrowly cylindrical, merging imperceptibly into the 3 mm long seta; it is deeply 8-ridged, the ridges continuing down as the edges of the seta when dry. The operculum is slightly convex, sometimes red round the rim, with an abrupt spike on top. The vaginula, at the base of the seta, is stiffly hairy. There is a peristome of 16 broad, pale teeth, grouped and initially fused, in pairs, which alternate with the capsule ridges. When the capsule first opens, the teeth remain loosely covering the capsule mouth, allowing gradual spore dispersal. After being wet for the first time, the hygroscopic mechanism is triggered and on drying out then and on subsequent occasions, the teeth bend backwards, splitting into single teeth—and sometimes partly split still further into half teeth—leaving the inner peristome of 8 hyaline hair-like processes still in position over the capsule mouth, partially restraining the spore mass. The teeth are finely papillose all over. The spores, 22–28 µm, are densely papillose with fine papillae. There is a bell-shaped calyptra, densely hairy with long erect hairs.

DISTRIBUTION: TAS, VIC; also in New Zealand.

ILLUSTRATIONS: Malta (1933), Figs 1–3.

Orthotrichum (especially the common *O. tasmanicum*) is the only genus likely to be confused with *Ulota*—the cushion growth form separates it from any species of *Macromitrium*. The distinction between *Ulota* and *Orthotrichum*, however, is a fine one. When the plants are fruiting, the broad teeth of the inner peristome in *O. tasmanicum* will separate it from *U. lutea*. When vegetative, the transverse thickenings of the marginal cells in the leaf-base, which are usually conspicuous in *Ulota*, will commonly be sufficient to identify it.

U. cochleata Vent.

This rare moss is found on the branches and twigs (commonest at 80–90 feet from the ground) of tall *Nothofagus* trees in Victoria (Ashton and McRae, 1970).

The leaves are usually more finely tapering than those of *lutea* and have 5 or more rows of hyaline cells at the basal margin (2–5 in *lutea*). Cells of the leaf base are long, narrow, very thick-walled and often porose.

The vaginula is naked without the stiff hairs found in *lutea* and the stomata are situated in an unusual position for the genus *Ulota*, in the middle and upper part of the capsule. In *lutea* they are only on the neck which is also longer and more tapered than in *cochleata*. The spores are papillose, 27–30 µm, often germinating to form a short filament in old capsules.

DISTRIBUTION: TAS, VIC; endemic.

ILLUSTRATIONS: Malta (1933) Fig. 1. (leaf)

Other species recorded for Australia are:

U. crassifolia (Hook.f. & Wils.) Hook.f. (*Muelleriella* in the Handbook). The Index's record from the region Austr. 1 refers presumably to Kerguelen Island, not Australia.

U. dixonii Malta (TAS)

U. laticiliata Malta (TAS)

U. membranata Malta (TAS)

U. viridis Vent. (TAS)

Zygodon Hook. & Tayl.

Predominantly a genus of epiphytes, this has a rather distinctive combination of characters: stems forming quite large turfs on bark of trees; dark fairly short leaves, corkscrewed part way round the stem when dry, in most species; distinctively shaped capsule tapering above and below, with evident neck and *c* 8 ribs when dry; glabrous, small calyptra, open down one side and falling off early; small, rounded hexagonal or squarish cells; and a nerve deeply sunk in a longitudinal furrow. The cells may be either papillose or smooth. Rhizoids are dimorphic, the main ones very wide and often thick-walled (18–36 μm) with much narrower lateral branches (4–6 μm). Filamentous (6–8 celled) gemmae occur on the stem among the upper leaves. The genus has been monographed by Malta (1923–1926).

This genus is closest to *Ulota* [57], *Orthotrichum* [53] and *Amphidium* [54]. From our species of *Ulota* and *Orthotrichum* it differs in having a calyptra which is side-split instead of campanulate; a growth form of turfs instead of small cushions; in having gemmae among the leaves; and in the usually plane leaf margins (except *Z. menziesii*) and almost parallel-sided leaves.

Most species of *Orthotrichum* and all species of *Ulota* have hairy calyptras. The habitat too is different, *Zygodon* growing predominantly on tree trunks while *Ulota* and *Orthotrichum* are commoner on branches and twigs, especially in montane regions or higher.

Amphidium differs in the growth form, the stems being very densely matted together in large turfs; in the habitat which is on damp shaded rock in montane to alpine regions; in the gymnostomous capsule on a very short seta; and in the absence of gemmae.

The three principal species in southern Australia are easily distinguished: *Z. intermedius* has papillose cells, *Z. menziesii* has smooth cells but leaf margins recurved below and converging at the apex to form a wide, blunt point. *Z. minutus* has smooth cells, plane margins, and abruptly pointed leaves.

Z. intermedius B.S.G.

This, the commonest species, forms large, dark green patches on trunks of trees in rain-forest, wet sclerophyll forest and fern gullies. It has a distinct preference for the underside of tree trunks which are leaning slightly.

The stems are commonly 5 mm tall, with the upper 3–4 mm green; densely rhizoidal below but not tightly compacted. Mature leaves are c 0·7–1·0 mm long × 0·2–0·25 wide, the single nerve failing below the apex. Cells in mid-leaf have irregularly rounded cavities c 7–8 μm diameter, sometimes slightly wider than long with one to several small papillae per cell; those at the very base of the leaf oblong (up to 2 × 1) but still thick-walled. The cells are usually in clear longitudinal rows and sometimes lateral rows.

The seta is tightly twisted (right hand) just below the capsule, slightly twisted (left hand) in the lower half; perichaetial leaves are not distinctive. The capsule mouth is crimson, with a row of peristome teeth often falling off early but the sticky spores remain in the capsule even when wide open. Spores are c 15–18 μm usually slightly oval, green, irregularly warted, with c 10 warts across the diameter.

DISTRIBUTION: TAS, WA, VIC, NSW; widely spread in the southern hemisphere.
ILLUSTRATIONS: Plate 45; Handbook, Plate 33.

Z. hookeri Hampe, recorded from TAS, VIC, NSW, is very similar, differing mainly in the slightly or conspicuously undulate and often slightly denticulate leaves and in the somewhat larger spores (20–24 μm according to the Handbook). It is illustrated in Allison and Child (1971), p. 75.

Z. reinwardtii (Hornsch.) Braun, from Tasmania (Rodway 1914) sounds very similar and combines some characters of *intermedius* and *hookeri*. These three species would well repay a joint investigation. If they proved to be

PLATE 45. *Zygodon intermedius* VIC—Fruiting plant × 30, leaf × 50, cells, including leaf margin, × 1000

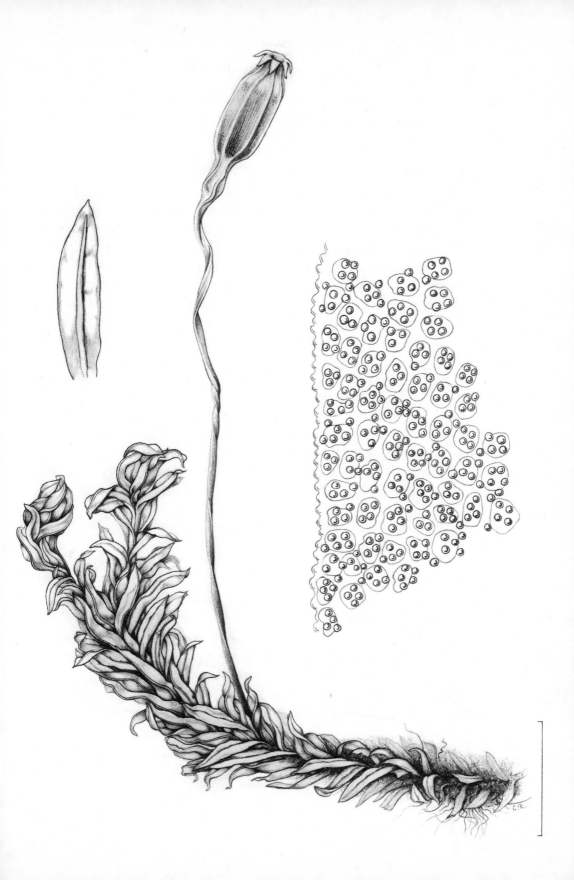

one, rather variable species, the correct name would be *Z. reinwardtii.*
Sainsbury (1955c, p. 19) feels that *Z. reinwardtii*, "at any rate in its typical
form, must be excluded from the Tasmanian flora".

Z. menziesii (Schwaegr.) Arnott

Smaller and much less common than *Z. intermedius*, occupying drier
habitats such as the trunks of bushes and trees in coastal scrub, this species
has shoots *c* 5 mm high, in very small, rather densely compacted, short
turfs, the green portion *c* 1·5–2·0 mm high. The leaves, 1·0–1·2 mm ×
0·2–0·3 mm, are twisted round the stem; the nerve failing below the apex;
the margins narrowly recurved in the lower half. Cells are extremely
variable in pattern, commonly thick-walled, especially at the ends. Many
cells in mid-leaf are rhomboidal, elongated diagonally out towards the
margins. The other cells in mid-leaf are squarish or oblong, *c* 6–10 μm wide,
in longitudinal and diagonal rows; those in the basal third of the leaf
elongated and enlarged, *c* 45–60 × 15–20 μm, rectangular-hexagonal, thin-
walled or thickened only on the end walls. There are no papillae.

The seta is twisted as in *intermedius*; the calyptra pale straw-coloured
and glabrous, lightly ribbed below; and the capsule mouth red.
DISTRIBUTION: TAS, WA, SA, VIC; also in New Zealand and South America.
ILLUSTRATIONS: Plate 46; Handbook, Plate 33; Brotherus (1924–5), Fig.
429; Malta (1924b), Figs 1–3, 9.

Z. rodwayi Broth. in Rodw., from Tasmania, sounds very similar but we
have not seen it.

Z. minutus C.Muell. & Hampe

Apparently this species occupies similar habitats to *Z. menziesii* from which
it is indistinguishable in leaf size and areolation; but the *nerve is shortly*
excurrent and the margins scarcely ever at all recurved. Stems are perhaps
slightly smaller than in *Z. menziesii*, *c* 1·5–2·0 mm, of which the upper
1 mm is green; the leaves not twisted round the stem.

The capsule, when green and unripe, has both the calyptra and *capsule*
mouth pale straw-coloured. The mouth darkens slightly when old but does

PLATE 46. *Zygodon menziesii* SA—Fruiting plant × 30, leaf (top right) × 50, cells
× 1000; *Z. minutus* VIC—Leaves (lower left) × 50

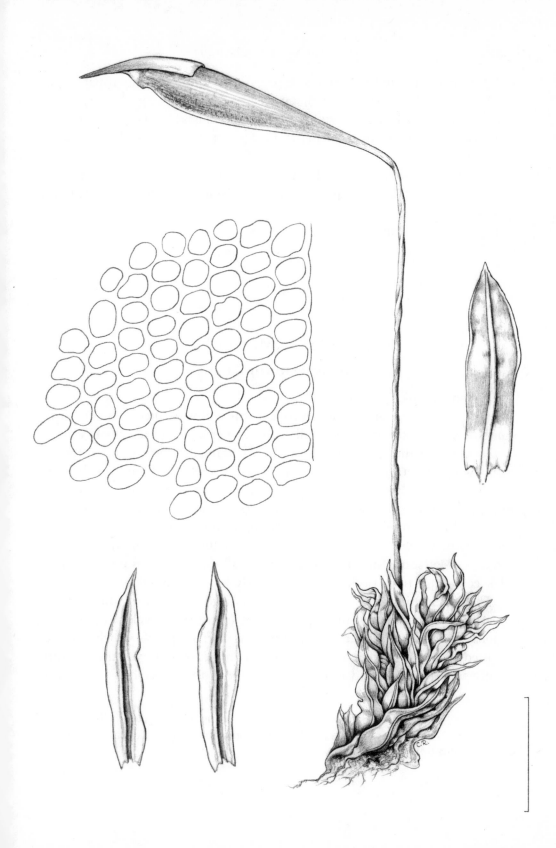

not turn red as in *menziesii*. Spores are green, 15–20 μm, very finely papillose, with *c* 20 papillae across the diameter.

DISTRIBUTION: TAS, WA, SA, VIC; also in New Zealand.

ILLUSTRATIONS: Plate 46; Malta (1924b), Fig. 13.

Z. obtusifolius Hook., a rare and quite distinct species known from Australia only in Tasmania, has very papillose cells and round-ended leaves. We have not seen it.

ILLUSTRATIONS: Malta (1924b), Figs 15–18.

GIGASPERMACEAE

Gigaspermum Lindb.

G. repens (Hook.) Lindb.

The plants, which are usually fertile, form a low white or silvery turf on bare soil mostly in low rainfall areas. The erect leafy stems (*c* 1–3 mm) are bud-like at the apex and arise from a perennial creeping underground stem which is practically leafless. They are sometimes found on the soil on limestone ledges where the stems may reach 1 cm high. Vegetative leaves are tiny, *c* 0·5 mm long, nerveless, delicate and pale green, very broad, almost circular, concave, with an abrupt long colourless hair-like point which often consists of only one or a few long cells; the margin entire except sometimes for some very small denticulations on the point. The cells range from isodiametric to slightly longer than wide, *c* 20–25 μm.

The perichaetial leaves are comparatively large, 1·0–1·5 mm long, closely overlapping and erect at first, spreading at maturity, nerveless, colourless, concave and broadly ovate, tapered to long narrow points which may be twisted and are often hair-like and reflexed on the lower leaves, the margins entire to very finely and distantly serrulate. In the lower half of the leaf

PLATE 47. (upper) *Goniomitrium enerve* VIC—Fruiting plants ×22, cells ×1000; (lower) *Gigaspermum repens* VIC—Fruiting plants (behind), male plants (in front) ×22, cells (lower left) ×1000

the cells are large and elongated, 20–25 µm broad, rectangular to rhomboidal and 4–6 × 1; in the upper half more regularly diamond-shaped *c* 5 × 1 and mostly thin-walled. The straw-coloured capsule is gymnostomous and hemispherical, *c* 1 mm wide and practically sessile with a wide flaring mouth when the lid is off. The operculum is broad and flat with a small apiculus, and the calyptra small and conical, covering only the top of it. The spores are very large, *c* 130 µm, brown, angled and coarsely granular. Male plants are small rosettes of broad, imbricate, pale green leaves which have long, hair-like strongly reflexed tips.

DISTRIBUTION: All states; TAS, WA, SA, VIC, NSW, ACT, QLD, ?NT; also recorded from New Zealand and S. Africa.

ILLUSTRATION: Plate 47; Brotherus (1924–5), Fig. 260.

This splendid plant was desribed by the Rev. W. W. Watts (1906) as "the lovely *Gigaspermum*" and, according to Willis (see *Bryologist* **74**, 531, 1971), was mistaken for and described in Black's "Flora of South Australia" as a minute flowering plant, *Trianthema humillima* F. von Muell. in the Aizoaceae. The wide open capsules containing the huge brown spores and nestling in the long silvery leaves make the plant readily recognized when mature.

G. subrepens C.Muell. (WA) is said to differ only in the more pointed leaves and should be considered synonymous until proved otherwise.

FUNARIACEAE

Funaria Hedw.

Bare soil is the customary habitat for the Australian species of this distinctive genus. The big, thin-walled cells and generally wide and concave leaves are shared with other genera, especially *Physcomitrium* [63] and *Tayloria* [65] but the *inflated balloon-like membranous calyptra*, with a long narrow beak on top is diagnostic. In the field the genus is usually recognizable at a glance by the combination of wide leaves, red-orange capsule, and the range of shapes of the capsule all of which are rather distinctive. The

stems are all individual and practically unbranched forming usually rather
open colonies. Where a double peristome is present, it has the unusual
feature of inner and outer teeth arranged in phase with each other, instead
of alternating. The common species fall into three groups according to
capsule shape: *hygrometrica* and *microstoma* have completely asymmetric
capsules with the mouth apparently almost in the side wall of the capsule;
glabra has the capsule twisted through 90° so that the plane of the mouth
is vertical; the others have the capsule erect and symmetrical. Vegetatively
they are scarcely to be identified, except by comparison with fruiting
material, but as they fruit freely this is seldom necessary. It would be most
interesting to examine the spores with the scanning electron microscope;
they appear to have quite specific markings but these are just beyond the
limits of comfortable resolution with the light microscope.

Key to species

1. Capsule asymmetric; in some species with the
   theca deeply grooved when dry   2
   Capsule erect and symmetrical, never grooved .   4

2. Capsule deeply grooved, mouth much smaller
   than maximum width of the capsule . . .   3
   Capsule wall smooth outside, mouth almost as
   wide as the capsule . . . .   *glabra*

3. Operculum 0·4–0·5 mm diameter. Spores 23–
   30 µm, papillose . . . .   *microstoma*
   Operculum 0·7–0·8 mm in diameter. Spores 10–
   12 µm, smooth . . . .   *hygrometrica*

4. Leaves narrow, lanceolate, at least in the upper
   half of stem . . . .   *producta*
   Leaves wide, ovate or oblong   5

5. Capsule appearing small on a long seta of up to
   c 1 cm . . . .   *gracilis*
                                                       (+ *cuspidata*)

   Capsule appearing large relative to the seta which
   is usually less than 5 mm . . . . . .   *apophysata*
                                                        (+*helmsii*)

F. hygrometrica Hedw.

This very common world-wide moss has been studied in much greater detail than any other and from many aspects (see e.g. Hoffman, 1964; Bopp and Böhrs, 1965; Proskauer, 1958) and has been the standard moss of elementary botanical textbooks for generations. Despite the amount of research already carried out on this species, new facts are still being discovered (Proskauer, 1958). It has the invaluable property (as a plant for teaching) of turning up regularly on forest and other soils after fire. Thus it commonly forms big and almost pure patches on the sites of old camp fires in forests, by roadsides, in waste places, nurseries, and on rubbish dumps, apparently associated with high potash concentrations. In appropriate habitats it will appear even in cities.

The short unbranched stems, *c* 0·5 cm tall, have a rosette of wide concave ovate leaves, *c* 4 × 1 mm, with a short or long point and usually widest at or above the middle, with a single nerve failing below the apex or excurrent. The cells are typical of the genus, wide and lax, hexagonal–squarish above and rectangular below, much narrower at the margins but not forming a conspicuous border since the cells are neither thickened nor distinctly coloured. When dry the leaves are very crumpled and slow to soak out.

Most, if not all, stems become fertile, first producing a terminal male cup, later overtopped by a female branch from below the cup. Sporophytes are usually very abundant indeed. The long seta, 1·5–2·0 cm long (the Handbook records it to 6 cm) is crooked down or even coiled in a circle at the top so that the capsule hangs downwards; when dry the mature seta is strongly twisted, untwisting when wet; hence the name *hygrometrica*. The capsule although erect when very young is pendent and completely asymmetric when ripe, *c* 3 × 1 mm, deeply grooved and with a mouth so displaced by curvature of the capsule as to appear to occupy a lateral position. The mouth is *c* 0·8 mm across, red-rimmed, with two sets of peristome teeth under a low domed operculum. The orange outer teeth are not radially straight but have a spiral (sigmoid) twist and are all joined at the tips by a small disc when undamaged. The usual peristome mechanism where the teeth flex outwards and inwards hygroscopically is thus translated into a movement of rotation, opening and closing the slits between adjacent teeth to allow spores to escape; the unusual coincidence between inner and outer peristomes makes this possible. In old capsules, if the teeth break free from each other, they bend in and out in the standard peristome movement. The outer teeth are papillose and have very strong transverse bars projecting into the capsule cavity and sticking out at the sides of the teeth. The inner peristome processes are similar in shape to the outer teeth but are hyaline.

When immature and still green the neck of the capsule is studded with white dots which are stomata (air pores), easily visible to the naked eye. When old, the capsules are a deep rusty brown. The spores are smooth, c 12 μm in diameter, but are extremely variable in size.

CHROMOSOME NUMBER: n=28 (NSW).

DISTRIBUTION: Throughout all Australia, including NT, and the rest of the world.

ILLUSTRATIONS: Watson (1955), Fig. 70; Allison and Child (1971), Plate 14; Troughton and Sampson (1973), Plates 39, 42–51; Brotherus (1924–5), Fig. 282.

With ripe capsules, this is as distinctive as a moss can be. *F. microstoma*, the only species with a similar capsule, has a mouth half the size and this gives it a very different appearance. With non-fruiting specimens, confusion is much more likely, e.g. with other Funarias and *Tayloria*. Consideration of the habitat and comparison with material of known identity are then the only hopes for identification.

F. microstoma Bruch ex Schimp. is rather similar to the last species but the leaf cells tend to be rectangular *throughout*, measuring c 150–200 × 60–80 μm in mid-leaf, the leaf apex ends in rather a long point, and the capsule is quite distinct. The seta tends to be shortish, somewhat less than 1 cm, and the capsule is shorter and relatively wider than in *hygrometrica* (c 1·7 × 0·8 mm) with a much smaller mouth, c 0·4 mm in diameter. The peristome teeth have much less pronounced transverse bars and the inner peristome is merely a low membrane or undetectable. The spores are much larger, 23–30 μm, and the surface rather sparsely ornamented with low, rounded papillae.

DISTRIBUTION: WA, SA, VIC; not given in the Index, but our material seems unmistakable; also in New Zealand, Europe and Asia.

ILLUSTRATIONS: Grout (1928–40), II, Plate 40.

This seems to be a salt-tolerant or salt-loving species, found both in seaside salt marshes and beside inland salt lakes. *F. salsicola* C.Muell. is likely to be synonymous according to the Handbook (p. 247).

F. glabra Tayl.

This species of dry roadsides is common enough, but seldom forms large patches so that it is not particularly conspicuous, but the capsule is unlike any other and even vegetatively it is more distinct than most. It also occurs, sometimes in conspicuous colonies, on wet ledges near streams and waterfalls.

The leaves, 2·0–2·7 mm × 1·0 mm are rather lanceolate, scarcely concave, with the nerve failing below the shortly pointed apex, occasionally very short. The margins are mostly lightly denticulate by projecting cell-ends and the marginal cells are much narrower, tending to form a rather distinct border, emphasized often by a slightly different colouration. We have found a form in which the leaves were rather strongly toothed with twinned teeth, as in *Rhizogonium bifarium*, presumably an abnormality.

The short, reddish-brown seta, 1·5–2·0 mm long, is straight and the capsule above, *c* 1·2 × 0·8 mm, is curved through a quarter circle like the deck ventilators on a ship so that its wide mouth, *c* 0·6 mm diameter, faces sideways. The peristome is double; the outer peristome red-orange with a very slight sigmoid twist, the outer surfaces of the teeth striate by oblique stripes formed by lines of papillae, the inner peristome paler and membranous, occasionally rudimentary. The capsule is smooth outside, apart from the basal third, the neck, which is rather crumpled when dry; but inside the capsule is heavily ridged longitudinally, as if lined with pleated fabric. The spores are dark brown, 26–30 μm across, covered with a mosaic of low irregular plate-like warts. The operculum is usually *flat* or even slightly concave but occasionally with a projecting nipple.

CHROMOSOME NUMBER: n = 52 (± 2) (NSW).

DISTRIBUTION: TAS, WA, SA, VIC, NSW, ACT, QLD, NT; also in New Zealand and New Caledonia.

ILLUSTRATIONS: Studies, Plate 9; Mueller (1864), Plate 1 (as *F. tasmanica*); Wilson (1859), Plate 175 (as *F. crispula*).

F. gracilis (Hook.f. & Wils.) Broth.

at its best, is well named. The small capsule, *c* 1·4 mm long, about half of which is neck, × 0·8 mm wide, with a flat operculum, is perched on top of a long graceful seta, 1–2 cm, very much like a glass thistle-funnel. The leaves are ovate, not usually particularly concave nor as wide as in some other species, c 1·0 × 0·4 mm, usually with a short projecting tip. The cells are slightly smaller than other species, c 30 × 15 µm above and 60 × 15–20 µm below, with the marginal cells not appreciably different.

The peristome teeth are orange and *straight* or nearly so, hyaline and papillose at the tip and with *sinuose* edges, densely ornamented with obliquely longitudinal striations, especially on the outer face, and sometimes partially split down the middle. The inner peristome is usually absent but a trace of membrane is sometimes detectable. The spores are 33–40 µm in diameter, smooth but with occasional irregular projections.

DISTRIBUTION: TAS, WA, SA, VIC, NSW, ACT, NT; also in New Zealand.

ILLUSTRATIONS: Wilson (1854), Plate 86 (as *Entosthodon*).

This is a plant of open woodland, dry sclerophyll forest, mallee, etc.

F. cuspidata Hook.f. & Wils. has an identical fruit but the leaves have the nerve clearly and usually far excurrent.

CHROMOSOME NUMBER: n = 26 (NSW).

DISTRIBUTION: TAS, WA, VIC, NSW, QLD; elsewhere? Also in New Zealand.

ILLUSTRATIONS: Studies, Plate 9; Wilson (1854), Plate 86.

The relationship of this species and *gracilis* is well worth further study. Plants with *apophysata*-like fruits but with peristomes are sometimes referred to *cuspidata*, where indeed they may belong, but are discussed under the next species.

F. apophysata (Tayl.) Broth. is a puzzling species or group of species which we do not yet fully understand. The leaves are wide and very concave, often nearly circular and usually with an abrupt almost hair-like point and a single nerve usually failing well below the apex, although the Handbook mentions a Tasmanian specimen where the nerve is excurrent and we have found the same in some NSW specimens. The cells are irregularly hexagonal above (sometimes throughout) 40–50 × 25–35 µm, and usually shortly rectangular below, 60–75 × 25 µm. The margins sometimes are crenulate with projecting cell ends.

Most typically the orange capsule is long and almost narrowly triangular in shape measuring c 2·3 × 0·9 mm of which ⅓ or ½ is neck, narrowed to the mouth (which is covered by a flat or occasionally slightly domed operculum) and deeply ridged just below it. The seta, c 1·4–2·0 mm, is

slightly shorter than the capsule. There is no peristome. The spores are dark brown, 27–36 μm, and are densely and very finely wrinkled or ridged.
DISTRIBUTION: TAS, WA, SA, VIC, NSW, ACT, QLD, NT; also in New Zealand.
ILLUSTRATIONS: Brotherus (1924–5), Fig. 274; Wilson (1854), Plate 86 (as *Physcomitrion*).

While this sounds a distinctive species, it is not always so. The seta can be much longer, giving young capsules much the appearance of *gracilis*. By contrast, specimens occur which are very similar in general appearance but have a distinct single peristome. Such specimens are most typical of mallee country and may, from the description, be *F. tateana* (C.Muell.) Watts & Whitelegge.

F. helmsii Broth. & Geh. has been recorded from the central regions of Australia (e.g. Willis, 1972b). It is chiefly distinct in the clearly toothed leaves.
DISTRIBUTION: WA, SA, VIC, NSW, NT; ?endemic.
ILLUSTRATIONS: ?

The relationship between this, *tateana*, and the *apophysata* and *gracilis* groups needs thorough investigation.

F. producta (Mitt.) Broth. is a very small plant, distinct vegetatively in the concave, ovate or oblong leaves tapered to a narrowly triangular or lanceolate (or even subulate) upper half. The leaves are *c* 1·2 mm long × 0·4 mm wide with a nerve extending almost to the tip. The cells are *c* 60 × 15 μm, narrowly rectangular or rhomboidal. The spores, 30–33 μm, are irregularly ridged and warted. It has no peristome.
DISTRIBUTION: TAS, WA, SA, VIC; also in New Zealand.
ILLUSTRATIONS: Handbook, Plate 38; Wilson (1859), Plate 175 (as *Entosthodon*).

The other species recorded from Australia are:
F. aristata Broth. (VIC, NSW)
F. bullata Broth. (VIC)
F. clavaeformis (C.Muell. & Hampe) Broth. (SA)
F. minuticaulis (Geh.) Watts & Whitel. (VIC)
F. papillata Hampe (QLD)
F. smithhurstii Broth. & Geh. (VIC, NSW, QLD)
F. squarrifolia Broth. (NSW)

F. subnuda Tayl. (=*F. glabra fide* Burges, 1952). (WA, SA, NSW) and

F. varia (Mitt.) Broth. (VIC), which are all known only from Australia. Most of them are certainly synonymous with other species but we have not yet studied them in detail.

F. perpusilla Broth. (VIC) is held by Sainsbury (1955e) to be a synonym of *producta*.

F. radians (Hedw.) C.Muell. we have never seen but, being a Hedwigian name, is very probably an earlier name for one or other of the species already mentioned.

Goniomitrium Hook.f. & Wils.

G. enerve Hook.f. & Wils.

These tiny, practically stemless plants grow singly or clustered together on bare earth. The leaves are few in a rosette, nerveless, pale green, very concave and broadly ovate, *c* 0·5–1·0 mm long, the upper often tapered to a point; the margin entire to minutely serrulate. The cells are large, shortly rectangular to 5-sided, 20–30 μm long × 15–20 μm wide, and thin-walled; more elongated (to 5 × 1), larger and rhomboidal in longer leaves.

The seta is thin and very short, the capsule gymnostomous (0·5–0·75 mm), showing above the leaves and shortly oval or globose, bright orange with a red-gold rim round the mouth which is smaller than the diameter of the capsule; the operculum is a small, low cone. The translucent calyptra which wholly covers the capsule is completely diagnostic of the genus, large and bell-shaped with 8 longitudinal flanges or pleats (which may be slightly serrate) below the short dark beak. The spores are big and bright yellow, 90–105 μm, the outer surface having a slightly projecting reticulum with large areolae (but not multi-cellular).

The tiny male plant is a pale green rosette of very concave broad leaves which end in a sudden short hyaline apiculus. The tips of the leaves do not usually turn outwards and are shorter than in the *Gigaspermum* [59] male plants.

DISTRIBUTION: WA, SA, VIC, NSW, ACT, ?QLD; endemic.

ILLUSTRATIONS: Plate 47; Wilson (1846), Plate 3.

A second species, G. *acuminatum* Hook. & Wils., has nerved leaves and the upper leaves are much longer with a narrow hyaline twisted tip reaching above the capsule which is larger (1 mm) and cup-shaped. This plant has a greater resemblance to *Gigaspermum* but the nerved leaves, the bright orange capsule and large bell-shaped pleated calpytra are completely diagnostic.

DISTRIBUTION: WA, NSW, QLD, NT, probably wider; endemic.

ILLUSTRATIONS: Wilson (1846), Plate 3.

Physcomitrella B.S.G.

P. readeri (C.Muell.) Stone & Scott, 1973

These ephemeral plants are small, 0·5–4·0 mm high or sometimes more by innovation, growing together in wide patches on mud or silt in recently flooded areas. The delicate bright green leaves are rather erect when moist but shrivelled when dry and do not recover readily; the upper larger (to 2 mm) and more crowded than the lower. The translucent leaves have a soft delicate texture and are oblanceolate, spathulate or obovate with a short apiculus, a narrow nerve failing far short of the apex and the margins plane and entire near the base, serrate above. The cells are very large and thin-walled, rectangular or rhomboidal to six-sided, 20–30 μm wide and mostly 2–3 × 1 but sometimes much larger in the lower part of the leaf and adjoining the nerve.

There is a very short seta, usually less than half the length of the cleisto-carpous capsule which is very variable in size, globular or oblong, apiculate, and dark brown to black at maturity. The stomata are at the base of the capsule and are not immersed; there is no sign of a line of dehiscence on the capsule. The small conical membranous calyptra has a short slit at the base and covers not much more than the apiculus of the capsule. The spores are dark rusty brown, 30–45 μm, spinulose with slightly hooked spines and the spores cling together in a mass when the capsule wall decays.

PLATE 48. (left) *Physcomitrella readeri* VIC—Fruiting plant (moist) × 37, cells × 1000; (right) *Acaulon integrifolium* VIC—Fruiting plants × 37, cells × 1000

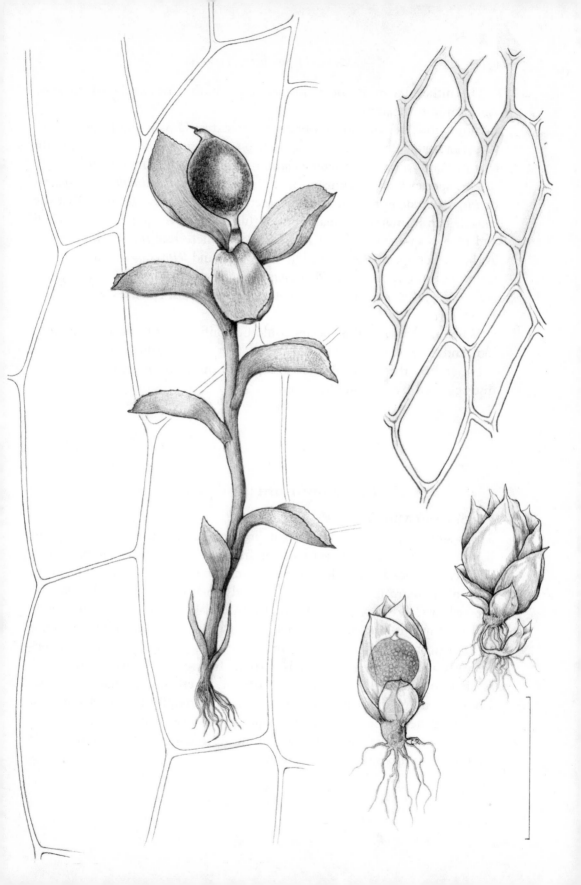

The antheridia are in small clusters of 4–5 in the leaf axils just below the archegonia at the apex.

DISTRIBUTION: SA, VIC, and probably other states; U.S.A. and Japan.

ILLUSTRATIONS: Plate 48; Studies, Plate 9 (as *Physcomitridium readeri*); Roth (1914), Plate 10 (as *Physcomitrella austro-patens*); Roth (1911), Plate 21.

This species was previously known either as *Physcomitrella austro-patens* or *Physcomitridium readeri*, a genus erected because it was thought that the capsule had immersed stomata, but as this is not so the genus cannot stand and reverts to a synonym of *Physcomitrella* where *readeri* takes precedence as the earlier specific epithet. The New Zealand "*Physcomitridium readeri*" is not the same species, but *Physcomitrella californica* in the U.S.A. and Japan is *P. readeri*; this has been checked at the British Museum and confirmed by Dr H. Crum.

When not fruiting this species could be confused with species of *Funaria* [60] and *Tayloria* [65]. The former is best distinguished by the longer nerve and by the terminal male discs when these are present; the latter is distinguished by the long leaf point.

Physcomitrium (Brid.) Fuernr.

P. conicum Mitt. in Hook.

These small plants form broad patches on moist alluvial soil or stream banks, with stems erect, 0·3–1·0 cm high, sparsely branched or unbranched. The leaves, 2·0–2·5 mm long, are oblong or obovate, shortly apiculate, with plane margins, sometimes shrivelled when dry. In the upper lamina the cells are shortly rectangular or rhomboid-hexagonal, 25–35 µm wide and 1–3 times as long, the marginal row narrower and protruding, forming an indistinct slightly serrated border; the cells of the lower lamina rectangular and longer. The nerve is narrow, failing just below the apex.

There is a relatively long seta, 0·5–1·0 cm; an erect, pear-shaped capsule, 1·0–1·8 mm long, with an annulus and a distinct apophysis, gymnostomous, wide-mouthed when empty; the operculum low convex with a short abrupt blunt beak; the calyptra symmetrical with a short inflated base cleft into 2–3 lobes, and a subulate apex.

DISTRIBUTION: TAS, VIC, NSW; also in New Zealand.

ILLUSTRATIONS: Handbook, Plate 38.

P. subserratum Hampe is synonymous according to Dixon (Studies).

There are many other spp. recorded for Australia, some of which may also be synonyms but none of which we know:

P. brisbanicum C.Muell. (NSW, QLD)

P. flaccidum Mitt. (NSW)

P. integrifolium Hampe & C.Muell. (VIC, NSW)

P. laxum (Hook.f. & Wils.) C.Muell.

P. minutulum C.Muell. (QLD)

P. nodulifolium Mitt. (QLD)

P. pyriforme (Hedw.) Hampe

EPHEMERACEAE
Ephemerum Hampe

E. cristatum Hook.f. & Wils.

The abundant green protonema, persistent and dichotomously branched, is the most readily recognizable feature of this extremely minute earth moss. The stem is less than half a millimetre high with a few leaves which are loosely arranged and arched inwards or spreading. As in *Bryobartramia*, however, the upper leaves keep growing as the capsule develops, so that the plant varies in appearance. The leaves are spathulate to narrowly lanceolate or oblanceolate, and only a few cells wide; above mid-leaf irregularly and deeply incised and dentate–ciliate with the teeth often recurved. The nerve is variable, sometimes inconspicuous but often cristate abaxially (as is the lamina) in the upper half, and sometimes excurrent from a truncate leaf apex. The cells are rectangular to irregularly rhomboidal, 10–20 μm wide, short or long, $2–9 \times 1$, firm to thick-walled.

The seta is extremely short and the foot rounded; the capsule tiny, cleistocarpous, rounded to shortly oval with a small apiculus, shining red to red-brown when mature and with no sign of a dehiscence line. The spores are very large, 40–80 μm, brown and coarsely warty. The calyptra is campanulate, delicate and torn at the base, covering down to the middle of the capsule.

DISTRIBUTION: TAS, WA, SA, VIC, NSW, QLD; endemic.

ILLUSTRATIONS: Plate 49; Hooker, W. J. (1845), Vol. 8, Plate 737; Roth (1911), Plate 22.

These plants may be found on humus-rich soil in red gum forests along the Murray River or in mallee scrub; along river banks and pond margins, or mixed with other tiny earth mosses on bare earth in clay-pans subject to periodical inundation.

Three species have been recorded from southern Australia—*cristatum*, *grosseciliatum* C.Muell. and *whiteleggei* Broth. & Geh.. The first two are probably the same moss and *cristatum* would have priority. The third, *whiteleggei*, has a line of dehiscence on the capsule and is not an *Ephemerum* (see *Eccremidium*, 16). There is also a further as yet undescribed species in WA, VIC, NSW; it is close to *recurvifolium* (Dicks.) Boul. of the northern hemisphere.

E. fimbriatum C.Muell. is a QLD plant.

SPLACHNACEAE

Includes: *Splachnobryum baileyi* Broth. (QLD)
S. *wattsii* Broth. (NSW)

Tayloria Hook.

This genus is the main Australian representative of the Spalchnaceae, a family of mosses characteristic of ground enriched by animal matter such as dung, dead birds, etc. and mostly distinguished by various enlargements of the sterile apophysis at the base of the capsule; sometimes this can be very conspicuous and brightly coloured. It has been shown for *Splachnum* and *Tetraplodon*, in the northern hemisphere, that the sticky spores are

PLATE 49. (top) *Ephemerum cristatum* VIC—Fruiting plant with persistent protonema × 50, cells from leaf margin (left) × 1000; (below) *Pottia drummondii* VIC— Fruiting plants × 30, cells (right) × 1000

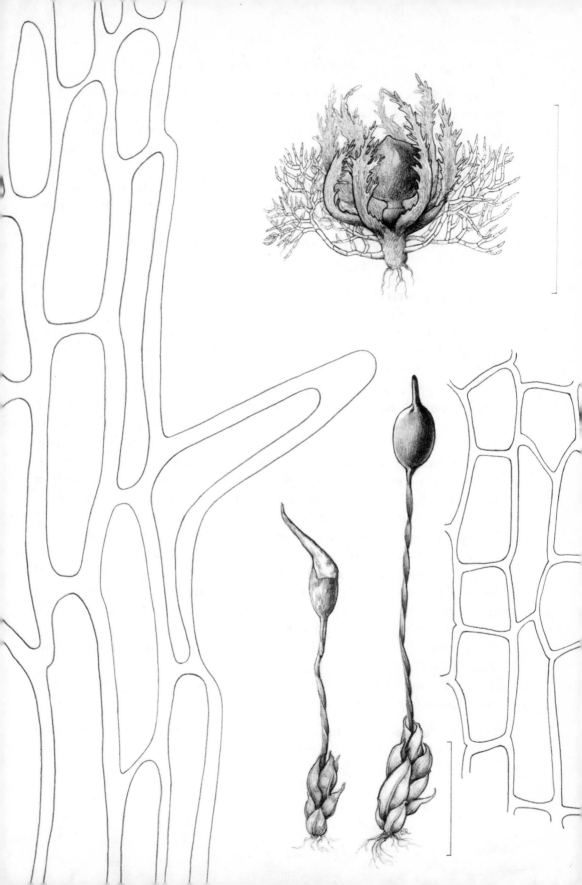

dispersed by flies, attracted by the smell of indole compounds excreted by the apophysis (see Erlanson, 1930 and references there). Whether the same is true of *Tayloria* is still unknown but the spores are certainly sticky and may well be dispersed in the same way as in *Tetraplodon*, the genus in which *Tayloria tasmanica* was first described. Vegetatively all species have very large, wide, thin-walled cells (Plate 50) rather similar to *Funaria* and its allies, and the stems carry a dense felt of rhizoids. The only common species is *T. octoblepharis*.

The Tasmanian species are discussed and illustrated by Willis (1950).

T. octoblepharis (Hook.) Mitt.

This is a variable species, especially in overall size and in the length of the seta, with several described varieties. It is not as closely restricted to animal substrates as many of its relatives, but is apparently found also on rotting wood or fabric, soil and even on rock. The shoots are usually in dense turfs, pale green, varying in height from 0·5 to 1·0 cm or more and bearing pinkish rhizoids, but not usually densely tomentose. Leaves are very variable in size, *c* 2–4 mm long, widest above the middle, and tapering to a long fine point (which may or may not consist of the nerve) especially in the uppermost leaves. The nerve is distinct, failing below the apex or excurrent; margins entire or slightly crenulate-denticulate by projecting cell-ends. The cells are very large, *c* 60 × 30 μm above mid-leaf but longer and laxer below, to *c* 150 μm long.

Capsules are very variable in size, colour and proportions, commonly *c* 2–3 mm long × 0·4 mm wide at the widest part. The theca varies from roughly isodiametric to 4 times as long as wide, forming from ¼ to ½ of the total length of the capsule; the apophysis shrivels as it matures, making the theca more conspicuous.

CHROMOSOME NUMBER: $n = 12$ (NSW).

DISTRIBUTION: TAS, WA, SA, VIC, NSW, ACT extending into QLD; also in New Zealand, ?New Guinea.

ILLUSTRATIONS: Plate 50 (inset); Handbook, Plate 33; Willis (1950); Allison and Child (1971), p. 78.

PLATE 50. *Tayloria gunnii* TAS—Fruiting plant × 7, cells including leaf margin × 1000; *T. octoblepharis* VIC (inset)—Capsules showing peristome and operculum × 7

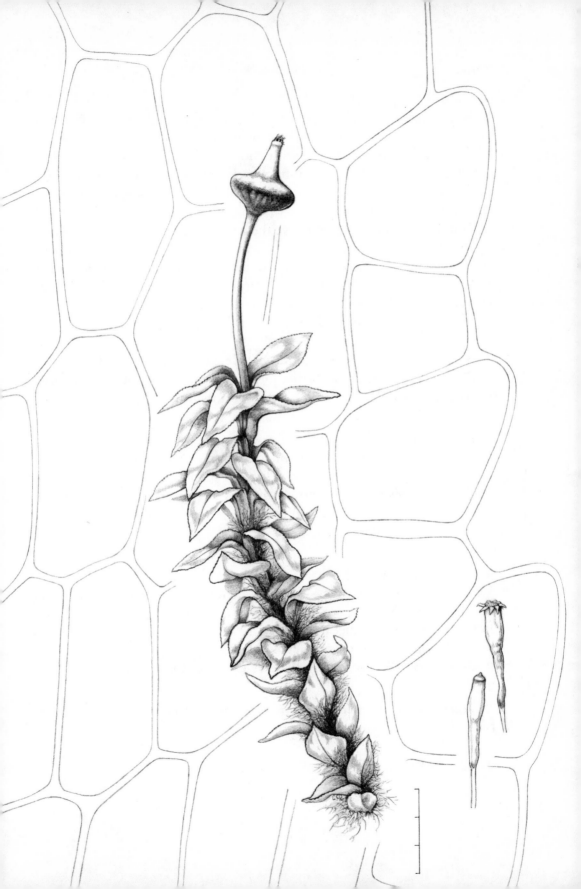

T. gunnii (Wils.) Willis

This rare Tasmanian species is very different from *octoblepharis*. The stems are quite tall—to 8 cm or more, of which the upper 0·5 cm is green and fresh—and densely tomentose with chocolate brown or crimson variably papillose rhizoids, but are not infrequently solitary, growing among other bryophytes and leaf litter on the floor of *Nothofagus cunninghamii* forest. The leaves are conspicuously and regularly denticulate in the upper half, nerved to above mid-leaf, and have the large wide cells typical of the genus.

The capsule is unique, with a very wide apophysis to *c* 2·5 mm diameter, surmounted by a narrow tapering theca, "reminiscent of a tall Welsh hat" (Willis, 1950 p. 32). The short red peristome teeth are erect and slightly curved in over the narrow capsule mouth.

DISTRIBUTION: Endemic to Tasmania.

ILLUSTRATIONS: Plate 50; Willis (1950).

When fruiting this species is unmistakable; vegetatively, the tall tomentose stems are rather similar to *Bryum pseudotriquetrum* [66] but the very large lax cells (*c* 100 µm long) at once distinguish them.

T. tasmanica (Hampe) Broth.

This species, also a Tasmanian endemic, has a similar capsule but with a much smaller, parsnip-shaped, ashy-grey apophysis. Vegetatively it resembles *octoblepharis*, with almost entire leaves and usually excurrent nerves. The rhizoids are variably papillose.

ILLUSTRATIONS: Willis (1950).

T. callophylla (C.Muell.) Mitt.

Vegetatively this resembles *T. gunnii*, with long tomentose stems and serrate leaves, but has a capsule like *T. octoblepharis* with the apophysis not much wider than the theca.

DISTRIBUTION: Tasmania and New Zealand.

ILLUSTRATIONS: Wilson (1854), Plate 87 (as *Eremodon robustus*).

T. henryae Dix., known only from Queensland, differs from *octoblepharis* only in the short, acute points of the leaves.

T. maidenii Broth. (NSW) we do not know.

BRYACEAE

Includes: *Anomobryum cymbifolium* (Lindb.) Broth. (QLD)
A. filescens Bart. is *Eccremidium pulchellum* [16, q.v.]

Bryum Hedw.

On first acquaintance the taxonomy of this infamous genus is a daunting prospect. Even in Australia, where *Bryum* has been far less intensively studied than in Europe, the Index lists 86 valid species, and another 6 in *Rhodobryum*, with at least as many again of *nomina nuda* and other illegitimate names. Fortunately Dixon, Sainsbury and, more recently, Ochi have studied the genus in detail and earned the gratitude of other bryologists for the comparative simplicity of the picture they have produced. Ochi (1970) reduces a great many species to synonymy, ending up with only 28, several of which are tentative so that we have been able, for practical purposes, to reduce his list still further; he has also provided useful illustrations of many of them. Other taxonomists have tended to credit the species with much less powers of variation and have recognized innumerable microspecies, analogous to those of *Hieracium*, *Rubus*, etc. in the flowering plants, based on very slight differences vegetatively or in the peristome. Both of these kinds of variation, the wide and the narrow, probably co-exist but no doubt much of the apparent variability is a mask for ignorance, and the most variable species such as *billardieri* and *pachytheca* are almost certain to be aggregates of several species as Syed (1973) has recently shown for *capillare*. In other groups such as *sauteri* and its allies, the species seem to be extremely constant in their rather minute differences. The kind of variation found in this huge genus is itself variable, and no single approach is adequate by itself. We cannot claim more than a cursory knowledge of the local species and have therefore had to accept for the most part the recent treatment by Ochi (1970), especially for distribution data and for species we have not seen.

All the species are fairly small, forming turfs or, occasionally, cushions; the stems rhizoidal below, occasionally even tomentose, with papillose rhizoids. The leaves of a few species are concave and obtuse, but most commonly they taper to a point with the nerve usually failing in the apex or excurrent. The leaves may be bordered or not, or intermediate or even variable in this respect. They are usually almost entire, and have a characteristic arrangement of rhomboid–hexagonal cells—often called bryoid areolation. The capsule is usually narrowly pear-shaped, pendulous or nearly so, with a conical operculum, and is terminal on the main stem with sympodial branches (subfloral innovations) from below the perichaetia. The peristome, at its most complete, retains the maximum amount of wall material of the cells from which it originates, giving the so-called *perfect peristome* in which the inner peristome has, between the processes, groups of 3 full length cilia which are *appendiculate* i.e. have transverse bars, the persistent remnants of cross walls. Some species have their cilia much less well developed, or "imperfect", and the degree of development has been widely used in this genus as a taxonomic character, although there are species in which it is very variable. Most species are dioicous (or usually so) but *pseudotriquetrum*, *sauteri* and *torquescens* are usually synoicous.

Three kinds of gemmae are found, and increasing importance is being attached to these in species discrimination; minute green axillary bulbils, as in *dichotomum* and *pachytheca*, brown or green axillary filaments, and subterranean rhizoid gemmae (tubers) which are tiny, round, red clusters of cells found on the rhizoids often as much as one or two cm below ground. The size and morphology of these have been extensively and very elegantly used by Crundwell and Nyholm (1964) and later by Whitehouse (1966) in elucidating the taxonomy of the *B. erythrocarpum* complex, and most recently by Syed (1973) with *capillare* agg.. Several of the European species which they recognized are to be found in New Zealand and Australia but the use of rhizoid gemmae as a taxonomic character is not easy. The plants have to be dug out with about 2 cm of soil below them and very carefully washed from the soil to avoid tearing the gemmae off. In general it is advisable to inspect the dry soil for gemmae, in a strong side-light, before washing. Even after gemmae have been found it can be hard to prove the connection back to the parent plant and the mixtures in which Bryums are often found are a snare for the unwary. With some practice and by taking especial care during collecting, the rhizoid gemmae are not difficult to observe. Sainsbury remarks that "scanty and imperfect material is seldom worth investigation, though attractive as an outlet for intellectual activity" (Handbook, p. 265) but, with the introduction of these new vegetative characters, that is now less true than formerly.

We are greatly indebted to A. C. Crundwell and J. T. Linzey for helpful comments on this genus.

Key to species

1. Leaves silvery and/or very obtuse   2
   Leaves neither silver-coloured nor obtuse . . .   3

2. Upper half of leaves hyaline. Not aquatic . . .   *argenteum*
   Leaves with a metallic sheen but not hyaline. Plants
       restricted to wet habitats . . . .   *blandum*

3. Leaves clearly bordered with 3 or more rows of cells   4
   Leaves unbordered or very narrowly bordered
       (1–2 rows)   10

4. Nerve excurrent in a long slender hair. Margins
       always entire   5
   Nerve failing, or excurrent in a short point.
       Margins sometimes toothed   6

5. Leaves twisted when dry. Capsule dark red . .   *capillare*
   Leaves scarcely altered when dry. Capsule pale
       brown . . . .   *caespiticium*

6. Margins toothed above. Nerve excurrent in a
       short, toothed point . . . .   *billardieri* etc.
   Margins entire or serrulate above. Nerve tip not
       toothed   7
                         (*B. rubens* may key out here)

7. Margins plane   8
   Margins recurved   9

8. Leaves usually less than 2 mm long. Axillary
       gemmae (bulbils) often present. Cells narrow,
       c 10–12 μm wide . . . .   *erythrocarpoides*
   Leaves usually more than 2 mm long. Axillary
       gemmae never present. Cells wide, *c* 15–20 μm .   *laevigatum*

9. Nerve excurrent in a distinct sharp point. Cells elongated parallel to nerve; cell walls thin. Synoicous . . . .   *pseudotriquetrum*

    Nerve failing below or in the apex, or excurrent in a very short, blunt point. Cells elongated obliquely; cell walls thick. Dioicous . . .   *laevigatum*

10. Nerve excurrent in a long slender hair. . . .   11

    Nerve failing, or excurrent in a short point . .   13

11. Leaves erect and appressed, not twisted or shrivelled when dry   12

    Leaves strongly contracted and twisted when dry   *capillare + torquescens*

12. Leaves denticulate above. Cells wide, *c* 15–18 μm, in mid-leaf. . . . .   *campylothecium*

    Leaves more or less entire. Cells narrow, *c* 8 μm wide . . . .   *australe*

13. Gemmae (bulbils) present in leaf axils   14

    Gemmae not in leaf axils   15

14. Gemmae usually very abundant and conspicuous, with rudimentary pointed leaves. Capsule neck slightly wrinkled but not differently coloured .   *dichotomum* (and mis-keyed *erythrocarpoides*)

    Gemmae usually inconspicuous, without leaves. Capsule neck strongly wrinkled, very distinct from the rest of the capsule in colour, size and shape . . . .   *pachytheca*

15. Cells in mid-leaf usually less than 10 μm wide; those in the leaf base squarish   16

    Cells in mid-leaf usually more than 10 μm wide; those in leaf base mostly shortly rectangular .   17

16. Capsule neck very distinct, swollen and wrinkled, not tapering. Leaf margins mostly plane . .   *pachytheca* (+ *coronatum*)

    Capsule neck tapering, not swollen; leaf margins recurved throughout . . . .   *crassum*

17. Plants robust, usually over 1 cm tall, often tomentose; usually in wet habitats . . . . .   (return to 7)

Plants small, less than 1 cm (usually *c* 0·5 cm), not
tomentose; usually in dryish habitats . . .   18

18. Margins plane even in upper leaves . . . .   *erythrocarpoides*
                                (sometimes *sauteri*)

Margins recurved, except sometimes in lower
leaves . . . .   19★

19. Rhizoid gemmae small, mostly well under 100 μm   *sauteri*

Rhizoid gemmae large, *c* 150 μm . . . . .   *micro–erythrocarpum*
                            (*tenuisetum*, *rubens* etc.)

★ Note: Plants in this group are best keyed out in Crundwell and Nyholm's (1964) key since it is likely that there are more species of erythrocarps present in Australia than have yet been found.

B. argenteum Hedw.

The mere handful of truly cosmopolitan mosses, found in all parts of the world, includes this as perhaps the most easily recognizable species and the commonest in cities. The greenish silver shoots, neatly packed together in a dense low turf, are common on damp earth in gutters, gaps in pavements, at the foot of walls, etc. even in the most heavily industrial cities, but are equally at home in other open habitats such as rocky crevices on mountain tops and even in hot and dry sandy areas such as the old Australian goldfields. It appears to be a nitrophilous species and also perhaps calcicolous, at least in Australasia.

The stems are always short, sometimes only 2–3 mm tall although usually more, with papillose rhizoids below. The leaves are concave and densely imbricated, commonly 0·7 × 0·3 mm on small stems, but rising to 1·2 × 0·4; varying from rounded and obtuse to obtuse with a sudden small point, or even tapering to a quite long acute apex. The nerve most commonly ends well below the apex but may extend into the apex and even be shortly excurrent (in the var. *lanatum*). The cells in the upper ½ of the leaf measure 40–90 × 12–18 μm, and are irregularly rhomboidal, hyaline and shining, with

quite thick walls, especially at the cell corners where there tend to be triangular corner thickenings. The Handbook says "thin-walled" but this
certainly does not hold good in Australia, although the thickness tends to be
exaggerated by collapse of the walls. In the lower ½ of the leaf the cells are
more nearly rectangular, rather thin-walled and green. There is no difference
in the marginal cells. Because the leaves are tightly imbricate the lower
green parts of the leaves tend to be completely covered and concealed by an
outer coat of dead glassy cells, which may have something to do with the
ability of this species to withstand the polluted air of industrial cities. Axillary
bulbils, rather like those in *dichotomum*, are commonly present and often
grow *in situ* to form fragile branches, easily knocked off even when quite
long, and helping to build the colony of moss into a dense turf.

The pendulous capsule is short and fat, dark brown when mature but
commonly with many stages of ripeness, from green to pale brown, present
together in the same colony. Spores measure 9–12 μm.

DISTRIBUTION: TAS, WA, SA, VIC, NSW, ACT, QLD, NT, Lord Howe; cosmopolitan.

ILLUSTRATIONS: Watson (1955), Fig. 82; Dixon (1924a), Plate 45(L).

On city pavements the silver-green patches of this moss can be confused
with nothing else, but in sandy mallee soils there may be some difficulty in
separating barren stems of the silvery-white *Gigaspermum repens* in which,
however, the leaves are nerveless; it is possible that there are species of
Plagiobryum still to be discovered in alpine areas and requiring capsules for
certain identification (the outer peristome is shorter than the inner).

B. blandum Hook.f. & Wils.

Despite the uncommonness of sporophytes this is one of the most easily
identified species in the genus, from a combination of colour, habitat, and
leaf shape. The stems are long and closely packed together into rounded
cushions or sometimes quite large mats in boggy trickles, small streams or
even waterfalls, sometimes submerged but more commonly just emergent.
The rhizoids are orange-brown to crimson-brown, mostly only sparsely
scaly–papillose or nearly smooth. The upper 2 cm or so, representing current
growth, is usually rose-tinted below, but sometimes darker red, dark or
silvery green above, shining with a metallic lustre, especially near the apex;
the colour is usually sufficient evidence for identification. Equally characteristic are the leaves which are concave, tongue-shaped, and *totally obtuse*,

274

1·6–2·3 × 0·8–1·0 mm, entire except for a little minute toothing at the very apex. There is a red nerve which tapers and fades out below the apex. The cells are narrowly rhomboidal and large, 80–150 × 12 μm, shorter near the apex, becoming minute and isodiametric at the very tip, shorter, wider and rectangular in the leaf base. A rather broad band of several marginal rows of narrower cells forms a border, but it is not very conspicuous.

We have not seen capsules in Australia. New Zealand plants have capsules *c* (2–) 3 mm long of which ⅓ is a conical shrunken neck, on a seta of 1·5–2·0 cm. The spores are smooth, *c* 25–32 μm.

DISTRIBUTION: TAS, VIC, NSW, ACT; also in New Zealand and Campbell Island.

ILLUSTRATIONS: Ochi (1968a), Fig. 1,2.

The habitat, completely obtuse leaves, red and metallic silver colouration, and long cells are sufficient to distinguish this species with little chance of error. *B. argenteum*, which can also have obtuse silvery leaves, is not aquatic and completely lacks any red tint except sometimes at the very base of the leaf; it is also much smaller, while *laevigatum*, which may be reddish but not silvery, usually has more acute leaves and much shorter cells.

B. capillare Hedw. sens. str.

It has recently been shown by Syed (1973) that *capillare* covers a number of distinct species of which three are recorded from Australia. There are also forms which are hard to place and which may represent further species.

To some extent this species combines the characters of *billardieri* and *campylothecium*. The obovate leaves, 1·2–2·0 × 0·5–0·8 mm, have the margins narrowly recurved and the nerve excurrent in a long, toothed hair-point, but they are soft-textured, evenly distributed along the stem, not in terminal rosettes, and tend to be shrunken and twisted when dry. In typical *capillare* the leaves are tightly cork-screwed round the stem but this form seems relatively less common in Australasia than in Europe. There is a fairly distinct, entire or serrulate border, 1–3 cells wide, sometimes yellowish. The rhizoid gemmae are brown to red-brown.

The sporophyte is rather small, the seta *c* 2 cm tall and the capsule *c* 3 mm long including the neck, usually light brown. The exothecial cells of the capsule wall are in irregular rows, those next the mouth smaller but not over-square. Spores are 10–13 μm in our material, sparsely wrinkled–papillose or almost smooth. It is a dioicous species; the synoicous specimens recorded in the Handbook are presumably *torquescens*.

DISTRIBUTION: TAS, WA, SA, VIC, ?NSW, ?ACT, ?QLD, ?Lord Howe; very widely distributed in both hemispheres.

ILLUSTRATIONS: Syed (1973), Figs 1–4; Ochi (1970), Fig. 31; Ochi (1969) Fig. 39, A–D (as *vino-viride*), Figs 40–42.

It tends to favour dry, often calcareous, rocks and soil.

B. immarginatum Broth., *plebejum* C.Muell. and *leptothecioides* Broth. & Watts of the Index are reduced to synonymy by Ochi (1970) and this viewpoint is provisionally accepted by Syed.

B. torquescens Bruch is very similar to *capillare* and has commonly been considered merely a synoicous form of that species. Syed (1973), however, has shown that it is specifically distinct. The main distinguishing features are: the leaves tend to be only slightly twisted round the stem (despite the specific epithet) not corkscrewed; the colouration generally very red or crimson— the tubers bright red instead of red-brown, the operculum crimson-brown instead of usually a light brown, the theca usually a beautiful deep bright red-brown. The exothecial cells of the capsule wall are in longitudinal rows, with 1–2 rows at the mouth much elongated transversely. Above all, it is usually *synoicous* which *capillare* never is. There are several other characters which distinguish the two species (see Syed, 1973).

DISTRIBUTION: TAS, WA, VIC, NSW, and probably elsewhere; the distribution is worldwide.

ILLUSTRATIONS: Syed (1973), Figs 23–25; Ochi (1970), Fig. 30 (as *synoicum*); Ochi (1969), Fig. 38 (as *pyrothecium*), Fig. 39 (as *philippianum*).

B. synoicum C.Muell. and *erythropyxis* C.Muell. are probably synonymous (Syed, 1973).

This appears to be the commonest of the *capillare* group in Australia and at its most typical is quite distinct from *capillare* sensu stricto but there are specimens not wholly consistent with either, and clearly the Australian plants need detailed investigation.

The commonest sources of confusion with both *capillare* and *torquescens* are *billardieri* and *campylothecium*, from both of which they differ in the soft texture of the leaves and in their even distribution along the stem; from the former also in the long hair-point and, from the latter, in the leaves being twisted when dry.

B. albo-limbatum (Hampe) Jaeg., also in the *capillare* group, is distinguished by its very large orange tubers (Syed gives 180–400 × 260–500 µm), toothed perichaetial leaves, and the presence of light brown papillose *filamentous gemmae* in the leaf axils.

DISTRIBUTION: WA, SA, QLD; endemic.

ILLUSTRATIONS: SYED (1973), Figs 21, 22; Ochi (1970), Fig. 30(H–J). We have not yet seen this species.

B. wallaceanum C.Muell. (VIC) and *leptothrix* C.Muell. (?QLD) are almost unknown but are perhaps related.

B. caespiticium Hedw.

The common little yellow *Bryum*, of dry sandy or silty soils in the mallee and similar habitats, is this species. It is commonly found with *Gigaspermum repens* and other small ephemeral dry ground mosses of inland Australia, and is probably a sand-binder capable of withstanding a certain amount of burial by wind-blown sand. The shoots, c 0·5–1·5 cm tall, vary from yellowish to reddish and have small ovate leaves, c 2 × 0·6–1·0 mm, tightly imbricate to form a small terminal rosette at ground level, unaltered when dry. There is a rather strong nerve, red or brown at the base where it is c 120 µm wide, and excurrent in a long slender slightly denticulate hair point, c 0·4–0·6 mm long. The cells are rhomboidal or obliquely rectangular, with c 2–3 marginal rows narrower and thick-walled and elongate, forming an entire or very slightly crenulate border which is distinct but often confined to the upper ½ of the leaf and is narrowly recurved.

The capsule is pendulous, c 2–3 mm long × 0·7 mm, pale brown or dark brown, cylindrical but expanded above to the wide mouth, and with an orange-brown operculum. The spores are 13–15 µm, green and smooth or very finely wrinkled.

DISTRIBUTION: TAS, ?SA, VIC, NSW, ACT; very widely distributed in both hemispheres and almost cosmopolitan.

ILLUSTRATIONS: Watson (1955), Fig. 81; Dixon (1924a), Plate 44(E).

The yellowish olive colour and dry sandy habitat are a good indication of this species. In the long excurrent nerve it resembles *capillare* and *campylo-thecium* but differs from *capillare* in having the leaves ovate, usually widest in or below the middle, instead of obovate, and unaltered when dry. *B.*

campylothecium differs in the denticulate leaf margin, thick cell walls and papillose spores.

B. billardieri Schwaegr. (including *truncorum* of the Handbook)

As a result of his recent investigation Ochi (1971) has concluded that *truncorum* is confined to Madagascar and Réunion. The Australian and New Zealand plants therefore all come under the broad heading of *billardieri* sensu lato. This is easily the most variable as well as the commonest species in our area and there seem to be no natural bounds to its variability. From species of *Rhodobryum* on the one hand and species like *campylothecium* and even *capillare* on the other, there is no very sharp distinction yet apparent. Quite possibly the synonymy should be even more extensive, incorporating still more of the described species under the same name, but more probably we have to deal with a group of microspecies whose recognition will have to await a thorough revision using new diagnostic characters as in the *capillare* group.

In general it is a relatively big species for this genus, with stems rarely less than 1 cm tall and occasionally 5 cm or more. The commonest form in Australia has the leaves *comose*, arranged in swollen terminal heads like miniature cabbages and packed densely together into quite extensive mats, but there are also variations which form quite loose and open colonies or where the leaves are of fairly even length and distribution along the stem. Below, there is usually some dark brown tomentum of warty–papillose rhizoids. The rather glossy leaves are of a dull olive or golden-olive colour and quite big, 2–4 (exceptionally to 7) × 0·9–2·0 mm, oblong or obovate, usually distinctly concave and very broad above with a short sharp small-toothed triangular *point* which is *bent back abaxially*. The nerve is pale and strong, shortly excurrent, and tapers sharply to a point. The cells are rhomboidal, sometimes thin-walled but more usually thick-walled, 45–60 × 15–20 µm in mid-leaf, more rectangular below, usually (especially the lower cells) porose; those at the margins much narrower and longer and thick-walled in 2–6 rows forming a distinct border which is often narrowly recurved and is toothed in the upper ½. The margin may be strong enough to form a conspicuous pale rib down the edge of the leaf, visible to the unaided eye, or

PLATE 51. *Bryum billardieri* VIC—Vegetative shoot (moist) and fruiting shoot (dry) × 11, cells × 1000

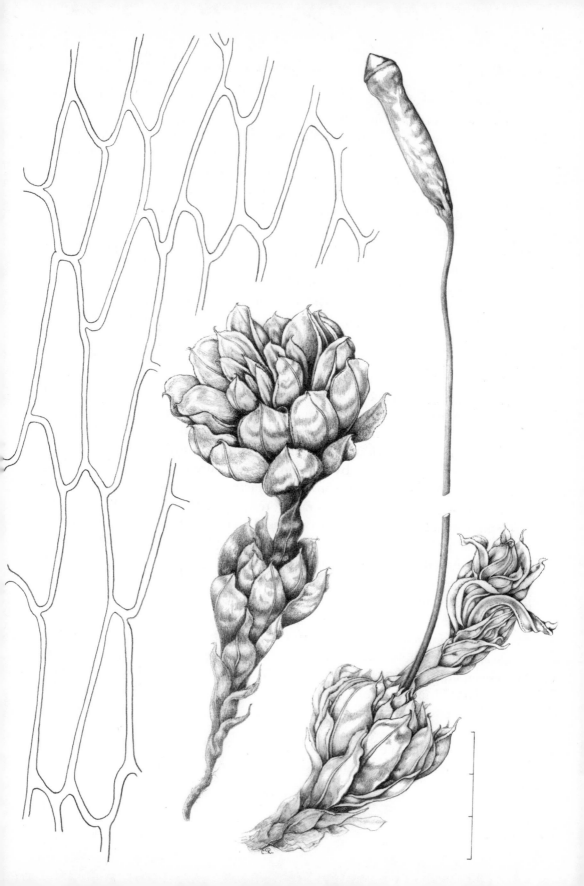

may be quite weak; there are all stages in between. When dry the leaves tend to be *shrunken and rather twisted*.

The species is dioicous; capsules are common and often strikingly large, but just as variable as the rest of the plant. Typically there is an orange-red to deep red seta, 1·5–4·0 cm tall or more, and a more or less pendulous cylindrical capsule, 3–5 mm long including a neck of about 1·0 × 0·8 mm diameter when dry, commonly pale brown or orange-brown but sometimes dark reddish brown. The spores are likewise very variable in size, in our material with a range of 10–21 μm, most of which can occur in a single capsule; they are green and sparsely to densely papillose. The outer face of each peristome tooth has conspicuously projecting ridges.

CHROMOSOME NUMBER: n= 10 (NSW, QLD); 11 (10+m) (TAS, NSW); 20 (NSW); 21 (20+m) (TAS). The variability by itself is sufficient to suggest a species-complex.

DISTRIBUTION: TAS, WA, SA, VIC, NSW, ACT, QLD, Lord Howe; also in Asia, S. and Central America, Africa, New Zealand and Oceania, according to Ochi, and also Europe according to the Index.

ILLUSTRATIONS: Plate 51; Allison and Child (1971), Plate 16.

Despite its variability this species is usually easily recognizable by the growth form of broad neat rosettes, the toothed border, leaves contracted when dry and especially by the slight recurvature of the nerve and leaf tip which seems to be a fairly constant character. It is found in a wide range of habitats from sandy mallee to rain-forest, but is typical of relatively moist humus-rich habitats, even in dry areas. The main difficulty is with *campylothecium* in which the nerve is excurrent in a rather long straight hair point, not recurved, and the leaves are not contracted when dry, and with *capillare* where the leaves are evenly spaced and not in terminal comose heads, and the nerve is excurrent in a long toothed point.

B. subfasciculatum (Hampe) Mitt., *subtomentosum* (Hampe) Mitt., *dilatato-marginatum* C.Muell., *abruptinervium* C.Muell., *flavifolium* C.Muell., *brunneidens* C.Muell., *leucothecium* C.Muell., *dobsonianum* C.Muell., *angeiothecium* C.Muell., and *Rhodobryum pohliaeopsis* (C.Muell.) Par. of the Index, are treated as synonyms by Ochi (1970).

B. robustum Hampe (NSW, QLD) seems to be very similar but has the nerve excurrent in a very short blunt stubby point, not at all tapering. It is endemic to Australia where it has a similar distribution to *billardieri*. We have not seen it. *B. forsythii* Broth. is considered synonymous by Ochi.

B. graeffeanum C.Muell. (NSW, QLD) (=*Rhodobryum subcrispatum* (C.Muell.) Par. =*R. olivaceum* Hampe—Ochi, 1970) and *B. leucocanthum* (Hampe) Mitt. (Lord Howe) are somewhat similar. Both are in the subgenus *Rhodobryum* which is characterized by small leaves on the creeping underground rhizoidal stems, an extension of the condition found in *billardieri*, and by very strongly toothed leaves.

B. erythrocarpoides C.Muell. & Hampe (=*curvicollum* of the Handbook —Ochi, 1970) has been much confused with related species. According to Ochi it differs in the usually quite distinct border of 2–3 rows even in the middle to basal parts of the leaf but this is not a reliable feature (Handbook, p. 270). It has green axillary gemmae of the *dichotomum* form, apparently not found in fruiting plants (Handbook; and Linzey, pers. comm.). The leaves are dimorphous: those on the main stems with excurrent nerves and an ill-defined border of 1–3 rows of cells, those on the innovations quite un-bordered and with the nerve failing below the apex or percurrent.

The capsule is commonly dark reddish brown with quite a long and *slightly curved neck*—at its most distinct this is quite diagnostic. The peristome is unusual: the cilia variably developed, from rudimentary to fully developed and appendiculate, even in the same capsule. The spores are 19–25 μm, finely papillose.

DISTRIBUTION: TAS, SA, VIC, NSW, QLD, Lord Howe; also in New Zealand, S. Africa and Oceania.

ILLUSTRATIONS: Studies, Plate 9; Ochi (1970), Figs 13–16.

Ochi reduces *curvicollum* Mitt., *diversinerve* Broth. & Watts, *filarium* Broth., *kiamae* Broth. and *subcurvicollum* Broth. to synonymy.

B. laevigatum Hook.f. & Wils. is rare in our area, but easily recognizable. It can be perhaps the biggest of our species, with thick shoots to 4 cm or more (the Handbook records it to 12 cm) × 2–4 mm wide, green above and brown below, matted together with dark brown extremely papillose rhizoids. The shining yellowish green elliptical leaves are rather concave, tough in texture, and only slightly twisted when dry, erect and appressed, rather bluntly pointed with a strong dark nerve usually reaching the apex or shortly excurrent. The margins are usually recurved and they, together with the nerve, form strong conspicuous ribs along the back of the leaf when dry. The cells are short and wide, up to *c* 50 × 15 μm, thick-walled and commonly porose, tending to be elongated *obliquely*, those at the margins usually narrowed to form a distinct strong border, but sometimes not.

The seta is *c* 3 cm long, with a large erect or horizontal or pendulous clavate capsule, *c* 2·5–5·0 mm long. The peristome teeth are hyaline-margined and have strong bars across the outer face, projecting at both sides. The outer face is finely papillose all over; the median line down each tooth is by no means as straight as the Handbook suggests, but varies from straight to zig-zag. In both operculum and exothecial cells of the capsule wall the cells are thick-walled with very pronounced corner-thickenings. There is an annulus of basically one row of high cells. In our material the spores measure from 12–19 μm. The species is consistently dioicous.

DISTRIBUTION: TAS, VIC, NSW, ACT; also in New Zealand, S. Africa and S. America.

ILLUSTRATIONS: Handbook, Plate 40; Allison and Child (1971), Plate 16.

The habitat is mainly bogs or marshy ground. It is a large and striking species, usually more robust than *pseudotriquetrum* with stiffer, thicker leaves and thicker cell walls (in both sporophyte and gametophyte) and unlikely to be confused with other species of the genus once it is known; only *billardieri* and *blandum* match it for size, but the former has the leaves more strongly toothed, much more twisted when dry, and usually comose while the latter has bigger cells, plane margins, different colouration and a more aquatic habitat. It is perhaps most like *Pleurophascum* in general appearance, but the cells there are quite different and there is no nerve.

B. crassinerve Hook.f. & Wils. is the same thing (Ochi, 1970).

B. pseudotriquetrum (Hedw.) Gaertn., Meyer & Scherb. is quite a robust plant of boggy ground up to 6 cm tall, usually characterized by tomentose stems (rhizoids very papillose), rather broad triangular–elliptic leaves with recurved margins, strong excurrent nerve which is red at the base, and leaves usually clearly bordered in 1–2 rows in the upper half and serrulate there. The leaves are thin-textured and characteristically shrivel when dry. The cells are elongated 50–90 × 15–19 μm and are scarecely thick-walled or porose.

The clavate capsule, 3–4 mm long, is pendulous from the top of a tall (*c* 4 cm) red seta. The peristome is rather similar to that of *laevigatum* with very finely papillose teeth, but a wide annulus of basically 2 rows of high cells, and one row of short cells. The exothecial and operculum cells are scarcely thick-walled, but have small corner-thickenings. The smooth spores measure 16–18 μm. The inflorescence is, at least usually, *synoicous*.

DISTRIBUTION: TAS, WA, VIC, NSW, ACT; throughout most of the world except Africa.

ILLUSTRATIONS: Ochi (1970), Figs 26–28; Watson (1955), Fig. 80.

It is a rare plant with perhaps a preference for upland regions. The wet habitat and tomentose stems will distinguish it from other similar species. Ochi gives *rubiginosum* Hook.f. & Wils., *austro-affine* Broth., and *sub-ventricosum* Broth. as synonyms.

B. creberrimum Tayl. (=*affine* of the Handbook) is treated as a synonym of *capillare* by Ochi (1969, 1970). Syed, however, states that it is a separate species, conspecific with *affine* Lindb., which is an illegitimate name. The type locality is Swan River, WA; we have seen no material. The Handbook describes *affine* as resembling *pseudotriquetrum* but having narrower more finely acuminate leaves and nerves further excurrent.

ILLUSTRATIONS: Ochi (1969), Fig. 38.

B. campylothecium Tayl.

At its best this species is easily recognizable; the stems are short, *c* 1 cm, with the leaves in terminal heads as in *billardieri*, 2·0–2·5 × 0·6–1·2 mm, *tightly imbricate* and *neither shrunken nor twisted* when dry (except when very young and succulent) with the nerve very broad and strong below and *excurrent* in *rather a long* fine, tapering erect *hair-point* which is smooth or denticulate. The margins are recurved, especially above, and usually denticulate at the leaf apex. The cells are 33–45 × 15–18 μm in mid-leaf, with rather thick walls which have sinuose and almost porose thickenings. The cells below are more oblong and those in the marginal 1–2 rows sometimes narrow and elongated forming a distinct border but in other leaves, even on the same stem, no border is detectable. The rhizoids, as in *billardieri*, are strongly warted-papillose.

The capsule is *c* 3·5 mm long, including a neck of *c* 1 mm, and is banana-shaped, pale brown and horizontal to pendulous, on a 2–3 cm seta. The spores are commonly 12–18 μm, rather densely papillose, but we have specimens with very large spores measuring 27–47 μm. The species is *dioicous*.

DISTRIBUTION: TAS, WA, SA, VIC, NSW, Lord Howe; also in New Zealand.
ILLUSTRATIONS: Ochi (1970), Fig. 32.

This species is characteristic of sandy soils by the sea or inland, e.g. in the mallee. It is like a small *billardieri* but differs in the long erect hair-point, tightly imbricate leaves, unaltered when dry, and in the weak and inconsistent border; there are forms, however, which are quite hard to place. *B. pallenticoma* C.Muell., *howeanum* Broth. & Watts, and *Rhodobryum peraristatum* Par. are all synonyms according to Ochi.

B. crassum Hook.f. & Wils.

This is quite a small species and not a conspicuous one. The leaves are clear green in the current growth, brown below, *c* 1·4–2·5 × 0·6 mm, oblong or lanceolate, rather narrow, with a fairly strong nerve, *c* 80 μm wide at the base, failing below the apex or shortly excurrent, rather long-excurrent on small stems. The leaves are erect, straight and closely imbricate, unaltered when dry. The cells in mid-leaf are rather narrowly rhomboidal, *elongated obliquely*, 30–60 × 8 μm, more rectangular below and almost square in the leaf base, *thick-walled* sometimes strongly so. The margins are narrowly recurved almost to the apex, crenulate and usually unbordered but sometimes narrowly bordered.

The seta, typically 2·0–2·5 cm long, is pale brown, bearing a small *conical* capsule, 2 mm long × 0·5 mm wide at the mouth where it is widest; the colour in our material is pale brown, although the Handbook gives it as purple-brown. There are 25–30 bars ("ventral lamellae" in the Handbook) across the inner surface of each tooth in the outer peristome.

DISTRIBUTION: TAS, VIC, NSW; also in New Zealand.
ILLUSTRATIONS: Handbook, Plate 40; Ochi (1970), Fig. 23.

This is a rather variable and not particularly distinctive species which we have found in dry sclerophyll forest; the recurved leaf margin, usual lack of a border, oblique thick-walled cells and narrow closely appressed leaves are the best distinguishing features. According to the Handbook (p. 276), "the areolation is characteristic, the strongly incrassate cell walls having their angles so much rounded that under a low power the cells appear to be oval". In Australian specimens these thickenings are not always as well marked as this suggests.

B. australe Hampe (=*appressifolium* of the Handbook *fide* Ochi, 1970) is rather similar to the preceding species but has a much longer-excurrent nerve, more finely tapering leaves, and cells elongated parallel to the nerve. The cells are moderately thick-walled but never as thick as they can be in *crassum* and not rounded-off at the corners.

The capsule is short and wide, c 2 mm long with a large operculum of c 1 mm, deep red-brown when mature, widest at or near the mouth when empty, giving a conical shape. There are about 40 transverse bars across the *inner* surface of each peristome tooth (25–30 in *crassum*). The smooth spores are 7–9 μm in diameter.

DISTRIBUTION: TAS, VIC, NSW; also in New Zealand.

ILLUSTRATIONS: Handbook, Plate 40; Ochi (1970), Fig. 21; Hampe (1844), Plate 26.

We have collected this plant in New Zealand, where it is common, but have not yet seen it in Australia. The description above fits New Zealand material.

B. dichotomum Hedw.

On damp and often clay soils and waste places this is a common species but is probably often overlooked as fruit does not seem to be common and the short erect shoots, typically 4–5 mm tall, are frequently rather sparse and mixed with other species, although it can form extensive low turfs over wet soil in the early stages of colonization of bare ground. The gemmae are the most distinctive feature, small bulbils c 0·3–1·0 × 0·1–0·3 mm like little green buds with minute pointed leaves at the apex, often so numerous in the axils of most leaves that the whole plant is swollen and opened up by them. The leafy points appear to distinguish them from the gemmae of *pachytheca*. The main stem leaves are widely ovate and concave, c 1·0–1·2 × 0·5–0·8 mm, with a strong yellowish nerve excurrent in a stout point. The cells in mid-leaf are rather widely rhomboidal, c 35–40 × 12–15 μm, those in the leaf bases almost square. The marginal row (or 2 rows) of cells are narrowed but scarcely enough to form a border and only in the upper ½ or less of the leaf.

The fruit has a corrugated neck as in *pachytheca*, but the capsule is pale to mid-brown, usually narrowly waisted below the mouth and with the *neck no darker* in colour than the rest of the capsule; this last feature, apparently trivial, gives the capsule a very different appearance from that of *pachytheca*.

CHROMOSOME NUMBER: n = 10 (NSW).

DISTRIBUTION: TAS, SA, VIC, NSW, ACT, QLD, Lord Howe; widely distributed in America and also in New Zealand.

ILLUSTRATIONS: Studies, Plate 9; Ochi (1970), Figs 5–7.

The gemmae are far more abundant and conspicuous than in any of our other species and this is the feature by which the species is usually recognized. We are not yet fully convinced that their leafy points are consistent.

B. pimpamae C.Muell., *brachytheciella* C.Muell., *subcupulatum* C.Muell. ex Rodw., *aequicollum* Broth. & Watts, *philonotoideum* Broth. & Watts of the Index are the same thing (Ochi, 1970).

B. pachytheca C.Muell. (= *B. bicolor* in Ochi, 1970)

Dixon's description of the capsule as being like an acorn in its cup is alone quite often sufficient to identify this common species of sandy and silty soils. The capsule is small, barrel-shaped, commonly from 0·8 × 0·8 to 1·6 × 1·2 mm excluding the operculum, deep crimson-brown in the wrinkled apophysis region, paler in the theca, and hanging from the curved tip of a short red seta, 0·5–1·0 cm tall, from which the capsule base flares out abruptly without tapering. Sometimes the capsule is rather longer, to 2 mm, of which the basal third is wrinkled and darker. The spores in our material are 10–13 µm and finely papillose but the Handbook gives 8 µm so there is likely to be some variation in spore size.

Apart from the capsule, the plant tends to be nondescript with golden-green to reddish lanceolate leaves, 0·6–1·0 × 0·3–0·4 mm, tapering to a narrow point and with a strong yellow or red nerve excurrent in a short hair or stout point which is sometimes hyaline at the tip. The cells in mid-leaf, c 21–45 × 9 µm, are rhomboidal to obliquely rectangular, and almost square in the leaf base. The walls are rather thick with a tendency to triangular corner thickenings at the cross-walls. There are rather inconspicuous obovate axillary gemmae, c 150 × 100 µm, which are yellowish at times, solid and not bud-like; they are best seen at the tips of new but full-sized branches. Ochi considers this a crucial distinction from the bud-like gemmae of *dichotomum* which have leaf rudiments at the tip, but it seems not impossible that one can develop into the other.

PLATE 52. *Bryum pachytheca* VIC—Fruiting shoot × 22, cells × 1000

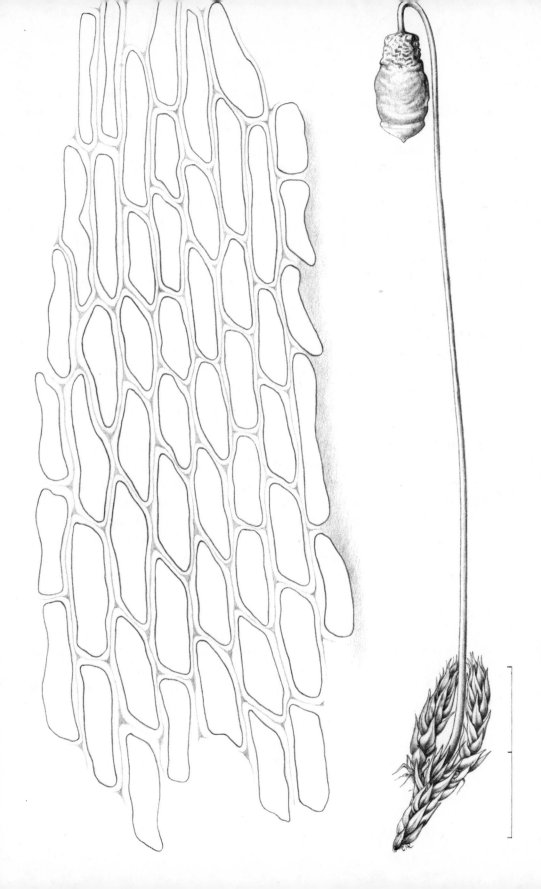

DISTRIBUTION: TAS, WA, SA, VIC, NSW, ACT, QLD, NT. The world distribution includes Europe, Africa, N. America, S.E. Asia and New Zealand but is hard to delimit until the full synonymy is established; at present Ochi includes *ovicarpum* Broth. and *balanoides* Tayl. ex Broth. of the Index list, in this species. Ochi (1973) now recognizes *bicolor* as a synonym for *dichotomum*, not for *pachytheca*.

ILLUSTRATIONS: Plate 52; Studies, Plate 9; Ochi (1970), Figs 8, 9 (as *bicolor*). *B. coronatum* Schwaegr., from WA, NSW, ACT, QLD, and Lord Howe, according to Ochi differs in the bud–like gemmae with pointed rudimentary leaves, and in the longer seta (1·5–2·0 cm), but we have some reservations about the constancy of this difference and have not yet examined authentic material of the species. It is illustrated in Ochi (1967).

B. sauteri B.S.G.

The groups of species allied to *erythrocarpum* which have been thoroughly investigated by Crundwell and Nyholm (1964) present particular difficulties taxonomically. The most powerful diagnostic features are the size and morphology of the rhizoid gemmae (tubers), minute reddish balls of cells borne on the rhizoids sometimes as much as a few centimetres below ground; once they have been found, identification of the species is usually easier, but it can be very hard to find them and still harder to be certain of which plants they are connected to, for mixtures of species are quite common. A further difficulty is that they are badly under-collected in our area, partly because of the formidable reputation of the genus and partly because, until the taxonomic effectiveness of the gemmae had been established, only fruiting material was usually collected. All species in this group have been much confused in the past. Our account of the erythrocarps is therefore tentative and based largely on our own rather meagre collections. *B. sauteri* has lanceolate leaves, 1·6–2·0 × 0·5–0·6 mm, with plane entire margins and with the nerve excurrent in a short, rather fine tapering point. Cells in mid–leaf are rhomboidal, 45–60 × 13–16 µm, those below more rectangular, the marginal 1–2 rows narrower forming a rather indefinite border which is most pronounced in the upper ½ of the leaf. The rhizoid gemmae are reddish brown, much the same colour as the papillose rhizoids, pear-shaped, and small, *c* 45–90 µm, and with the cells rounded but not protuberant.

The capsule is pale brown and pendulous, 1·3–3·0 mm long, swollen in the middle and slightly narrowed just below the mouth. The spores are green,

15–18 µm, and very finely papillose. This is one of the few *synoicous* Bryums in our area.

DISTRIBUTION: VIC, NSW (det. Linzey); also in Europe and New Zealand.
ILLUSTRATIONS: Crundwell and Nyholm (1964), Fig. 5; Whitehouse (1966), Fig. 4 (tuber).

This seems to have a preference for bare, rather heavy clay soils or very wet silt. It agrees with Ochi's (1970) account of *B. chrysoneuron* C.Muell. of which the type material is unfortunately missing, probably destroyed in the bombing of Berlin, and which is recorded from TAS, WA, NSW, ACT, QLD. *B. sauteri* is the earlier name. Ochi has reduced *suberythrocarpum* C.Muell., *leptopelma* C.Muell., *lonchoneuron* C.Muell., *microthecium* C.Muell., and *wattsii* Broth. to synonymy under *chrysoneuron*.

B. micro-erythrocarpum C.Muell. & Kindb. (1892)

There is quite a close resemblance here to the preceding species but this is a plant of peaty soil in our experience, with large red (or sometimes rather pale) rhizoid gemmae, *c* 150–200 µm in diameter; the leaves are reddish tinted and slightly denticulate above. It is dioicous; the capsule is relatively stout, and dark red–brown in colour on a reddish seta.

DISTRIBUTION: TAS, VIC, NSW (det. Linzey), QLD; also in Europe, N. America and New Zealand. This species is the same as *B. tryonii* Broth. (1893), but material of it has often also been identified as *chrysoneuron*.
ILLUSTRATIONS: Crundwell and Nyholm (1964), Fig. 7; Ochi (1970), Fig. 12 (as *tryonii*); Whitehouse (1966), Fig. 4 (tuber).

B. rubens Mitt. has large red gemmae, 150 µm diameter or more, plentiful

at the base of the stem, instead of on long rhizoids. The leaves are rather similar to those of *micro-erythrocarpum* but clearly narrowly bordered except on poorly developed stems.

DISTRIBUTION: VIC, but probably much wider; Europe, Russia, India, Japan and New Zealand.
ILLUSTRATIONS: Crundwell and Nyholm (1964), Fig. 9.

We have only found this species a few times, associated with *Pottia truncata*, but it will undoubtedly prove to be much more widespread in Australia.

B. tenuisetum Limpr. has recently been recognized from the high plains of Victoria and the identification was kindly confirmed by Crundwell. It is very similar to *micro-erythrocarpum* except for the spherical tubers, 50–200 μm in diameter, which are exquisitely coloured, a golden- rather than lemon-yellow, with red cross-walls. No doubt the species will turn up elsewhere. DISTRIBUTION: VIC, but probably other States also; Europe and N. America. ILLUSTRATIONS: Crundwell and Nyholm (1964), Fig. 6.

The other species recorded by Ochi (1970) from Australia are all very rare and much in need of further collecting and investigation:
B. angustirete Kindb. (NSW)
B. altisetum C.Muell. (VIC, NSW) which is very close to it.
B. calodictyon Broth. (NSW)
B. cellulare Hook. in Schwaegr. (NSW, QLD)
B. cheelii Broth. (NSW)
B. inaequale Tayl. (?TAS, WA, ?SA, ?VIC)
B. nitens Hook. (northern NSW, QLD)
The remaining species in the Index, not discussed by Ochi are:
B. aeruginosum C.Muell. (NSW)
B. amblyacis C.Muell. (VIC)
B. bauerlenii C.Muell. (NSW)
B. bateae C.Muell. (NSW)
B. brachycladulum C.Muell. (NSW)
B. chlororhodon C.Muell. (VIC)
B. cupulatum C.Muell. (TAS, SA, VIC, NSW)
B. limbifolium Broth. & Watts (Lord Howe)
B. minutissimum C.Muell. (QLD)
B. roseodens C.Muell. in Geheeb (NSW)
and *Rhodobryum humipetens* (C.Muell.) Par. (QLD)
We know nothing of any of them.
B. chrysophyllum Ochi (=*B. laxifolium* Besch., an illegitimate name) has recently been added to the Australian list (Ochi, 1973).

Brachymenium Schwaegr.

Vegetatively, there is little or nothing to separate this genus from *Bryum*. They share the same kind of short, rosette-like shoots, ovate leaves and rhomboidal cells. The capsule, however, is orientated between about 45° below and above the horizontal, sometimes even more erect, and has a rather characteristic shape, but the most distinctive feature is the inner peristome which consists of very narrow hair-like processes with almost no cilia between.

B. preissianum (Hampe) Jaeg.

The usual habitats of this widespread species are on calcareous rock, or less commonly clay banks in moderately dry areas, but it is nowhere common. Only when fruiting is it at all noticeable. The measurements given here are for plants on dry rock; on clay soil or moister habitats the plants may be larger. The stems, c 3 mm tall, have dark olive-brown leaves, c 0·8 × 0·3 mm, erect, straight, and appressed to the stem when dry. The nerve is strong, yellow or brown, excurrent in a stout short point. The upper cells are rhomboidal, c 24–40 × 9 µm, those below somewhat shorter, and irregularly rectangular, those in the leaf base almost square; at the margins, which are quite entire and plane or gently recurved, the cells are much narrower, forming a rather indistinct border.

The capsule, c 2·0 × 0·8 mm on a seta of c 1 cm, is almost oval in outline, widest at or slightly below the middle, of a bright dark orange-brown, inclined or sometimes almost erect. There is a high (c 0·4–0·5 mm) conical operculum, slightly hooked at the top, covering a double peristome. The outer teeth, c 0·4 mm long, are papillose above and almost smooth below; the inner peristome of similar height has very fine hair-like processes, hyaline and papillose from a high papillose membrane about half the height of the peristome. Because of the papillae the inner peristome appears to be sticky and the small smooth spores, c 10 µm diameter, stick to it readily. This species is dioicous.

CHROMOSOME NUMBER: n=22 (TAS)

DISTRIBUTION: TAS, WA, SA, VIC, NSW, QLD; also in New Zealand.

ILLUSTRATIONS: Handbook, Plate 39; Ochi (1970), Fig. 4; Hampe (1844), Plate 25 (as *Bryum*).

Only the shape and structure of the sporophyte, and to some extent the habitat, are helpful in separating this from species of *Bryum*. The peristome is the conclusive feature. *B. lanceolatum* Hook.f. & Wils., endemic to Tasmania, is not mentioned by either Sainsbury (1955d) or Rodway but is discussed and illustrated by Ochi (1970 p. 9). It is autoicous and has large leaves, *c* 4 mm long or more.

We know nothing of *B. klotzschii* (Schwaegr.) Par. which the Index records from S. America and Australia. The remaining species in the Index are tropical:

B. indicum (Doz. & Molk.) Bosch & Lac. (QLD). Also from S.E. Asia, S. America and Oceania.

B. semperlaxum (C.Muell.) Jaeg. and *B. wattsii* Broth. appear to be endemic to Queensland.

Leptobryum (B.S.G.) Wils.

L. pyriforme (Hedw.) Wils.

It would be difficult, nowadays, to tell whether this species is native or introduced. It grows, almost throughout the world, wherever plants are grown in pots in greenhouses. Flower pots are by far its commonest habitat, although it has also been recorded in the field. The very narrow leaves and large, pear-shaped capsules, are quite distinctive.

The stems form open tall turfs, 1–3 cm high, with very slender, distant, U-channelled leaves, 2–5 mm long (the apical leaves are by far the longest forming a conspicuous coma), spreading almost at right angles from a short base which sheaths the stem. The cells are long and narrowly rectangular or rhomboidal in the leaf base, *c* 90 × 9 μm, with either square or pointed ends; those above are very much shorter, especially at the margins, *c* 2 × 1. Rhizoids are densely finely papillose, brown to crimson or violet in colour. Red-brown, ovoid multicellular gemmae (tubers), becoming almost black with age, are sometimes present in abundance in the axils of the lower leaves and on underground rhizoids. They have an intensely pitted surface. Deciduous shoots, as in *Campylopus*, sometimes also occur.

There is a moderately long seta, 1–2 cm, with a half-pendulous pear-shaped capsule, 1·5–2·0 × 0·8 mm; the basal half is a rather narrow, wrinkled neck, the upper half a plump, glossy urn, smooth when mature but becoming ribbed when old. The peristome is double, and perfect as in *Bryum*, and the smooth green spores measure 12–18 μm. The outermost cells of the capsule wall are thin-walled.

CHROMOSOME NUMBER: n=22 (NSW).

DISTRIBUTION: TAS, SA, VIC, NSW; elsewhere? Almost cosmopolitan.

ILLUSTRATIONS: Watson (1955), Fig. 75; Dixon (1924a), Plate 40; Brotherus (1924–5), Fig. 326; Whitehouse (1966), Fig. 3 (tuber).

The leaf shape and even the cells are not wholly inconsistent with *Ditrichum* [14], which the plant rather resembles, but no *Ditrichum* has rhizoids either papillose or violet. Capsules, too, are the rule and their characteristic shape, as well as the perfect peristome, will prevent any mis-identification. Any fine-leaved moss in a flower pot stands a very good chance of being this species. The only other common species in the same habitat is *Funaria hygrometrica* [60] which has wide leaves and is altogether different.

L. sericeum Kindb. (TAS) is likely to be synonymous but we have not seen it.

Leptostomum R.Br.

Despite the isodiametric, thick-walled cells this genus is generally held to be closely related to *Bryum*, although commonly put in a family of its own. Andrews (1951) has argued for its retention in the Bryaceae. The stems are matted together by dense tomentum into quite large pads and the large capsules have only a rudimentary peristome. The leaf margins are quite strongly recurved and the nerve is excurrent in a long irregular hair-point. The plants are mainly epiphytic.

L. inclinans R.Br. forms pads of usually bright pale green shoots, bright brown below where they are matted together by rhizoids into quite a hard corky pad, commonly 2–3 cm thick. The rhizoids are dimorphous: thick ones with a coarsely papillose-scaly surface, and much finer ones which are almost smooth. The leaves, 1·5–2·0 × 0·4–0·7 mm, are almost oval in outline, very obtuse at the tip and with the margins recurved except at the apex. When dry the leaves are closely appressed to the stem, sometimes with a helical twist round it. The nerve is sunk in a rather open channel and is excurrent in a smooth or faintly denticulate unbranched point, commonly

0·4 mm long but sometimes exceedingly long, to 2 mm or more. At the leaf apex the margins are usually slightly toothed. The cells are smooth and iso-diametric throughout, small and thick-walled, *c* 10–12 μm across, but very variable in shape.

The capsules are slender and streamlined and very long (5–6 mm), often on long setas (2–5 cm) which tend to project nearly horizontally from the tree trunks where the plant grows, and from which the capsules dangle like fish on the ends of trout rods. The capsule is widest in the upper half, very gradually narrowed to the seta, and has a rounded-conical operculum. When dry the neck is distinctly narrowed and wrinkled, usually more than ¼ of the capsule length. The peristome is merely a low papillose membrane, round the rather small capsule mouth. The spores are characteristically very coarsely warted-papillose.

CHROMOSOME NUMBER: n=6, 12 (NSW).

DISTRIBUTION: TAS, VIC, NSW, ACT, QLD; also in New Zealand.

ILLUSTRATIONS: Allison and Child (1971), Plate 17 and p. 86; Hooker, W. J. (1818–20), II, Plate 168 (as *Gymnostomum*).

This plant, especially where the stems are sparse and not matted into a hard pad, is perhaps likeliest to be mistaken for *Leptotheca gaudichaudii* [75] which can have similar colouration and similar cells; but it differs from that species in the long flexuose hair-point and much more densely leafy stems, especially near the apex. This is reflected in the overlap of leaves, each leaf being covered to about ¾ by the leaf below it. In *Leptotheca* the overlap is less than ½ and the leaves tend to be arched, giving a chain-like (catenulate) appearance to the stem.

L. macrocarpum (Hedw.) Pyl. from New Zealand, Lord Howe and Norfolk Island, is wholly distinct in the *branched* hair-points and finely papillose spores. It has been recorded from TAS, NSW and QLD (Watts & Whitelegge, 1906) but possibly in error.

L. erectum R.Br., recorded from VIC, NSW, QLD and New Guinea, has an almost erect instead of hanging capsule, and leaves rather strongly twisted round the stem. It is clearly very close to *L. inclinans*.

We know nothing of *L. depile* C.Muell. (VIC).

Mielichhoferia Nees & Hornsch.

This genus is separated from *Pohlia* on the grounds that it lacks an outer peristome and has sporophytes basal instead of terminal on the stem. But the outer peristome varies from completely absent to equal in size to the inner one and the fruit is terminal on a short basal lateral branch, sometimes scarcely distinguishable from the condition in *Pohlia* so that the grounds for retaining this as a separate genus are shaky. The whole question of the relationship between *Pohlia*, *Mielichhoferia* and *Mniobryum* could profitably be re-opened. The species in this genus are well known "copper mosses", indicators of mineral-rich soils, in the northern hemisphere but they do not seem to show the same feature in Australia as far as one can judge without actual chemical analysis.

M. bryoides (Harv.) Wijk & Marg. (=*ecklonii* of the Handbook)
This is one of the commonest plants of damp earth banks in upland dry sclerophyll forest. Where there is a seepage through gravelly soil extensive patches often develop, looking rather like *Pohlia wahlenbergii* because of the reddish stems and rather glaucous leaves. The stems are short, usually *c* 1 cm tall, tomentose at the very base with papillose rhizoids and with lustrous leaves, *c* 1·0–1·5 × 0·2–0·3 mm, which are narrowly triangular, appearing glaucous in the field because of a very fine wrinkling of the cuticle. The nerve is strong, failing below the leaf apex. The long narrow cells, *c* 100–140 × 6–7 μm, are mostly slightly sinuous and project at the margins, especially near the apex, to make them crenulate-denticulate.

The capsule is horizontal to slightly pendulous—rarely nearly erect—seeming very large on the rather short seta of *c* 1–2 cm; it measures 2–3 mm long, sometimes more, and is almost cylindrical but tapered at the base to a conspicuous wrinkled neck and slightly narrowed to the mouth before dehiscence. The outer peristome is absent or sometimes present but each tooth fades out above instead of continuing to a triangular point. The appearance is very much that of wax teeth melted by heat at the tips. They are hyaline and closely transversely barred, but not thick and coloured. The inner peristome consists of 16 linear processes from a basal membrane, usually densely papillose and occasionally with projecting transverse bars, touching together at the tips to form an open basket over the capsule. The height of the basal membrane varies greatly. The spores range in size from

15–21 μm and have a surface either very finely wrinkled or finely papillose.
DISTRIBUTION: TAS, SA, VIC, NSW, ACT; also in New Zealand, South Africa, Madagascar.
ILLUSTRATIONS: Plate 53; Studies, Plate 9; Brotherus (1924–5), Fig. 302.

The main difficulty with this species is to separate it from *Pohlia* [72]. Vegetatively we can find no consistent difference—both genera have papillose rhizoids and very finely wrinkled leaf surfaces—but antheridia and archegonia are usually present on evident basal branches in *Mielichhoferia*, terminal on main shoots in *Pohlia*. The short seta and imperfect outer peristome will identify *Mielichhoferia* when in fruit.

M. australis Hampe (VIC) differs in having a basal membrane too low to be visible over the rim of the capsule; in view of the variability of the peristome in the genus it does not seem safe to consider this a distinct species without better evidence.

There is nothing in the type description of *M. forsythii* Broth. (NSW) to separate it from *M. australis*. *M. turgens* Broth. (NSW), however, has an outer peristome unusually fully developed, as tall as the inner; it could be a good species.

Orthodontium Schwaegr.

This small genus has been revised by Meijer (1952) who recognizes a total of only 8 species, half of which are found in Australia. They are little plants with long, very narrow leaves, single nerved to near the apex, and long, narrow cells. The capsule is moderately erect, with a double peristome of which the inner usually consists of long hair-like processes which may be longer than the outer teeth. The grooving of the capsule which has long been considered an important character is shown by Meijer to occur in most species and so has little weight. Vegetatively the species seem to be almost indistinguishable but fruit is common; even so the differences between *australe* and *lineare* are those

PLATE 53. *Mielichhoferia bryoides* VIC—Fruiting plants (centre) × 7, with enlargements of vegetative and very young fruiting shoots (bottom right) and of ripe capsule (top left) × 15, cells × 1000

of size only, but the size ranges overlap and Sainsbury (Handbook, p. 256) was sceptical of the distinction. In practice, except possibly in NSW, only one species (or pair of species) is at all common. It would be small loss to bryology if the whole genus were treated as a single, moderately variable species.

O. lineare Schwaegr.

The short stems, *c* 0·5 cm, and fine silky yellow-green leaves give this common species something of the look of *Ditrichum* [14], but the leaves are more spreading and flexuose, never falcate-secund, and the cells and peristome are quite different. Unlike most Bryaceae it has smooth rhizoids. The leaves are flat and ribbon-like, tapering steadily from the base to a rather blunt point, commonly 3·5–4·0 mm long and only 0·4 mm or less wide, but up to 6 mm long according to Meijer. The cells are very narrowly rhomboidal, *c* 75 × 15 μm above, longer below, and wider near the leaf base where they reach 150 × 21 μm in our material; those in the leaf base are rather thin-walled, tending to collapse. The margins are plane and practically entire.

The capsule is very variable in size and shape, *c* 2·0–2·5 mm long, mostly cylindrical, commonly grooved when dry and mature but sometimes smooth, with a short beaked operculum. The peristome is characteristic; the outer teeth are broad and up to 150 μm long, slightly papillose; the inner with equally long or longer hair-like processes, usually from a basal membrane. The spores are very variable in size, commonly 15–20 μm, and finely papillose.

CHROMOSOME NUMBER: n = 22 (TAS, NSW).

DISTRIBUTION: TAS, WA, SA, VIC, NSW, ACT; also in S. Africa and New Zealand and, according to Meijer, introduced into Europe.

ILLUSTRATIONS: Meijer (1952).

This is a common species in dry and wet sclerophyll forest, especially in upland or montane areas, frequently on rotting fibrous bark but with a special preference for charred logs. Meijer recognizes 3 subspecies of which only ssp. *sulcatum* occurs in Australia.

O. *australe* Hook.f. & Wils. is described with 2 subspecies in Australia, ssp. *australe* found only in TAS and ssp. *robustiusculum* in TAS and VIC. According to Meijer (p. 42) "the differences between O. *lineare* ssp. *sulcatum* and O. *australe robustiusculum* are mainly in the dimensions of the sporogonia, setae etc.". These measurements seem to be: capsule length 0·8–2·5 and 2·5–3·2 mm respectively; seta length 4–13 (average 8) mm and 11–25 (average 20) mm respectively, although there are some discrepancies in the data given. Fruit of *australe* which we have seen is certainly considerably larger than the common *lineare*, but Meijer's data show that the other subspecies are intermediate in size and we have to agree with Sainsbury that the distinctions are not convincing, at least on paper. The earlier name is *lineare*.

O. pallens (Hook.f. & Wils.) Broth. has a slightly pear-shaped capsule with a very thin transparent inner peristome membrane without any processes, and spores 20–25 µm. Vegetatively it is similar to *lineare*.
DISTRIBUTION: TAS, WA, NSW; endemic.
ILLUSTRATIONS: Meijer (1952).

O. inflatum (Mitt.) Par. has very narrow leaves, 3·0–5·0 × 0·2–0·3 mm maximum dimensions, a short broad pear-shaped capsule only 1·0–1·2 mm long × 0·6–0·7 mm broad, and big spores, 20–30 µm.
DISTRIBUTION: VIC, NSW; endemic. Perhaps not uncommon in NSW?
ILLUSTRATIONS: Meijer (1952).

Pohlia Hedw.

Pohlia is one of those troublesome genera where the species, once known, are usually recognizable with a hand lens but where it is hard to find unambiguous microscopic characters to discriminate between them. A satisfactory world-wide treatment of the genus, preferably based on new vegetative characters, is badly needed and, until that is accomplished and the precise relationships established between species of northern and southern hemispheres, most of the names used in Australia must be suspect; *P. nutans* especially is a widespread and polymorphic species, possibly just a taxonomic pigeon-hole, to which many of our species could be referred. The trouble is that some Australian species differ slightly from their northern counterparts: *nutans* has clearly long-decurrent leaves, *wahlenbergii* has, at least sometimes, secund leaves and an annulate capsule. Whether these features are significant can be assessed only by a complete and thorough revision of the genus. One is tempted to feel that the Australian species list should read: *P.* (?)*nutans sensu lato, cruda, tenuifolia, cf. wahlenbergii, cf. bulbifera.*

The striking metallic sheen of *cruda* is an optical interference effect brought about by a minute wrinkling of the cuticle. The same feature can be detected microscopically but to a lesser extent in *wahlenbergii*, where it causes the glaucous colouration, and to a very small extent in some of the other species. The rhizoids in all the species we have studied are papillose. The cells in *Pohlia* tend to be narrower than in *Bryum* with rather more square ends; this in turn allows the cells to be in longitudinal files down the leaf, instead of staggered, giving a different overall impression.

By far the commonest species, or group of species, is *P. "nutans"*.

Key to principal species

1. Plants with red stems. Leaves (at least the apical
ones) glaucous or with a metallic iridescent sheen   2
Stems green or black. Leaves shining but not
glaucous and without a metallic sheen . . .   *nutans et al.*

2. Leaves very narrow, usually slightly falcate, rather
glaucous and only slightly lustrous . . . .   *wahlenbergii*
Leaves widely ovate, symmetrical, very lustrous .   *cruda*

P. nutans (Hedw.) Lindb.

Most collections of *Pohlia* in Australia will turn out to be this species. It is very variable in size and colour and in the capsule, and it grows in a wide range of terrestrial habitats, either wet or dry, gravelly or peaty; most commonly it occurs mixed with other bryophytes but sometimes forms pure stands. Very commonly, isolated stems are scattered through other mosses and may produce a deceptively large amount of tomentum of papillose rhizoids, easily attributable to the wrong species.

The shoots vary from 0·5–2·0 cm in height and are green or black but not red, cornose above, sparsely foliate and with abundant papillose rhizoids below. The leaves are decurrent, green or yellow, variable in size but averaging perhaps 1·0–2·5 mm × 0·7 mm, narrowly lanceolate or ovate, with a conspicuous orange-brown nerve which is strong at the base and either ends well below the apex or may be slightly excurrent. The margins are usually plane, unbordered, entire or slightly crenulate above, and even denticulate near the apex; the comal leaves narrower and more evidently toothed. The cells vary from narrowly rectangular to rhomboidal, typically 50–100 × 9–12 μm. The cuticle is smooth or nearly so.

The inflorescence is (usually) paroicous, with antheridia just outside the female bracts.

The pendulous capsule is very variable in shape and size, basically oblong but sometimes expanded in the middle when fully mature or shrunk behind the mouth when immature; when old it is sometimes distinctly conical with a rather triangular outline. Usually it hangs on the end of a long or very long (2–4 cm) seta from which it often drops off when old, leaving the seta standing. The capsule, 3–5 × 1·0–1·5 mm, has a wide annulus.

DISTRIBUTION: TAS, VIC, NSW, ACT, (?SA, WA); almost world-wide.

ILLUSTRATIONS: Watson (1955), Fig. 76; Dixon (1924a), Plate 40(I).

This species is most likely to be confused with *Mielichhoferia* [70, q.v.] from which it is most easily separated when fruiting and with *Brachymenium* [67] which has a basal, not terminal, sporophyte and an areolation very much as in *Bryum*.

P. novae-seelandiae Dix. differs from *P. nutans* in the much narrower capsule, downwardly curved rather than pendulous from a curved seta, *c* 3·5–5·0 mm × 0·7–1·0 mm wide, almost cylindric in the upper two thirds. The cells too are rather narrower, averaging about 70 × 8 μm. It has only been found in the Victorian Alps and in New Zealand.

P. clavaeformis (Hampe) Broth. is vegetatively rather similar to *P. nutans* and has a pendulous capsule of somewhat similar shape but the cilia of the inner peristome are almost absent whereas in *nutans* they are quite conspicuous. It sounds much the same as *P. nutanti-polymorpha* (C.Muell.) Broth. from New Zealand and may well turn out to be a form of *nutans*.

P. mielichhoferia (C.Muell.) Broth. resembles a *Mielichhoferia* in external appearance. It has broad, triangular to ovate leaves, *c* 1·4 × 0·4 mm, tending to cluster at the tips of branches forming "comal tufts". It resembles *P. nutans* in being not at all iridescent but has a distinctive more or less erect capsule, narrowly pear-shaped and almost symmetrical. The cells are rather wider, not unlike a *Bryum*, *c* 45–60 × 9–12 μm, those below wider and more rectangular. It seems to be confined to VIC and NSW.

P. tenuifolia (Jaeg.) Broth. differs from all the other *Pohlia* species in the very narrow, almost subulate, leaves. It resembles *Orthodontium* vegetatively but differs in the papillose, instead of smooth, peristome teeth, in the papillose

301

rhizoids and in the much smaller capsule. It is confined to NSW and New Zealand.

ILLUSTRATIONS: Handbook, Plate 39.

P. cuspidata Bartr., known only from WA, is a minute plant 2–3 mm tall, distinguished by glossy yellowish leaves in which the nerve is excurrent as a short point. Bartram (1951, p. 468) states that there are "linear, sinuose propagula" on the sterile stems, which might suggest a form of *P. proligera* (Kindb.) Lindb. In such a small plant, which could be dwarfed when growing on limestone rock, the excurrence of the nerve seems a weak character on which to found a species.

What appears to be *P. bulbifera* (Warnst.) Warnst. has been collected on Mt Kosciusko by Weber and McVean. It is characterized by top-shaped axillary gemmae. It is otherwise known only from the northern hemisphere.

P. wahlenbergii (Web. & Mohr) Andrews in Grout (= *albicans* of the Handbook)

At its most luxuriant this is a very striking and attractive plant forming dense masses with bright red stems and whitish or even bluish-white contrasting leaves, probably corresponding to the variety *glacialis* (Brid.) Broth. It seems to be an upland or even subalpine plant of wet roadside banks and ditches in forest but is seen at its best in exposed boggy ditches. The red stems vary greatly in height from 1 to 8 or even more cm, of which the terminal 1–2 cm is fresh growth and rather pink in colour. The leaves tend to be evenly distributed up the stem, not in comal tufts, and to be slightly secund. They are quite small, *c* 1·0–1·5 mm × 0·3 mm, triangular–ovate to narrowly lanceolate, clearly decurrent, and often slightly asymmetric, being bent *sideways* at the apex. There is a strong red nerve ending well below the apex.

The cells are rhomboidal, rather like those of *Bryum*, *c* 90–190 × 12–15 μm, rather longer and narrower than in northern hemisphere material; the marginal 1 or 2 rows narrow, not forming a distinct border but with the tips projecting to give a crenulate or denticulate margin at least in the upper half of the leaf. The leaves are more glaucous and less metallic than in *P. cruda*.

On New Zealand plants the capsule is small, short and fat, widely pear-shaped and pendulous on a curved seta. The stomata in the capsule wall are immersed and there is said to be no annulus below the operculum. These features of the capsule are sometimes held to justify a transfer to another genus, as *Mniobryum wahlenbergii* (Web. & Mohr) Jenn. We prefer the more conservative treatment, especially since the very similar species, *P. cruda*, remains in the genus *Pohlia* and since the annulus is said not to be consistently absent, as it should be for *Mniobryum*. We have seen no fruiting material from Australia.

DISTRIBUTION: TAS, VIC, NSW; almost cosmopolitan in montane and alpine regions.

ILLUSTRATIONS: Dixon (1924a), Plate 41(G).

This species is often confused with *Mielichhoferia* [70, q.v.] and we do not yet know any way of telling them apart vegetatively, although the basal sporophyte of the latter is conclusive.

P. tasmanica (Broth.) Dix. is a very similar plant kept doubtfully distinct by Sainsbury on account of its narrower leaves and longer narrower cells (200 × 14 μm). However Victorian material shows a variety of cell size and leaf shape spanning the range between New Zealand *wahlenbergii* and Tasmanian *tasmanica* and we consider them to be conspecific unless contrary evidence comes to light.

P. cruda (Hedw.) Lindb.

In a strong light this little moss appears almost lurid. The red stems, about 2 cm high, contrast with the wide, ovate lustrous yellowish-green leaves which have a very strong sheen, as in *wahlenbergii* but much more pronounced; they become reddish with age, lower down the stem. They vary in size from 2·0–2·5 × *c* 0·9 mm, with a strong red nerve failing well below the apex, and narrowly rhomboidal cells, 100–225 × 9–12 μm often with a slight sigmoid twist at the ends.

We have seen no fruiting plants. The capsule is said to have an annulus.

DISTRIBUTION: TAS, VIC, NSW; also in New Zealand and widely distributed in both hemispheres.

ILLUSTRATIONS: Dixon (1924a), Plate 40(H); Grout (1928–40), II, Plate 74.

This is an exclusively alpine or sub-alpine moss of damp soil, flushes, rock crevices etc. It can be mistaken for no other species once the lustre of the leaves is recognized.

MNIACEAE

Mnium Hedw.

M. rostratum Schrad. (=*longirostrum* of the Handbook)
This is a rare species which might easily be mistaken for a *Distichophyllum* because of the very big round leaves, strong border, and single nerve. The leaf arrangement is almost distichous on the prostrate, rhizoidal vegetative stems, with rather distant leaves; on the erect stems, which culminate in perichaetia, the leaves are spirally arranged. The leaves are mostly oval, *c* 3 × 2 mm (the perichaetial leaves up to 7 × 3 mm) with a single nerve reaching the apex and joining with the border there. The rectangular to rounded–hexagonal cells are *c* 30 × 15–30 µm in smaller leaves, up to 50–60 × 25–50 µm in large leaves. It is a very rare species in Australia, in damp rock crevices by waterfalls or on the ground in wet forest.
DISTRIBUTION: TAS?, VIC, NSW, QLD; also in New Zealand and almost world-wide.
ILLUSTRATIONS: Handbook, Plate 42; Allison and Child (1971), Plate 17 and p. 87.

The Handbook treats the specific epithet *rostratum* as dating from 1791, and hence invalid since the starting point of moss nomenclature is with Hedwig in 1801; but the Index gives the correct date for Schrader's name as 1802, and it is therefore valid. Koponen, in a recent monograph (1968), has transferred this species to a segregate genus, *Plagiomnium*, as *P. rostratum* (Schrad.) Koponen, but the grounds for separation do not seem to us to be particularly strong. The record for Tasmania is based on a misidentification (Sainsbury 1955d, p. 35), but it is not unlikely to occur there.
M. rotundifolium Bartr. (QLD) is unknown to us (see Willis, 1955d).

MEESIACEAE

Meesia Hedw.

M. muelleri C.Muell. & Hampe
This is a rare subalpine or alpine bog plant from the high plains of VIC and NSW. The shoots, in dense deep cushions, are up to 9 cm tall or more,

of which 1–2 cm is current growth and the rest is matted together with *crimson*, densely papillose–warted *rhizoids*. The narrowly lanceolate leaves, 1·5–2·0 × 0·4–0·5 mm, are widest shortly above the base, broadly channelled, erect and flexuose when dry, and not much altered when wet; very slow to soak out when moistened. There is a very broad strong nerve in the leaf base, *c* 150 μm, tearing off a strip of stem-tissue when the leaf is pulled off, and tapering markedly above and failing below the apex. The apex is generally very obtuse, but sometimes acute, the margins entire except for a slight crenulation at the very tip. The cells are shortly rectangular through-out, those above *c* 20–30 × 10 μm, those in the leaf base *c* 40–60 × 15–20 μm; all cells fairly thin-walled.

The fruit is carried on a very long red flexuose seta *c* 4 cm tall; the capsule short and thick, erect at the base but strongly curved so that the peristome faces horizontally. The neck is often distinct and swollen and the operculum conical with a small apiculus. There is a double peristome of unusual form; the outer teeth short and almost square, the inner processes long and slender, both peristomes yellow or brown and shining. The inner teeth are half erect and separate when dry, but lie close together and almost flat over the capsule mouth when moist, apparently forced there by leverage of the short outer teeth at their bases; only the outer teeth seem to be hygroscopic. The peristome teeth are inserted well below the capsule rim leaving a saucer-like depression when the peristome is folded-in and flat. The spores are green or brown, 40–55 μm, intensely sculptured with fine papillae and ridges.

DISTRIBUTION: VIC, NSW, ACT: also in New Zealand.

ILLUSTRATIONS: ?

The crimson rhizoids, obtuse leaves and the habitat are sufficient to suggest the species even in the absence of the very characteristic capsules.

M. triquetra (L.) Aongstr., predominantly a northern hemisphere species, has been found, together with the preceding species, in NSW (Willis 1955d, p. 76) and VIC. It is similar in habitat but quite distinct in the triquetrous leaf arrangement and distinctly toothed leaf tips.

ILLUSTRATIONS: Brotherus (1924–5), Fig. 389.

AULACOMNIACEAE
Leptotheca Schwaegr.

L. gaudichaudii Schwaegr.

The neat, pointed stems of this common species, c 1·0–1·5 cm tall, are usually massed together to form extensive open turfs, bound together rather loosely underneath by tomentum. Perhaps most characteristically epiphytic in rain-forest (particularly on tree ferns) or wet sclerophyll forest, it is also common enough on rock faces, roots, soil etc. The colour is pale yellowish green or yellowish brown, often with a slightly glaucous sheen.

The leaves, (0·5–)1·0–1·5 mm long × c 0·4 mm wide, are widely lanceolate and mostly acute and slightly toothed at the apex where the strong nerve is excurrent in a short tapering, spike-like point. The leaves are rather distant, each overlapping only c ½ of the leaf above; when dry they are arched with the points inwards giving a chain-like (catenulate) appearance. The margins are plane or recurved. The cells are very variable and irregular in shape, but are thick-walled and roughly isodiametric, c 9 µm across. Brown filamentous gemmae (6–12 or more cells long) are often found on the upper part of the stem in the leaf axils.

The capsule is very narrowly cylindrical and erect on a long seta.

CHROMOSOME NUMBER: $n = 10, 20$ (TAS).

DISTRIBUTION: TAS, VIC, NSW, ACT, QLD; also in S. America and New Zealand.

ILLUSTRATIONS: Handbook, Plate 41.

The shape of the leaf and hair-point is quite distinctive, but until that is well known there is a possibility of confusion with several other species. *Rhizogonium mnioides* [77] is distinguished by the leaves being strongly curled when dry, not just arched. *Hymenodon pilifer* [79], which is also superficially similar and grows on tree ferns and can have somewhat similar colouration, is immediately distinct under the microscope by the short nerve. *L. wattsii* Card. (TAS) is unknown to us, as also to Sainsbury (1955d).

Aulacomnium palustre (Hedw.) Schwaegr., a rare plant of upland bogs, is related to *Leptotheca*. The erect shoots are characteristic; yellowish green in colour, with chocolate-brown tomentum enveloping the stem for most of its length. The cells are strongly singly papillose and the margins revolute throughout, almost to the apex. Usually the nerve is sinuose in the upper half and curiously buckled even when moist.

This species is probably overlooked and may be commoner than its rather few records indicate. It has recently been found (I.G.S.) in Victoria. DISTRIBUTION: TAS, VIC, NSW; also in New Zealand and widely distributed in the northern hemisphere.

ILLUSTRATIONS: Handbook, Plate 41; Watson (1955), Fig. 90; Brotherus (1924–5), Fig. 388.

MITTENIACEAE

Mittenia Lindb.

M. plumula (Mitt.) Lindb.

The stems, which are usually less than 1 cm high, may be unbranched or with a few branches from the base and are inclined at an angle to the substratum; either gregarious or scattered, and frequently mixed with small *Fissidens* [10] species to which they sometimes show a superficial resemblance in size and habit.

Sterile shoots are flattened, the leaves frequently distichous with the lines of insertion of the long decurrent leaf bases running lengthwise down the stem; but there is often a third irregular row of leaves appressed to the stem. The leaves are asymmetric, usually less than 1 mm long, oval to oblong with the apex shortly pointed or rounded; the lower leaves nerveless, the upper with a single nerve ending above mid-leaf. The cells in mid-leaf are approximately isodiametric, 22–29 μm. Male shoots may show the distichous habit of the sterile shoots in the lower region but at the apex the antheridia are surrounded by longer radially arranged leaves with oblique to transverse insertions. There are two other types of shoot, the leaves of which are radially arranged; either for the length of the stem or else with the stem almost bare except for the terminal cluster of longer leaves (to 2 mm long) surrounding antheridia or archegonia.

The seta is 2–3 mm and the cylindrical capsule about 1 mm long with a slender tapering operculum almost the same length. There is a short, narrowly conical calyptra closely covering the apex of the operculum and falling with it as a rule. The particularly beautiful peristome consists of 16 outer teeth which are slender and whip-like, curling with moisture changes, and an inner peristome of about 32 processes which are bent inwards to form a dome over the green spores. The detailed anatomy of the plant has been investigated and illustrated by Stone (1961a, b).

The protonemal stage of this remarkable moss is of particular interest as, like the northern hemisphere cave moss *Schistostega*, it has two phases according to the available light (Stone, 1961b). In caves, wombat holes and cavities where the light is unilateral, the protonema is apparently luminous and exhibits a striking green lustre. (The protonema of *Schistostega* is thought to be the origin of the tales of goblin gold in caves in Europe.) This effect is produced by the lens-like cells (c 15–20 μm), which are highly convex on the under-surface, and spread out in a plane at right angles to the light. A few sterile, but usually no fertile gametophores develop in these situations and the protonema may persist for many months, reproducing itself vegetatively. Where there is a higher light intensity, on shaded banks and on clay adhering to the bases of fallen forest trees, the protonema stage is composed of cells of the usual cylindrical type and fertile gametophores develop in profusion.

DISTRIBUTION: TAS, WA, VIC, NSW, ACT, QLD and also New Zealand.

ILLUSTRATIONS: Handbook, Plate 43; Stone (1961a, 1961b); Goebel (1906) Fig. 58; Brotherus (1924–5), Fig. 373.

It is not uncommon, particularly in the ranges east of Melbourne but has been found near the coast on Wilson's Promontory, in alpine regions, caves in the west of Victoria and recently for the first time in Western Australia. Not yet recorded for SA.

RHIZOGONIACEAE

Includes: *Bryobrothera crenulata* (Broth. & Par.) Thér.

Rhizogonium Brid.

Like their closest relations, *Goniobryum*, *Mesochaete* and *Hymenodon*, these are almost exclusively plants of wet forests. They are characterized by small rounded–hexagonal cells and by sporophytes borne on specialized female branches on the lower half of the stem. There are two quite distinct leaf arrangements, distichous and spiral, but the structures of both leaves and sporophytes in the two groups are very similar. Several of the species have the rather unusual characteristic of twinned teeth on the leaf margin, as in the genus *Atrichum* [5]. All species are densely gregarious and tend to form extensive patches, matted together by rhizoids at the very base.

Key to species

1. Leaves distichous, at least on the branches . . . 2
Leaves not distichous 5

2. Stems dendroid, branched above. Leaves distichously arranged on the branches only . . . *bifarium*
Stems not dendroid, usually unbranched. Leaves
distichous throughout 3

3. Nerve excurrent in a long or short point . . . 4
Nerve failing below apex . . . . *distichum*

4. Leaves clearly bordered with narrow cells. . . *pennatum*
Leaves unbordered . . . . *novae-hollandiae*

5. Leaves much curled when dry, similar in length
all down the stem . . . . *mnioides*
Leaves no more than flexuose when dry. Shoots
tassel-like, with short sparse leaves below and a
brush of long leaves above . . . . . . *parramattense*

R. bifarium (Hook.) Schimp.

This species is immediately distinguishable, even in the field, from others in Australia by the dendroid habit; short stems, *c* 0·5–1·0 cm tall, bearing small, triangular leaves spirally arranged, and with a 1 cm crown of 6–9 spreading, distichously-leaved branches above. The branch leaves are distichously inserted, widely ovate, and long decurrent on both sides, *c* 2·7–4·0 × 1·0–1·4 mm, strongly nerved right to the apex with the nerve toothed at the back above and slightly excurrent. The nerve is asymmetrically placed so that the outer half of the leaf is much narrower than the inner.

The margins are bistratose in the distal half, with twinned teeth. The cells are uniform throughout, rounded–hexagonal, c 10–15 μm across.

The short swollen male branches (1–2 mm long) are borne just at the base of the spreading crown of vegetative branches and there is often also a single one at the base of the stem. The female branches, with long slender bracts, are borne on different plants but in a similar position at the top of the stem. When mature, the short plump asymmetric capsule, 2 × 1 mm, hangs downwards from the tip of a red seta, up to 2 cm long. Several capsules can be present on the one plant.

DISTRIBUTION: TAS, VIC, NSW, ACT; also in New Zealand and S.E. Asia.

ILLUSTRATIONS: Plate 54; Handbook, Plate 44; Brotherus (1924–5), Fig. 376.

It is an uncommon species, on rotting wood in wet forests; unmistakable once the dendroid habit is noticed.

R. distichum (Sw.) Brid.

The unbranched dark green or yellowish shoots of this species, c 1 cm long or more, have 12 or more pairs of distichous, ovate or oblong leaves, 1·5–2·0 × 0·8–1·0 mm wide, sometimes smaller towards the base of the stem, coarsely toothed at the apex and with the nerve failing well below the apex and not toothed on the back. The cells are roughly hexagonal but irregular in shape and size, c 12–21 μm across, not elongated. The leaves tend to be slightly secund, pointing towards one side of the stem.

DISTRIBUTION: TAS, VIC, NSW, QLD; also in New Zealand and S.E. Asia.

ILLUSTRATIONS: Brotherus (1924–5), Fig. 376.

One of three rather similar distichous Rhizogoniums, this species is distinguished from the others by the short nerve and the lack of border. It grows on rotting stumps and trunks in wet forest. *R. graeffeanum* (C.Muell.) Jaeg. (NSW) is very similar if not identical.

R. novae-hollandiae (Brid.) Brid.

This species is similar to *distichum* but has the nerve excurrent in a short point and the leaf margins bluntly or strongly toothed or crenulate in the upper half. The marginal cells in 1–2 rows, in part of the leaf, tend to be

PLATE 54. *Rhizogonium bifarium* VIC—Female plant with young sporophytes (above left) and male plant (below) with antheridial buds in centre of crown and one at base of stem, both dry × 7; Moist plant (top right) × 7, leaf × 50, cells × 1000

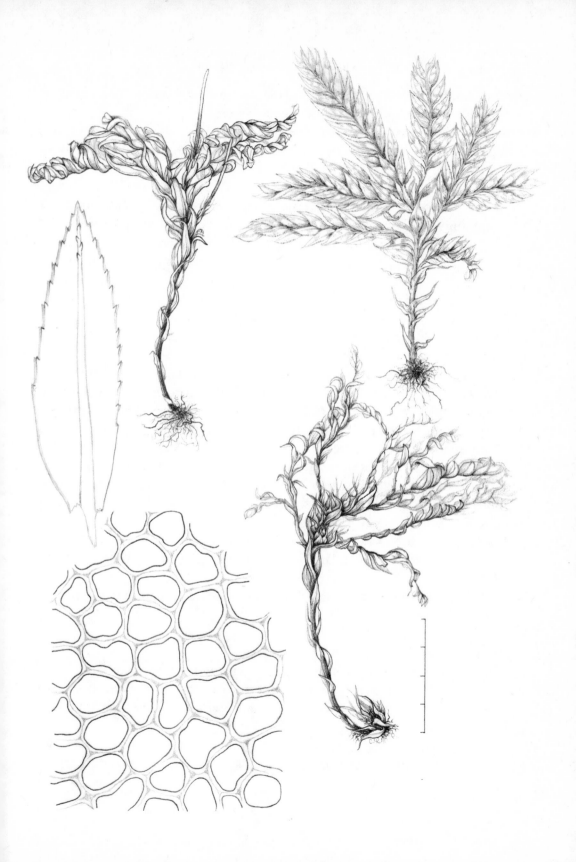

considerably narrower but not forming a distinct border. The cuticle is *minutely striate* in all the material we have seen.

CHROMOSOME NUMBER: n=5 (TAS).

DISTRIBUTION: TAS, VIC, NSW; also in New Zealand, S.E. Asia and South America.

ILLUSTRATIONS: ?

The habitat is rather similar to other *Rhizogonium* species but most commonly is on living trunks of trees or tree ferns, rather than rotting wood. The striolation of the cuticle is a character which does not seem to have been mentioned before; it needs to be checked on a wide range of specimens.

R. pennatum Hook.f. & Wils.

The Australian plants of this species seem to belong to the var. *aristatum* (Hampe) Dix., characterized by the long–excurrent nerve. This is a larger plant than the two foregoing species, with the fronds 2–3 cm long or more, and 3–4 mm wide. The leaves are narrowly lanceolate, $1\cdot0-1\cdot5 \times 0\cdot4$ mm, rather widely spaced. There is a strong rib-like border, *c* 2 cells thick, of narrow, thick-walled cells, toothed near the apex. The nerve is strong, excurrent in a rather long mucro. The cells in the leaf base tend to be elongated, to *c* $2-3 \times 1$, especially near the nerve.

The female branches occur at the base of the stem, and are very similar to those of *parramattense* with very slender yellow bracts forming a narrow bud-like branch at the base of the stem, usually almost buried in rhizoids and substratum.

DISTRIBUTION: TAS, NSW; also in New Zealand.

ILLUSTRATIONS: Plate 55; Brotherus (1924–5), Fig. 376; Wilson (1854), Plate 92.

This very handsome species with long fronds is perhaps more common on rocks and soil in forest than as an epiphyte.

R. mnioides (Hook.) Wils.

The shoots here, 2 cm high or more, are evenly foliate throughout, not at all tassel–like, with leaves of fairly uniform size (up to $4 \times 0\cdot5$ mm). Above the densely tomentose lower parts of the stem, the leaves become very

PLATE 55. *Rhizogonium pennatum* TAS—Vegetative plant (left) and female plant with reproductive branches at ground level × 7, leaf × 50, cells × 1000

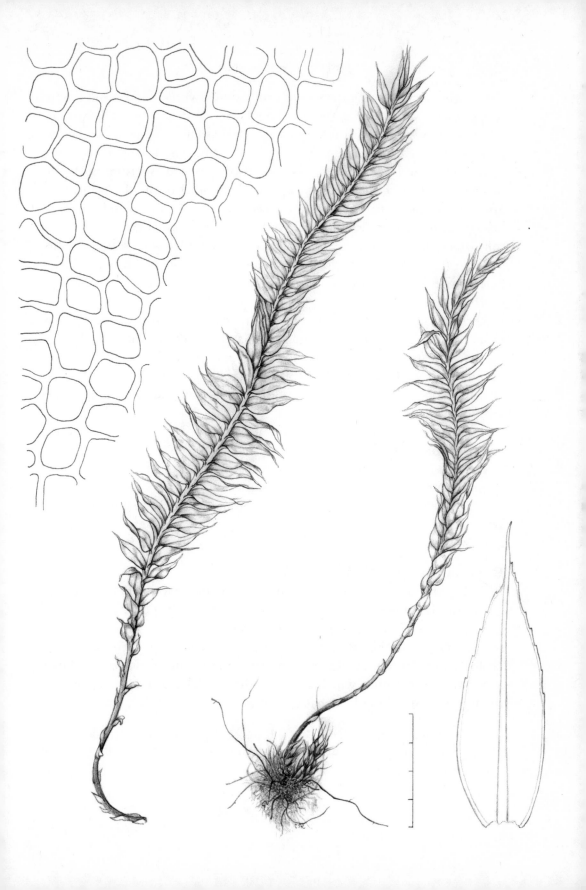

curled when dry, making the shoots distinctively woolly below and crisped above. The leaves are decurrent with very narrow strips of cells descending from the nerve and both margins of each leaf as far as the next leaf below, giving a sharply angled appearance to the stem. The leaves have a very strong nerve and a long very narrow lamina which is bistratose in the outermost cell row and bears conspicuous twinned teeth; the cells in the upper part of the leaf are typical of *Rhizogonium*, thick-walled and rounded–hexagonal, *c* 9-12 μm in diameter, but in the leaf base there is a patch of cells, especially next the nerve, which are rectangular with rounded ends, *c* 2-3 × 1 (up to 27 × 9 μm) forming an almost hyaline area in the leaf base, best seen under very low magnification. The marginal cells at the base are nearly square.

The plants are dioicous; the female branches have long slender bracts, more or less hidden by the foliage leaves, and occur at about ⅓ of the way up the stem from the tomentose base. Several female branches can occur on one stem, each with a small tuft of rhizoids at its base. The antheridia are said to be terminal at the apex of main stems, but we have not yet seen them.

DISTRIBUTION: TAS, VIC, NSW, ACT, QLD; also in New Zealand and S. America.
ILLUSTRATIONS: Plate 56; Handbook, Plate 44.

This is a common enough species in wet forest, but easy to mistake for *parramattense* (q.v.), especially when moist.

R. brevifolium Broth. is like a very short-leaved *mnioides* but the leaves are not crisped when dry. It is confined to QLD and S. America.

CHROMOSOME NUMBER: n=6 (QLD).

R. parramattense (C.Muell.) Reichdt.

Probably the most robust of the Australian species of *Rhizogonium* this is very similar to *mnioides* in leaf arrangement and in the general appearance of the plant when moist. Its most characteristic feature is in the tassel-like stems, almost bare below with small narrowly triangular stem leaves and no tomentum except at the extreme base, and a flail of big leaves above, 4-7 mm × 0·5-0·6 mm, which are often slightly falcate. The appearance of the frond is something like that of a horse's tail. When dry, the leaves are

PLATE 56. *Rhizogonium mnioides* VIC—Female plant ×7, with female branch projecting from the right of the shoot below half-way down, leaf × 50, cells × 1000

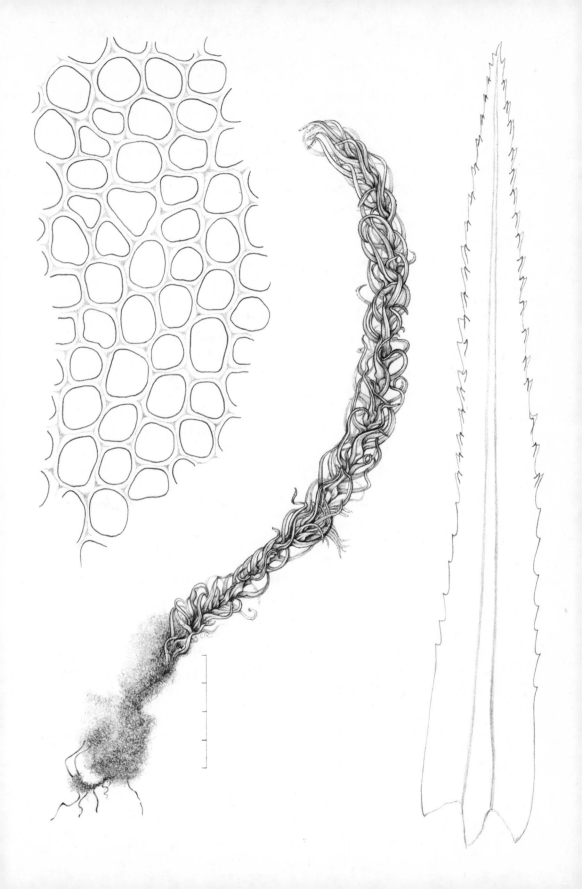

slightly twisted and flexuose but not at all crisped as in *mnioides*. The cells are isodiametric throughout, hexagonal or rounded, *c* 9–13 μm in diameter, rather irregular in size and shape but, if anything, slightly shorter in the leaf base and even wider than long, and with thinner walls. There is a very strong nerve, broad at the base (to 150 or even 200 μm) reaching right to the apex or even excurrent, and toothed on the back in the upper half. The leaf margins above are bistratose and strongly toothed with twinned teeth. As in *mnioides* the leaf is decurrent with narrow strips of cells reaching the leaf below.

The species is dioicous; the male branches are short, fat buds, *c* 1·0–1·5 mm, usually with abruptly pointed bracts, axillary or nearly axillary towards the base of the stem, sometimes with up to 6 or more adjacent buds. The female branches, on different stems, are found nearer the base of the stem and are quite different in shape; long and narrow (*c* 4–5 mm) with very long and slim perichaetial leaves exposed and easily visible because of the small leaves on that part of the stem. The capsule is borne on a long seta of *c* 3–4 cm, and is short and plump (*c* 2·5–3·0 × 1·0–1·5 mm) with a wide mouth; the operculum has a long, slightly curved beak about half the length of the capsule. The spores are smooth and *c* 13–18 μm in diameter.
CHROMOSOME NUMBER: n=6 (NSW).
DISTRIBUTION: TAS, VIC, NSW, QLD, Lord Howe; endemic to Australia.
ILLUSTRATIONS: Plate 57

All the Australian material we have seen is dioicous and therefore cannot be referred to the synoicous *R. spiniforme* (Hedw.) Brid., but Dixon (1942) and Bartram (1952) record *spiniforme* from north Queensland, presumably correctly, and it has also been recorded from NSW. We have not seen the Tasmanian plants which Sainsbury (1955d, p. 23) attributed to *spiniforme* but presumably they were sterile. Most likely the specimen was *parramattense*. At first sight very similar to *mnioides* this species can be quite easily distinguished from it by the tassel-like fronds with small leaves below, not tomentose except at the very base, the lack of elongated cells in the leaf base, the leaves flexuose but not crisped when dry, and by the position of the male flowers on basal branches instead of being terminal.

PLATE 57. *Rhizogonium parramattense* NSW—Female plant with sporophyte (right), base of male plant with antheridial branches at ground level (left) ×7, leaf ×50, cells ×1000

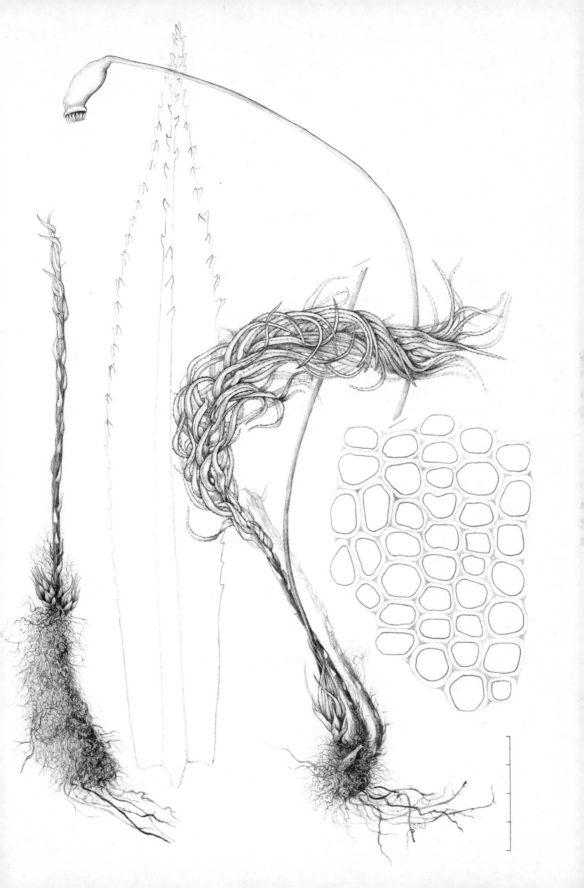

Of *R. alpestre* C.Muell. (TAS), the remaining species recorded for Australia in the *Index*, we have seen no material. It is not mentioned by either Rodway or Sainsbury (1955d). If it is a *Rhizogonium* it is near to *distichum* in structure, but Watts and Whitelegge (1906, p. 146) suggest it may be a *Porotrichum*.

Goniobryum Lindb.

G. subbasilare (Hook.) Lindb.

This species is rather similar to a *Rhizogonium* in general appearance but has quite different cells. The stems, 1–2 cm high, bear dark green, narrowly lanceolate leaves arranged sometimes distichously, sometimes in roughly 3 ranks, most commonly not in ranks but rather complanate. The leaves generally have twinned teeth (sometimes single) along most of the margins especially at the apex. The cells throughout are very large, wide and thin-walled, rectangular, hexagonal or rhomboidal, *c* 50–150 × 25 μm in mid-leaf, collapsing when dry so that the leaves shrivel. The chloroplasts tend to aggregate at the ends of the cells, giving a characteristic chequered appearance to the leaf when dry. The marginal 1–2 cell rows are narrower but do not form a distinct border.

The female branches are basal among the rhizoids, as in *Rhizogonium*, and the capsule is similar but not contracted below the mouth.

DISTRIBUTION: TAS, VIC, NSW; also in New Zealand, Pacific Oceania and South America.

ILLUSTRATIONS: Plate 58; Brotherus (1924–5), Fig. 379.

It is not uncommon in alpine bogs and wet forests, especially on decaying logs or earth, but rather easily overlooked because of its sombre colour and undistinguished shape. The stems and nerves are sometimes red when old.

Hymenodon Hook.f. & Wils.

H. pilifer Hook.f. & Wils.

This elegant moss forms large patches of pale glaucous green delicate stems, 1·0–1·5 cm long, usually projecting outwards and downwards from the

PLATE 58. *Goniobryum subbasilare* VIC—Fertile plant × 7, cells × 1000

trunks of tree ferns in wet forest. The base of the stem is anchored by a tomentum of bright chestnut-coloured papillose rhizoids. The oval leaves (0·3–0·5 × 0·2–0·25 mm) are arranged regularly round the stem but some-times appear to be in 2 or 3 ranks. They are distinguished by a long hair-point (0·2–0·4 mm) springing abruptly from the leaf tip, but with the nerve usually failing below the apex. The margins are plane, crenulate throughout and sometimes denticulate at the apex; the mamillose cells, squarish–hexagonal, *c* 9 µm, not elongated at the leaf base. The cuticle is *minutely wrinkled*, causing the glaucous sheen of the leaves.

The female branches are basal, with long narrow bracts, hidden among the rhizoids. The seta (1·0–1·5 cm long) hangs outwards and downwards, carrying a broadly cylindrical capsule with no outer peristome but the inner peristome teeth touching at the apex to form a high, rounded–conical, basket through which the spores are sifted. The separate male plants are said to be dwarf, as in *Leucobryum* [28], but we have not seen them.

DISTRIBUTION: TAS, VIC, NSW; also in New Zealand and Pacific Oceania.

ILLUSTRATIONS: Plate 59; Handbook, Plate 43; Brotherus (1924–5), Fig. 375.

This species might be mistaken for a *Rhizogonium* [77] at first glance, or for *Leptotheca* [75] which is similar in size and colour, but the hair-points which are projections of the leaves, not excurrent nerves, are quite diagnostic.

Mesochaete Lindb.

M. undulata Lindb.

The robust stems of this unmistakable species (*c* 2 × 0·6–0·8 cm) form dense swards on river banks, rotting trees and other damp habitats in wet forest in the warmer regions of eastern Australia, just reaching into easternmost Victoria. The leaves are in 4 rows and strongly complanate so as to appear distichous and extremely densely imbricated. The oblong–ovate leaves, which are strikingly undulate when dry, measure *c* 2·5–4·0 × 1·5 mm, and are divided by the strong nerve into 2 very unequal halves, the lowermost narrow, the uppermost greatly widened at the base. There is a very strong

PLATE 59. *Hymenodon pilifer* VIC—Pendent fertile plant × 15, leaf × 50, cells × 1000

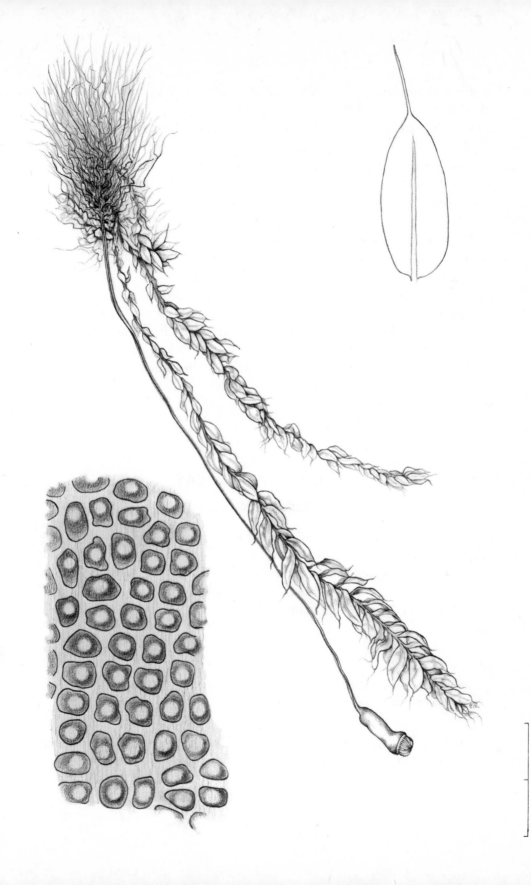

border of narrow thick-walled cells, bearing strong multicellular teeth. Not uncommonly the apical leaves are deciduous, falling off to leave a black spike of stem at the shoot tip: presumably the fallen leaves act as gemmae. The cells are square or hexagonal, *c* 10 µm across, and thick-walled, becoming slightly narrower and more elongate in the leaf base.

The capsule is produced laterally from near the middle of the stem, on a stout red seta. The capsule itself is massive, 5–6 mm long, curved, dark, and deeply grooved even when moist.

CHROMOSOME NUMBER: n = 10 (NSW).

DISTRIBUTION: VIC, NSW, QLD, Lord Howe. Also SA according to Watts and Whitelegge (1906, p. 149); endemic to Australia.

ILLUSTRATIONS: Plate 60; Brotherus (1924–5), Fig. 380.

This is a subtropical species reaching its southernmost limit in Victoria. Because of the leaf arrangement, shape and border it is unmistakable, even when the distinctive fruit is absent.

M. grandiretis Dix. (QLD) differs only in the cells which are twice as large, 20–25 µm across.

BARTRAMIACEAE

Bartramia Hedw.

Unlike its near relatives this genus does not have clusters of branches below the inflorescences. Apart from one group of species (to which *B. stricta* belongs) the genus is distinct vegetatively in having leaves with a conspicuous sheathing base, a very narrow, long subulate lamina and small almost isodiametric cells. The sporophyte is generally similar to that of related genera.

Key to species

1. Leaves with a distinct sheathing base	2
Leaves without sheath	4

PLATE 60. *Mesochaete undulata* NSW—Fruiting plant × 7, basal cells (below) upper cells of leaf including marginal tooth and with border left blank × 1000

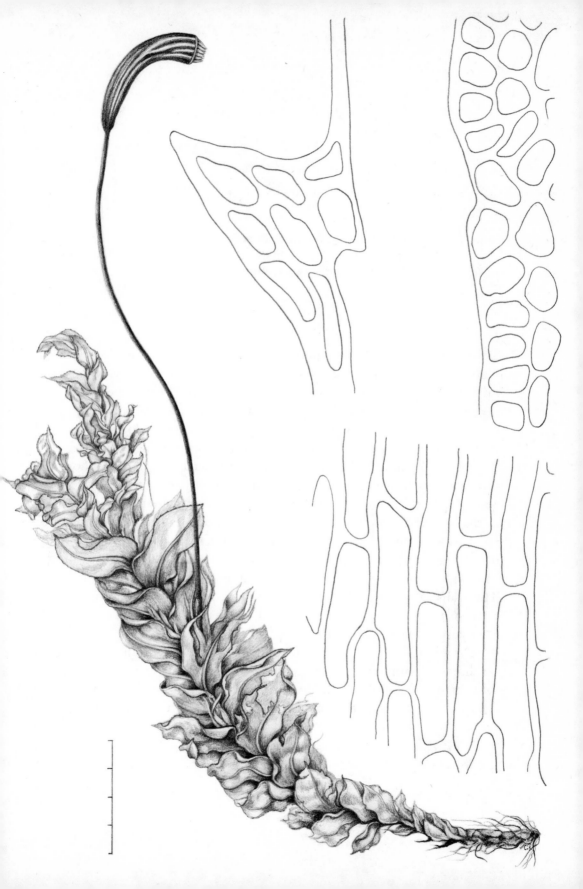

2. Cells of the limb isodiametric . . . . . .   *hallerana*
    Cells of the limb elongated, 2 × 1 or longer . .   3

3. Capsule gymnostomous. Plants less than 1 cm
    high. Leaves not wide-spreading. . . . .   *Bartramia* cf.
                                                      hampeana

    Capsule peristomate. Plants more than 1 cm tall.
    Leaves often spreading, sometimes erect . .   *papillata*

4. Capsule peristomate. Leaves strictly erect, brittle,
    closely appressed to stem when dry . . . .   *stricta*
    Capsule gymnostomous. Leaves spreading, more
    flexuous . . . .   *Bartramia* cf.
                                                      compacta

B. hallerana Hedw.

This is a handsome species with rather tall stems, to 8 cm or more, densely matted into very soft cushions or tall turfs by chocolate-brown tomentum in the preceding year's growth, i.e. from c 1 cm below the apex. The rhizoids are densely papillose. The leaves have a short sheathing base, c 1 mm long by 0·8 mm wide, quickly narrowed to a long arched and twisted subula 5–6 mm long × 0·2–0·3 mm wide, tapering to a long, very fine, toothed point. The nerve is conspicuous, occupying most of the upper subula and both nerve and margins are entire in the leaf base but toothed above it, the teeth sometimes twinned. The margins are narrowly recurved in the base and sometimes above it. In the subula the cells are squarish or rounded–hexagonal, c 7–10 μm diameter, thick-walled, and with single dome-shaped papillae; those of the leaf base narrowly rectangular, mostly smooth, with thin (or sometimes thick) rather sinuose walls, c 50–60 × 8 μm. The walls in the sheathing base are usually, but not always, thinner than those of the lamina. The cells at the top of the sheath are still elongate but have papillae at the cell ends.

Sporophytes are very characteristic; the capsules are almost *immersed*, on short, 3–4 mm setas, overtopped by the perichaetial bracts and usually, when mature, completely hidden by growth of branches from below. The capsule, c 1·0–1·5 mm, is asymmetrically placed on the seta with the mouth

PLATE 61. *Bartramia hallerana* TAS—Fruiting plant × 7, leaf × 50, basal and upper cells × 1000

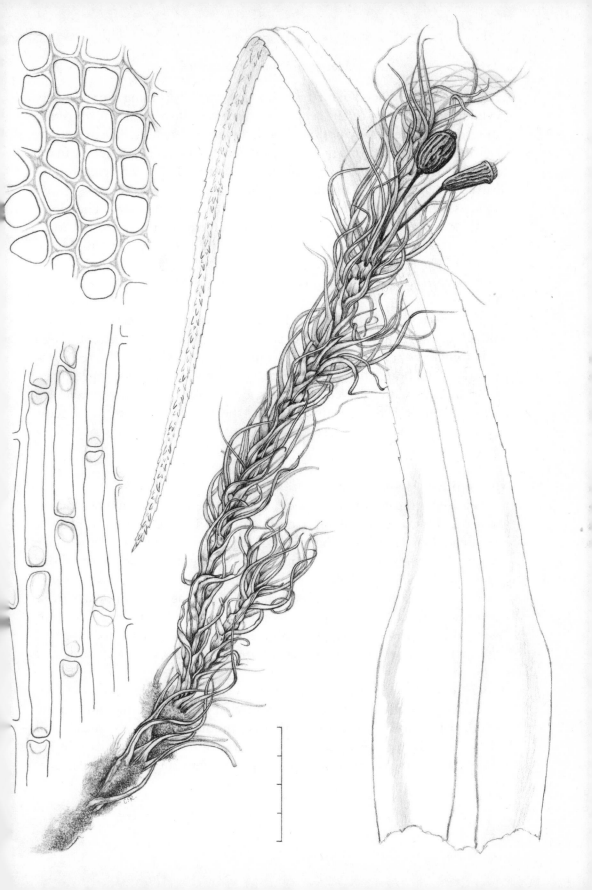

obliquely erect. When young it is almost spherical, becoming more broadly cylindrical when mature, and deeply grooved even when moist (*c* 16 grooves), the grooves evident from both outside and inside the capsule. The outermost cells of the capsule wall are *square* and not very thick-walled, especially in the grooves. There is a double peristome, the outer teeth deep brown, strongly barred and very densely and finely papillose; the inner processes pale yellow and less densely papillose, split down the middle as in *Philonotis*. The spores are dark brown, 20–25 µm, strikingly ornamented with high large warts.

DISTRIBUTION: TAS, VIC, NSW; also throughout Europe, Asia and in Africa, N. & S. America and New Zealand.

ILLUSTRATIONS: Plate 61; Dixon (1924a), Plate 39; Allison and Child (1971), Plate 18.

This species is commonest on rocks and cliffs in damp forest, but can also occur as an epiphyte. It is not evidently calcicolous in Australia, although it is so in Europe. If the immersed capsules are present the species is un-mistakable (although they can be quite hard to notice) and even vegetatively the tall stems with very curled leaves are distinctive. We are grateful to A. C. Crundwell for pointing out that the usual spelling of the specific epithet, *halleriana*, is incorrect.

B. papillata Hook.f. & Wils.

This is by far the commonest species and is quite frequent on well drained soils of river-banks and damp parts of forests. The stems are quite short, *c* 1–3 cm, loosely matted together by papillose rhizoids. The leaves are erect, spreading or flexuose, sometimes very slightly falcate at the shoot apex, bright blue-green above and pale brown below. There is a strong, hyaline–whitish sheathing base, typically 0·8 mm long × 0·3 mm wide, from which emerges a spreading glaucous-green limb, *c* 2–3 mm long × 0·1 mm wide, the leaf bases overlapping to give a smooth whitish plated cover to the stem from which the contrasting green leaves spread outwards and upwards.

There is a rather broad, single nerve which is not very well defined, even less in the subula where it apparently takes up about one third of the width below and more above. The margins are entire in the sheath and closely serrulate in the subula, especially near the tip. The subula cells are shortly rectangular, *c* 20–40 × 6 µm, papillose with single domed papillae.

The leaves are 2 cells thick above, obscuring the cell outlines. In the leaf base the cells are longer, rectangular–rhomboid, c 50–70 × 12–15 µm, thin-walled except sometimes for the short end-walls which may be thickened. The abrupt narrowing from leaf base to subula causes conspicuous shoulders in the leaf outline where the cells are shortly rectangular and thin-walled.

The seta, c 1–2 cm long or more, carries a symmetrical, spherical capsule 1–2 mm across, deeply 16-grooved when mature and with a rather small mouth c 0·6 mm across. When dry the capsule may contract to a curved cylindrical shape—possibly squeezing the spores out, although we have not observed this. The peristome teeth are smooth or very faintly papillose, with transverse bars and *sinuose* margins. The inner peristome is little more than a low membrane with irregular projecting tooth-like lobes. The whole peristome lies flat, or even concave in the middle, below the rim of the capsule mouth. The bright brown spores, c 26 µm diameter, are densely warted, the warts almost coalescing to form areolae. There is a flat operculum with a stubby nipple in the middle.

The male inflorescences are terminal on the stem and have perigonial leaves longer and wider than vegetative leaves, enclosing the antheridia and paraphyses which do not form a striking disc as in *Breutelia*.

CHROMOSOME NUMBER: n=8 (NSW).

DISTRIBUTION: TAS, WA, SA, VIC, NSW, ACT; also in New Zealand.

ILLUSTRATIONS: Handbook, Plate 45; Wilson (1854), Plate 86.

B. cf. hampeana

The leaf shape and papillae are usually sufficient to distinguish the preceding species from all others except perhaps a gymnostomous species which was first described from the Grampians, Victoria, as *Glyphocarpa erecta* Hampe. This plant has also featured in both *Bartramia* and *Bartramidula*, but is a *Bartramia* and may be conspecific with *B. hampeana* C.Muell. from S. Africa. Otherwise the correct name would be *B. hampei* (Mitt.). The stems are erect, less than 1 cm high with few branches; the leaves have a shining yellowish or white sheath with distinct shoulders and a limb which is not usually wide-spreading. The plants are dioicous, the seta is c 5–7 mm long and the *gymnostomous* capsule erect (occasionally inclined) and furrowed when dry, with a short flat-topped conical lid.

DISTRIBUTION: TAS, WA, VIC, NSW.

ILLUSTRATIONS: ?

B. stricta Brid.

This species grows on dry earth banks and is possibly overlooked because of its resemblance to *Campylopus* spp. [22] and the apparent rarity of fruit. The stems, which may be sparsely branched, are densely leafy with strict brittle leaves which are appressed to the stem when dry, slightly spreading when wet. The leaves are up to 3 mm long, wide at the base, with a straight, brown insertion, and narrow gradually to the pointed apex giving a narrowly triangular shape. The margins are plane or weakly recurved at the leaf base, more strongly above and with 1–2 rows of very thick-walled sharp spines. The cells are mostly oblong, thick-walled, and, except for those at the base, have small projecting papillae at the cell ends. The cells in the upper part of the leaf are *c* 6 μm wide and 4–7 × 1; in mid-leaf, *c* 10–12 μm wide and 2–4 × 1; below, thinner-walled, *c* 10 μm and 3–6 × 1 except for a marginal alar band of 2–3 rows of more or less quadrate cells, *c* 13–20 μm, creating a slight fold for a short distance up from the base on each side near the margin. The nerve is about 60–70 μm wide below, 30 μm above, forming a strong abaxial ridge, rough with projecting cell-ends and occasional spines and often excurrent in a saw-edged golden point. The outer stem cells are small and thick-walled unlike those of *Philonotis* [85, q.v.]. The rhizoids are smooth.

The perichaetial leaves are not distinct except that they lack the alar quadrate cells. There is a straight seta *c* 1·0–1·5 cm long, and an erect capsule, *c* 1·0–1·5 mm, globose when wet and more or less oblong and deeply grooved when dry, with a convex lid and single peristome of finely papillose, lanceolate, fragile teeth inserted below the rim. The outer cells of the capsule are small, 5–6-sided, isodiametric and often collenchymatously thickened; the spores are dark red-brown with large rounded warts, 30–40 μm. The plants are synoicous with antheridia, archegonia and paraphyses mixed.

DISTRIBUTION: TAS, WA, SA, VIC, NSW, QLD; also in Europe, round the Mediterranean, N. & S. America.

ILLUSTRATIONS: Dixon (1924a), Plate 38; Grout (1928–40), II, Plate 67.

The alar quadrate cells of the leaf would separate this plant from sterile *Conostomum* species.

B. cf. compacta

A similar moss from WA with softer, more spreading leaves appears to have a close affinity with *Bartramia compacta* Hornsch. of S. Africa and vegetatively is certainly more like a *Bartramia* than a *Bartramidula*. The globose capsules we have only seen in the immature state except for one old capsule which showed furrows.

The stems are less than 1 cm long; the leaves, up to 3 mm long, are spreading and have no sheath at the base. The variably recurved margin is spiny except at the leaf base and the usually excurrent nerve is strong and adaxially rough. The leaf cells are almost smooth or with slightly projecting ends.

The plants are synoicous; the seta is about 1·3 cm long and the gymnostomous capsule has a low conical lid and stomata confined to the base of the capsule. The isodiametric outer cells of the capsule have collenchymatously thickened walls.

B. timmioides C.Muell. (NSW) is unknown to us.

Bartramidula B.S.G.

B. pusilla (Hook.f. & Wils.) Par.

These plants, which form loose pale or glaucous green patches on intermittently moist soil, have weak primary stems, sparsely rhizoidal and straggling, often with irregular divergent branches *c* 1 cm long. The leaves are tiny, *c* 0·5 mm long, narrow, linear–or triangular–lanceolate, only a few cells across even in the widest part (including the cells over the nerve), more or less erect when dry, usually spreading or sometimes squarrose when wet; the margins varying from almost entire (with slight projections at cell junctions) to obscurely dentate; the nerve weak, finishing short of the apex. The cells are shortly rectangular, firm-walled to somewhat incrassate, slightly angular at the margins, not usually papillose.

The slender seta, 5–10 mm long, is curled or arcuate or straight; the thin-walled capsule globose or shortly oblong with a short neck, warty or wrinkled when dry, pendulous or erect, gymnostomous and usually with a small mouth. The scattered stomata extend to more than half way up the capsule. The operculum is convex with an umbo; the calyptra *c* 1 mm long, side-split, subulate, only covering the apex of the capsule. The spores are large, 48–65 μm, papillose.

DISTRIBUTION: TAS, WA, SA, VIC, NSW; endemic.
ILLUSTRATIONS: Wilson (1859), Plate 174.

Conostomum curvirostre is not greatly dissimilar, but lacks divergent branches and has the leaves nerved to the apex or beyond. The curved beak on the operculum is also distinctive.

B. hampei Mitt. is much more robust, with leaves 2–5 mm long. First recorded from Mt William, VIC, and by Rodway from TAS. We consider it to be a *Bartramia* [81, q.v.].

Breutelia [B.S.G.] Schimp.

This genus of large ground mosses is usually thought to be characteristic of wet banks and rocks. At least in Australia, however, *B. affinis* is often baked dry on exposed rocks and roadsides although always where it will be soaked in some seasons. The leaves of all species are plicate and very densely packed on the stems; the elongate cells papillose with single papillae next to or overlying the junctions of cell-ends. Most rhizoids too are finely papillose in all species. All our species are dioicous and the male plants have splendid, usually bright brown or orange terminal "flowers" (perigonia). At their most distinct the species are very different from each other but, in atypical specimens, they can be surprisingly hard to identify. The most useful guiding rules are that, in Australia, *B. affinis* is overwhelmingly the commonest and the only one to tolerate dry conditions; *B. elongata* has not yet been found outside Tasmania and is by far the largest plant; *B. pendula* seems to be restricted to upland habitats. The cells forming the stem surface are similar to those of *Philonotis* [85, q.v.].

Key to species
  1. Leaves plicate throughout. Always in wet habi-
        tats. Stems usually little branched   2
      Leaves plicate at base only. Often in dry habitats.
        Stems usually densely branched at the shoot tips   *affinis*

2.   Very big plant, not obviously tomentose. Leaves
        usually 5–6 mm long, falcate and secund. Upper
        cells linear, porose.   .   .   .   .   *elongata*
      Plant smaller, tomentose. Leaves 2–4 mm long,
        seldom falcate and secund. Upper cells irregular,
        shorter, scarcely porose .   .   .   .   *pendula*

B. affinis (Hook.) Mitt.

By far the commonest species in Australia, this is very variable in robustness. Shoots measure from 3 to 10 cm long (of which the terminal 1–2 cm is current year's growth) and 1–2 mm wide when dry; the tips often firmly pointed. Both male and female plants are richly branched by sub-floral innovations (whorls of side-shoots from immediately below the inflorescence). The leaves are tightly pressed to the stem and not uncommonly slightly secund, spreading to one side of the shoot. The stems are usually densely tomentose with chocolate brown, papillose rhizoids. The leaves, usually yellowish green, are *c* 3 mm long, mostly with a wide ovate plicate base tapering to a long fine point much of which is taken up by the nerve; the plications are confined to the base. The margins are narrowly revolute throughout most of the leaf and are usually denticulate in the upper half.

The cells are very varied in shape and size, varying from narrowly oblong to almost square or quite irregular, but with rather thick walls and rounded ends; those above measure 10–20 μm long × 7–8 μm wide, those in the leaf base more elongate, up to *c* 30 μm. At the leaf base the 6 or more outermost rows of cells are squarish and rather thick-walled, forming a distinct marginal/alar group. The cells almost throughout, except for alar cells and the adaxial surface of the nerve, are conspicuously papillose, usually at the ends of the cells, with single, high papillae, and the underlying cell cavity often widened there. The papillae are scattered and occur on about half the total number of cells. All cells are thick-walled but scarcely porose.

Sporophytes are quite common. The capsule, *c* 2 mm long, is globular when fresh, widely cylindrical and deeply grooved when dry, with a low conical operculum and a double peristome.

CHROMOSOME NUMBER: n=6 (TAS, NSW).

DISTRIBUTION: TAS, WA, SA, VIC, NSW, ACT; missing from QLD? Also in New Zealand.

ILLUSTRATIONS: Brotherus (1924–5), Figs 411, 412; Hooker, W. J. (1818–20), II, Plate 176.

This is a very common dull-yellowish moss on soil and rock in a wide variety of habitats. The pointed shoot tips usually separate it easily from *Philonotis tenuis* [85] with which it often grows.

B. pendula (Smith) Mitt.

This is a very similar plant to *B. affinis* but is usually slightly more robust. It seems to be confined to mountainous regions, and elsewhere is not common. Superficially very like *affinis* it can be distinguished by the leaves which are plicate throughout, instead of only at the base, and by the alar cells which are fewer, larger and in 3 (sometimes to 6) rows, and *thin-walled*, collapsing when dry, since they lack the firm thick walls of *affinis*. The capsule is larger.

CHROMOSOME NUMBER: $n=6$ (NSW).

DISTRIBUTION: TAS, WA, VIC, NSW, ACT; also in S. America, S. Africa and New Zealand.

ILLUSTRATIONS: Handbook, Plate 45; Studies, Plate 9 (including *B. sieberi*); Allison and Child (1971), Plate 19 and p. 93.

B. elongata (Hook.f. & Wils.) Mitt.

One of the most splendid of all mosses, at its biggest and best this is quite unmistakable; with long, sparsely branched or unbranched thick shoots not unlike a *Lycopodium* but with densely packed, deeply striate leaves which are falcate and secund especially in female plants. Despite Sainsbury's comment to the contrary (Handbook, p. 309), the stems are often tomentose although the tomentum is usually covered by the leaves. In small specimens it may be hard to separate this species from *pendula* and there is no single character that provides an unambiguous separation. The alar cells are much the same as in *pendula* but the other cells throughout the leaf tend to be thicker-walled and are porose, which is probably the best distinguishing feature. The leaves tend to be slightly arched back in a continuous curve when moist, whereas in *pendula* they tend to be more abruptly bent back above the base (squarrose), but this difference is not always detectable.

PLATE 62. *Breutelia elongata* TAS—Male shoot (left) and female shoot with fruit × 4, cells × 1000

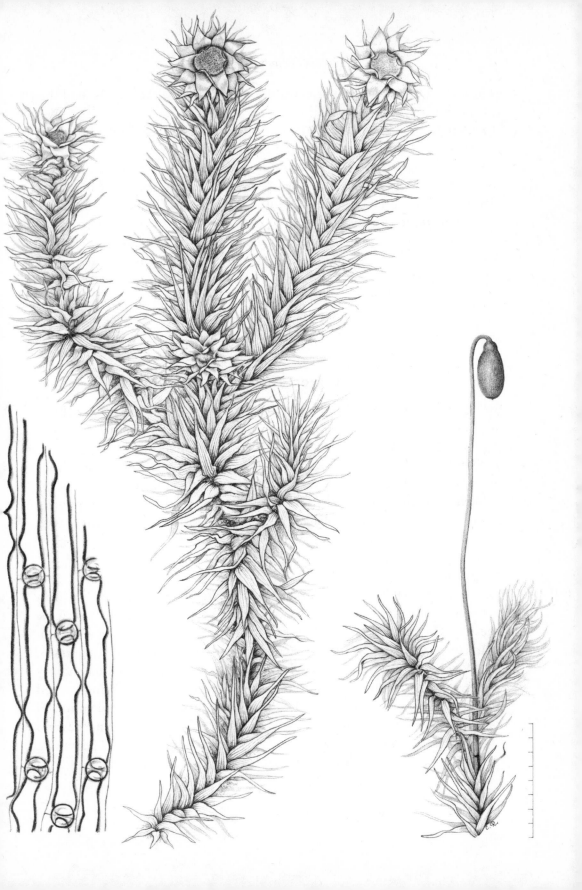

The leaf arrangement, too, is generally much more falcate and secund than in either of the preceding species.

DISTRIBUTION: Recorded only from TAS; also in New Zealand.

ILLUSTRATIONS: Plate 62; Handbook, Plate 45.

B. crassa (Hook.f. & Wils.) Jaeg. has been recorded from TAS but Rodway's Tasmanian material, referred to this species, was found by Sainsbury (1955e, p. 38) to be *elongata*.

We have not seen material of *B. fusco-aurea* Broth. (NSW). The other species given for Australia are all attributable to C.Müller and may be treated with some suspicion: *B. baeuerlenii*, *B. campbelliana*, *B. leptodontoides*, *B. lonchopelma*, *B. pseudo-philonotis*, *B. witherheadii*.

Conostomum Swartz

Peaty and stony soils in montane to alpine regions are the usual habitats for this predominantly southern hemisphere genus. The relationship to *Bartramia* and *Breutelia* is evident in the almost spherical capsules and in the characteristically papillose cells with a single papilla at the end of each cell. It is distinctive in having the leaves arranged more or less in 5 ranks, the beaked operculum and the curious peristome in which the teeth, if present, are joined at the tips. The foliage too tends to be distinctively coloured with glaucous blue laminas and golden nerves.

C. pusillum Hook.f. & Wils.

The pale brown spherical capsules which are often produced in great abundance are the most conspicuous feature of this common subalpine species. The narrow erect shoots, usually less than 1 cm tall × 0·4 mm diameter, are rather closely packed together into low turfs. The leaves, *c* 1·2 × 0·2 mm, are very narrowly triangular and finely pointed, tightly appressed to the stem in 5 ranks which tend to be obscured by being twisted round the stem. The nerve is well defined, *c* 60 μm across (up to 90 μm) and occupying about ⅓ (sometimes to ½) of the leaf base, extending beyond the apex as a hair-point which is denticulate on the back. The young growth is bluish, with nerves and hair-points golden; the old leaves are a dull golden colour. The cells are narrow, obliquely rectangular or rhomboidal, c 30–45 × 9 μm, with c 6–10 (rarely more) rows of cells on either side of the nerve in the leaf base. They are smooth in the lower half of the leaf and papillose in the upper half or less, with a single papilla on the end of each cell.

The globular capsule, c 1·5–2·0 mm long × 1·0–1·5 mm broad, on a seta c 1·0–2·1 cm long, is deeply 8–10 grooved, and can vary from erect to pendulous. The rather small mouth, 0·3–0·4 mm diameter, has a raised reddish rim projecting well above the point of insertion of the single peristome. There are 16 red or brown teeth, often hyaline near the base, and all bound together at the tips into a hyaline point. When moist the teeth are straight and erect forming a complete cone over the capsule mouth. On drying out they pull backwards drawing the tip down into the capsule mouth and causing gaps to appear between the teeth. The hyaline base to the teeth, presumably unthickened and therefore more flexible, may assist this movement. There is an interesting parallel here to the movement found in *Funaria hygrometrica* [60]. The operculum has a short and stiff curved conical beak, covered when young by a smooth membranous side-split calyptra. The spores are 53–60 × 40–50 μm, coarsely papillose.

CHROMOSOME NUMBER: $n=8$ (NSW).

DISTRIBUTION: TAS, VIC, NSW, ?ACT; also in New Zealand.

ILLUSTRATIONS: Studies, Plate 9; Wilson (1854), Plate 86.

By far the commonest of the Australian species of this genus, *C. pusillum* is liable to confusion with related genera also having spherical capsules, but is easily separated from small *Breutelia* [83] by the smooth, not papillose rhizoids, and from *Philonotis tenuis* [85] most easily by the size of the plant and by the intact stem epidermis.

C. pentastichum (Brid.) Lindb. (*australe* of the Handbook) differs in the much longer stems, of a bright glaucous colour, strongly 5-ranked leaf arrangement, and wide diffuse nerve.

DISTRIBUTION: TAS, ACT; also in S. Africa, S. America and New Zealand.

ILLUSTRATIONS: Handbook, Plate 45.

C. curvirostre (Mitt.) Mitt. is similar to *pusillum* but even smaller. It is distinguished mainly by the fruit which is not grooved and has no peristome, but a long curved beak. The surface cells of the capsule have the walls evenly thickened all round. The spores are smaller: $45-48 \times 32-35$ µm.

DISTRIBUTION: VIC, NSW; also in New Zealand, although not given by the Index.

ILLUSTRATIONS: Brotherus (1924–5), Fig. 400.

Bartramidula pusilla [82] is gymnostomous like *C. curvirostre* but the capsule has unevenly thickened surface cells and tends to be warted rather than smooth, with an almost flat operculum; the nerve stops short of the leaf apex instead of continuing beyond it.

Philonotis Brid.

The cell-tip papillae and the capsules, globular when young, deeply grooved when mature, and with a flattish peristome, evidently relate this genus to *Conostomum*, *Breutelia* etc. from which it differs most obviously in the leaf shape and structure. *Conostomum* differs in the 5-ranked leaf arrangement, and joined peristome teeth, *Breutelia* in the plicate leaves and *Bartramia* in the sheathing leaf bases. Of these genera *Philonotis* and *Breutelia* have the stem with a set of branches emerging from just below the inflorescence (subfloral innovations), and also have an outer stem layer of large, hyaline cells which break down early leaving a curiously rough surface. In Australia *Philonotis* is not nearly as characteristic of bogs, springs, marshes and other permanently wet habitats as it is in other countries.

P. tenuis (Tayl.) Reichdt.

Roadsides, river banks, quarries and channels subject to intermittent flooding are typical habitats of this common species. The tall reddish stems (usually more than 2 cm) are slender, usually well under 1 mm wide except for the spreading leaf tips, and densely matted together into large soft masses by smooth or nearly smooth chocolate-coloured rhizoids. The leaves, c 0·8–1·6 × 0·2–0·5 mm, are narrowly triangular, each tapering to a long fine point made up mostly of excurrent nerve. The margins are usually recurved and strongly denticulate (sometimes with twinned teeth); the leaves spread outwards through the mat of tomentum which usually conceals the stems and are commonly slightly curved and falcate at the stem apex. They are usually a bright pale green or yellowish green with a golden sheen. The upper lamina cells are narrowly rectangular, c 40 × 5 μm, with rounded ends, those in the leaf bases wider and shorter, c 20–30 × 15 μm. The margins, back of nerve and back of upper lamina are (usually at least) papillose–denticulate with projecting cell tips, the uppermost end of each cell standing out to form a low papilla. In addition, all cell surfaces are very finely wrinkled, giving the characteristic pale lustre.

The perigonial bracts, which surround the male inflorescence, are very widely sheathing with a small abrupt spike-like limb, but the inflorescence itself is small and often inconspicuous, not forming a wide disc as in *P. scabrifolia*.

The capsules are on quite long setas, 2 cm or more, persisting for more than one year and eventually overtopped by the branches from below the female inflorescences. The capsules, c 2·5 × 1·5 mm, are plump, asymmetric, globular when young and forming a thick, curved, strongly 16-ribbed cylinder when mature and dry, with a mouth c 1 mm across. The double peristome is flat, sunk below the rim of the mouth when wet. The outer teeth bear strong transverse ribs on the outer face, while the inner peristome is hyaline, also transversely barred but papillose as well, the processes deeply *split down the middle* so that the peristome is divided into lobes, each consisting of the halves of two adjacent teeth. This structure is revealed by the rather short stubby cilia which mark the spaces between processes and which here occur in the middle of the peristome lobes. Because of this structure the openings in outer and inner peristomes coincide instead of alternating. This is like an imitation of the *Funaria hygrometrica* peristome where the teeth and processes are genuinely opposite. The in-and-out hygroscopic movement of the peristome is very slight but, because of the structure, is sufficient to cause narrow slits to appear between the teeth, through which the spores can sift out. The spores are round, c 18–24 μm, brown and densely papillose.

Sainsbury (1952, p. 74) records a form with masses of minute lateral branches which act as propagules, and we have seen the same form in Victoria.

DISTRIBUTION: TAS, WA, SA, VIC, NSW, ACT, QLD, NT, Lord Howe; also in Africa and New Zealand.

ILLUSTRATIONS: Plate 63; Handbook, Plate 46.

There are several mosses of similar size which share the same sort of habitat, the same red stems and pale glistening leaves. Vegetatively *Pohlia wahlenbergii* [72] is distinguished most effectively by the strongly papillose rhizoids and smooth wide leaf cells of a different shape; *Conostomum pusillum* [84], which can on occasion be surprisingly hard to distinguish, sometimes differs in the long terminal cell of the leaf, but more consistently in the intact stem epidermis. *Breutelia* spp. [83] differ in the papillose rhizoids and plicate, much larger leaves where the lamina is *c* 48 or more cells wide on each side of the nerve, compared with *c* 12 cells in *P. tenuis*.

P. pseudomollis (C.Muell.) Jaeg. (NSW, QLD, Lord Howe) is said by Dixon (1942, p. 32) to be "very doubtfully distinct from *P. tenuis*".

P. pyriformis (R.Br.ter.) Wijk & Marg. (=*australis* of the Handbook) is a very similar species but much less common and probably confined to very wet habitats. It differs from *tenuis* in the clearly falcate-secund leaves at the stem apices, rather larger leaves, 1·5–2·0 × 0·4–0·6 mm, longer stems, and plane leaf margins. The nerve is less excurrent than is common in *tenuis* but this is a very variable feature as, indeed, are all the features in which the two species are said to be distinct; in New Zealand these differences appear to be more reliable than in Australia. A much more complete distinction exists in the male plants which have conspicuous terminal discs enclosed by the male bracts.

DISTRIBUTION: TAS (doubtful), VIC; also in New Zealand.

ILLUSTRATIONS: Handbook, Plate 46; Allison and Child (1971), Plate 19 and p. 94.

P. scabrifolia (Hook.f. & Wils.) Braithw.

There are few mosses as easily recognized as this species, with its small dendroid stems and distinctive shade of light bluish green. It is not uncommon on damp earth banks in wet sclerophyll forest or rain-forest in upland regions to 5000 ft (1650 m). The glaucous colour is usually very beautiful and striking, especially in shade, contrasting with the bright brown tomentum at the base of the stem and with the golden colour of the branch tips, but sometimes this moss is predominantly a dirty fawn colour. However, some bluish shoots can always be found. The stems are very slender, 1–3 cm tall,

PLATE 63. *Philonotis tenuis* VIC—Fruiting plant × 7, leaf × 50, cells × 1000

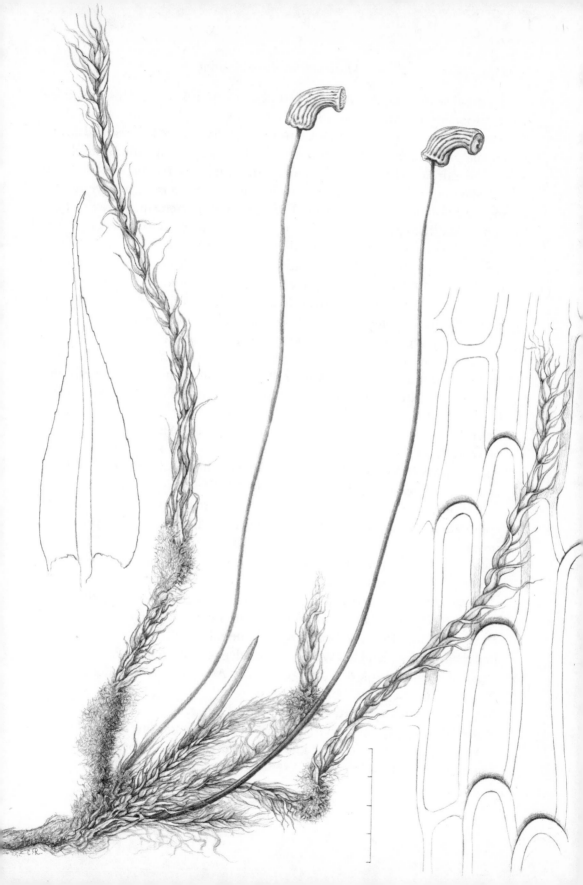

dendroid with 1–2 cm of unbranched stem surmounted by a densely pinnate (occasionally bi-pinnate) portion. The stems are quite white with a thick waxy bloom. The dendroid structure is usually rather less marked in the female plants, and vegetative stems may be branched at ground level, not at all dendroid. The stem leaves, measuring typically 1·4–1·6 × 0·4–0·5 mm and tapering to a fine point, are much larger than the branch leaves which measure only 0·6–0·8 × 0·2 mm. The nerve is excurrent in a short or long denticulate hair-point. The cells are short, rectangular–oval, c 15 × 8 μm, each cell with a median or terminal *high* domed papilla, half the cells having these abaxially and the other half adaxially, the two kinds of cell completely intermixed. Unlike *tenuis* the rhizoids too are papillose. The male bracts are very long (c 3 mm), wide and *golden*, spreading to form a very conspicuous cup enclosing a conspicuous disk of antheridia and paraphyses. The cells of the bracts are much more elongated, papillose at the ends.

The female bracts are similar in length but much narrower and more erect. The capsule structure is basically as in *tenuis* but the inner peristome is almost smooth instead of evidently papillose, and the spores kidney-shaped (12 × 20 μm).

DISTRIBUTION: TAS, WA, SA, VIC, NSW, ACT; also in S. Africa, S. America and New Zealand.

ILLUSTRATIONS: Matteri (1968, p. 200); Handbook, Plate 46; Brotherus (1924–5), Fig. 406; Mueller (1864), Plate 8 (as *Bartramia catenulata*).

The only troublesome forms of this species are those where the glaucous colouration is reduced or lacking but, once this is realized, they present no problems.

One would suspect, from the name alone, that *glaucescens* (Hornsch.) Broth. might be the same species but we have seen no material.

Because of the great plasticity of species in this genus, it is likely that many of the other species in the Australian list are merely synonyms:

P. austro-falcata Broth. & Watts (NSW; endemic).

P. dicranellacea (C.Muell.) Watts & Whitelegge (NSW; endemic).

P. hastata (Dub.) Wijk & Marg. (QLD; S.E. Asia, Africa, S. America).

P. jardinii (Besch.) Par. (Lord Howe).

P. longiseta (Michx.) Britt.

P. pallida (Hampe) Jaeg. (?VIC, QLD); said to occur in New Zealand (Index Muscorum) but we do not know the source of the record which is not mentioned in the Handbook.

P. slateri (Hampe) Jaeg. (NSW, QLD; endemic).
P. tortifolia (C.Muell.) Watts & Whitelegge (NSW, QLD; endemic).

HYPNODENDRACEAE

Hypnodendron (C.Muell.) Lindb.

The big, conspicuously dendroid stems of this genus make these among the most striking mosses in the Australian flora. Erect, unbranched or scarcely branched shoots and flat pinnate fronds are common enough in the field, but the great majority of shoots have a terminal cluster of radiating branches like an umbrella. This genus and its relatives have recently been thoroughly investigated and revised by Touw (1971); we agree with his opinions as far as our material goes, and have used his taxonomic treatment in preference to that in the Index. All Australian material attributed to *arcuatum* (Hedw.) Lindb. ex Mitt. has proved to be either *spininervium* or *vitiense*.

Key to species

1. Stems densely tomentose, more or less throughout. Leaves spreading widely; branch leaves not at all complanate . . . . *comosum*
   Stems tomentose only at base, apart from scattered tufts. Leaves erect or somewhat spreading; branch leaves usually at least partly complanate   2

2. Stem leaves usually green; attachment to the stem U- or V-shaped; leaves spreading from the stem at an angle, especially at the basal corners . . *spininervium*
   Stem leaves colourless; attachment to the stem straight and horizontal; leaves closely appressed to the stem, at least at the basal angles . . . *vitiense*

H. spininervium (Hook.) Jaeg. subsp. *archeri* (Mitt.) Touw

The erect shoots, typically *c* 4 cm tall, tomentose at the base and with a 3 cm wide crown of 6 or so branches, form dense miniature forests in the soft wet soils and rotting logs of stream beds in rain-forest and fern gullies. The stem leaves, 1–2 mm long, are usually green, slightly narrowed at the very base and rather concave in the lower half so that they stand clear of the stem with the tips spreading slightly. The line of insertion on the stem is a shallow V- or U-shape, matching the concave leaf base. The margins below are reflexed. The branch leaves are usually much larger, *c* 1·5–3·0 × 0·5–1·0 mm, complanately arranged and sometimes almost in 3 ranks, with an *upper* row of smaller leaves. The cells are smooth or papillose, long and narrow, *c* 40–60 × 7 μm, shorter and almost square at the angles, and often with rather orange walls but not always forming a conspicuous alar group. The margins and back of the nerve are strongly toothed, sometimes with twinned teeth.

Sexual branches near the centre of the frond replace vegetative branches in the pinnate system; they are scarcely larger than the adjacent branch leaves. The capsules are up to 8 per frond, cylindrical, slightly curved, and deeply grooved, held horizontally on tall red setas.

DISTRIBUTION: TAS, VIC, QLD; also in New Zealand.

ILLUSTRATIONS: Plate 64; Touw (1971), Fig. 20; Brotherus (1924–5), Fig. 384.

H. vitiense Mitt. subsp. *australe* Touw

This species is very similar to *spininervium* but is usually bigger and with a thicker stem; we have found it up to 10 cm tall. It differs principally in the stem leaves which are usually almost colourless, flat, tightly pressed to the stem especially below mid-leaf, inserted on the stem in a straight line, and not narrowed at the base.

CHROMOSOME NUMBER: n=9, 2n=18 (NSW).

DISTRIBUTION: TAS, VIC, NSW, QLD. This subspecies is endemic to Australia but subsp. *vitiense*, which differs slightly in the stem leaf shape and the toothing of branch leaves, is found from S.E. Asia to New Guinea, and in the Pacific Islands; doubtfully recorded from QLD.

ILLUSTRATIONS: Touw (1971), Fig. 23.

This species and the previous one grow intermixed, at least in Victoria, and for this reason, coupled with the small and restricted vegetative differences,

PLATE 64. *Hypnodendron spininervium* VIC—Small fruiting plant × 4 (part of stem removed to reduce size), leaf × 50, cells × 1000. Older fruiting plant, inset, life size

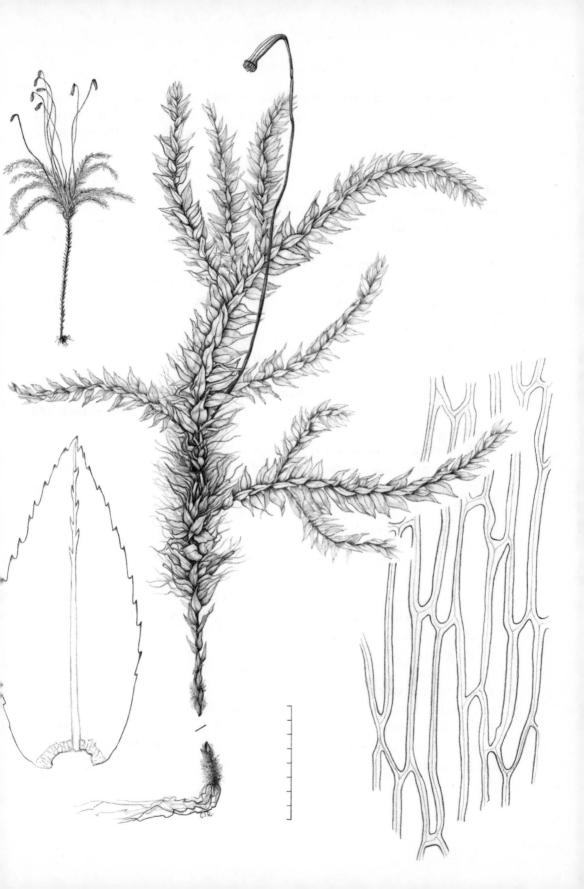

the distinctions between them look unconvincing on paper; but the separation, at least in herbarium specimens, is easy and unambiguous. An interesting problem is how far the two species can be separated ecologically. At present we have no data on this.

H. comosum (Labill.) Mitt. (=*Mniodendron comosum* of the Handbook; *M. dendroides* of the Index)
The dendroid fronds of this species have a yellowish tint which is quite different from the dull green fronds of the previous two species. The erect stems are densely tomentose especially at the base, and have widely spreading narrow, plicate leaves. The branch leaves are similar in shape, plicate, and arranged all round the axis, not at all complanate, with the nerve excurrent in a rather short point.

The dark red-orange, deeply grooved capsules are usually numerous; Touw records up to 35 on a single frond.
DISTRIBUTION: TAS, VIC, NSW; also in New Zealand.
ILLUSTRATIONS: Frontispiece; Handbook, Plate 48; Touw (1971), Fig. 36; Labillardière (1806–7), Fig. 253; Brotherus (1924–5), Fig. 386; Allison and Child (1971), Plate 21.

This is not as common as the previous species but grows in similar habitats and sometimes mixed with it. The tomentum and the shape, plication and arrangement of the branch leaves, and usually the stem leaves, are diagnostic features.

H. comatulum (Broth.) Touw, endemic to QLD, has relatively wide leaves with the nerves not excurrent, and pale capsules.

H. colensoi (Hook.f. & Wils.) Mitt., which is like a small *comosum*, is unique in having crystals in the surface cells at the base of the nerve. It is a New Zealand species doubtfully recorded from TAS and NSW, where it might be worth hunting for.

Braithwaitea sulcata (Hook.) Jaeg. is a magnificent epiphytic moss, one of the handsomest of all, with big, usually bipinnate fronds and short obtuse branch leaves, which give a rounded smooth texture to the branches quite different from *Hypnodendron*. There are sometimes filamentous gemmae arising from rhizoids among the branch leaves.

DISTRIBUTION: NSW, QLD, Lord Howe; also in New Zealand and New Caledonia. There are several records from VIC and even one from TAS, but these seem improbable and require investigation.

ILLUSTRATIONS: Handbook, Plate 47; Touw (1971), Fig. 42; Brotherus (1924–5), Fig. 382.

The record of *Sciadocladus* (=*Hypnodendron menziesii* (Hook.) Par.) for TAS is doubtful according to Touw.

SPIRIDENTACEAE

Includes: *Spiridens muelleri* Hampe (Lord Howe)

RHACOPILACEAE

Includes: *Powellia australis* (Hampe) Broth. (QLD)

Rhacopilum P. Beauv.

For the spelling of this name the same considerations apply as to *Rhacomitrium* [12, q.v.].

R. convolutaceum (C.Muell.) Reichdt. (=*strumiferum* of the Handbook) The prostrate, pinnately branched stems of this common species form dense or open mats on rocks, logs and earth in wet or dry sclerophyll forest, deep green in moister, shadier habitats, and yellowish in drier and sunnier sites. The leaves are arranged in 4 ranks along the stem, 2 lateral with the leaves spreading sideways and 2 dorsal with the leaves alternating almost in a single line. The stems are mostly densely tomentose with tufts of dark brown rhizoids arising near the base of each lateral leaf.

The lateral leaves are oblong–ovate and commonly measure $1 \cdot 0$–$1 \cdot 2$ $\times 0 \cdot 5$–$0 \cdot 6$ mm, with a strong nerve excurrent in a short smooth hair-point

0·2–0·3 mm long. The dorsal leaves are more triangular–ovate, slightly smaller *c* 0·8 × 0·5–0·6 mm, with the nerve excurrent in a long hair-point of 0·4–0·5 mm. When moist, the leaves are spreading and slightly channelled, the margins gently rolled inwards to give a flat U-shape in the upper half of the leaf. When dry, the leaves are curled and somewhat spirally contorted. The cells in mid-leaf are shortly and obliquely rhomboidal near the margins, more isodiametric and irregular away from the margins, *c* 9 μm wide; those below are slightly more rectangular but not greatly different. Most cells are papillose with a single domed papilla; sometimes mamillate with a small nipple on top of the papilla. The margins are usually denticulate throughout by projecting cell ends, especially near the apex, but there is no distinct border.

Sporophytes arise from short specialized side branches from which the seta emerges, 1–2 cm long, flexuous, yellow when young and red when mature, surrounded at the base by a cluster of sheathing perichaetial leaves *c* 1·5 mm long with nerves excurrent in long hair-points and cells larger than in ordinary vegetative leaves. Smooth hair-like paraphyses sometimes project beyond the bracts, sometimes not. The capsule is *c* 2·5 mm long, horizontal, downwardly curved and *deeply grooved*, capped by an operculum with a long beak of *c* 0·5–1·0 mm. The calyptra is membranous and side-split, *c* 3 mm long with a few long, sparse or sometimes dense hairs at the base. There is a double peristome set obliquely at the tip of the capsule so that it is directed almost downwards; the outer teeth *c* 0·4 mm long, very finely and densely cross-striate on the outer face, barred on the inner face, hyaline and papillose in the upper half and split at the slender tips. The inner peristome is united as a smooth membrane for half its height; the other half is papillose and with long cilia in groups of 3, between the processes, with nodules or cross bars. With hygroscopic movement, the outer peristome teeth penetrate past the inner peristome, and on drying out pull the inner peristome open, allowing the spores to fall out. These are green, 12–13 μm, and very finely wrinkled.

CHROMOSOME NUMBER: n = 10, 2n = 20 (QLD).

DISTRIBUTION: TAS, WA, SA, VIC, NSW, ACT, QLD, Lord Howe; also in New Zealand.

ILLUSTRATIONS: Handbook, Plate 49; Wilson (1854), Plate 92 (as *australe*).

This species may be the same as the more widespread, American and South African *R. tomentosum* (Hedw.) Brid., according to Weber (unpub.).

The type description of *strumiferum* (C.Muell., 1851, p. 563) says: "From all related species distinguished at a glance by the big struma of the capsule, extremely hairy calyptra, and hairy perichaetia". The struma (or Adam's Apple) in the capsule neck is a variable feature, sometimes present and sometimes not. The same is true of the hairiness of the calyptra and the projection of the hairs from the perichaetia. There seems no adequate reason for maintaining both species and *convolutaceum* is the earlier name.

R. purpurascens Hampe, which differs from *convolutaceum* in having rather bigger cells in the leaf base, is also most probably conspecific.

R. crinitum Hampe (WA) we do not know.

CRYPHAEACEAE

Includes: *Forsstroemia australis* (C.Muell.) Par. (NSW, QLD)
F. subproducta (C.Muell.) Broth. (NSW, QLD)

Cryphaea Mohr

The Australasian species, with the exception of *tasmanica*, are rather similar plants, with almost buried prostrate primary stems bearing erect slender, fawn-coloured secondary shoots, irregularly Y-branched, forming loose tufts on branches or twigs in humid forest. All species fruit abundantly with capsules almost sessile on specialized short lateral branches, and almost completely hidden by the long perichaetial bracts. There is a double peristome, the outer of wide teeth with transverse bars, the inner of isolated slender processes, keeled up the middle, both sets consisting of 16 segments, and usually papillose. There is a conical operculum and a symmetrical calyptra, split in a frill round the base. The cells are short and pointed. Of the species recorded from our area, *dilatata* and *ovalifolia* are included in *Cyptodon* in the Index. That genus is said to differ in having the fruiting branches arranged on two sides of a shoot instead of all on one side, but that seems to us a growth form of little significance and we have retained the species in *Cryphaea*.

Key to species

1. Leaves tapering to a long point.   2
    Leaves obtuse or with short broad points . . .   3

2. Capsule without annulus, or annulus rudimentary   *exannulata*
    Capsule with distinct well-developed annulus   .   *tenella*

3. Plants aquatic, usually dark green . . . . .   *tasmanica*
    Plants epiphytic, golden when dry . . . . .   *dilatata*

C. exannulata Dix. & Sainsb.

The pale brownish yellow secondary shoots of this plant are very slender, *c* 4–5 cm long but only 0·8 mm wide when dry, rather sparsely pinnately branched with the branches more erect than divergent. The tiny leaves are tightly pressed to the stem, rather deeply and broadly V-keeled, ovate, tapering to a long, quite broad point; 1-nerved to the apex and entire-margined, measuring *c* 1·2–1·4 × 0·4–0·5 mm. The cells are rhomboidal, rectangular or obliquely rectangular, *c* 15–20 × 9 μm, those at the angles in a rather large group square or over-square, with rounded ends and triangular corner thickenings filling in the corners. The immersed capsules have the peristome inserted well below the red rim of the capsule mouth; the outer teeth *c* 150 μm tall, papillose in the upper half. An annulus is lacking but the cells of theca and operculum adjoining the line of dehiscence are large and thick-walled, easily confused with one. The spores are 18–20 μm.

DISTRIBUTION: VIC, NSW, QLD; also in New Zealand.

ILLUSTRATIONS: Plate 65.

This is a rare species of rain-forests, perhaps more widely distributed than is yet known.

C. tenella Hornsch. ex C.Muell. is very similar vegetatively to *exannulata* but has a well developed annulus and peristome teeth almost twice as long. The spores are given as 24–28 μm (Handbook, p. 328).

DISTRIBUTION: TAS(?), NSW(?); also in New Zealand. We have not yet seen genuine Australian material of this species. Apparently it has often been mistakenly recorded.

ILLUSTRATIONS: Handbook, Plate 50; Allison and Child (1971), Plate 22.

PLATE 65. *Cryphaea exannulata* NSW—Fruiting shoot with background impression of rest of plant, × 7, cells × 1000

C. dilatata Hook.f. & Wils. has a more robust appearance than either of the foregoing. The erect shoots are not much longer, 2–3 (up to 6) cm tall, quite densely pinnately branched either in one plane or with all branches secund, pointing towards the same side of the plant. The leaves, however, are ovate, concave and moderately cucullate at the broad obtuse apex, and measure 0·8–1·5 × 0·3–0·8 mm (the Handbook gives them appreciably bigger). They are nerved almost to the apex. The cells in mid-leaf are diamond-shaped, c 20 × 9 μm, those near the margins nearly square, those below longer, to 40 μm; those at the angles in c 6 rows square or over-square, forming a conspicuous alar group.

There is no annulus in this species either.

DISTRIBUTION: SA, VIC, NSW, QLD; also in New Zealand.

ILLUSTRATIONS: Brotherus (1924–5), Figs 490, 491 (as *Cyptodon*); Wilson (1854), Plate 88.

This species differs from *tenella* and *exannulata* in the much less tapering, more concave leaves which give the stems quite a different appearance. It occurs in extensive patches on tree roots and low branches along rocky river-banks.

C. tasmanica Mitt. in Hook.f. & Wils.

Still more robust is this species, very different in appearance from the others except for the numerous immersed capsules on closely clustered reproductive branches. The main shoots are prostrate, to c 13 cm, sparsely branched, the branches prostrate or sometimes erect. The stem is black, with usually very dark green leaves and densely beset with short erect branchlets, mostly fertile, c 2–4 mm long. The stem leaves are c 2·0–2·3 × 1·4–1·6 mm, broadly ovate with rather wide points, slightly plicate and undulate, 1-nerved to the apex. The margins are slightly denticulate with minute teeth, one tooth per marginal cell. In the upper half of the leaf the cells are rhomboidal or rectangular or obliquely rectangular, c 10–13 μm across; those in the lower ¼–½ of the leaf are much longer, to 50 μm except for a broad band of cells at the margins which remain isodiametric and small, and are more or less quadrate, c 14 × 14 μm; those right at the corners are often orange and slightly enlarged to form an alar group.

PLATE 66. *Cryphaea tasmanica* TAS—Fruiting plant (below) × 4 and enlargement of portion with capsules × 15 (above), cells × 1000

The perichaetial bracts are long and finely pointed. The capsules, slightly over 1 mm long, are ribbed and have an operculum with a short fine conical beak. There is a membranous 1-sided calyptra, tightly fitting over the operculum at maturity, and a conspicuous annulus. The peristome is double, very pale; the outer surface of the outer teeth clearly marked out with a double row of cell outlines, the surfaces and margins of the teeth sparsely papillose with high papillae; the inner face with transverse bars. The inner peristome is of isolated processes, almost as high as the outer, joined by a very low basal membrane. Each process is papillose along the edge and is folded lengthwise into a series of slit-like longitudinal pockets, sometimes perforated. The spores are finely papillose, green and *c* 18 μm across.

DISTRIBUTION: TAS, VIC, ACT; also in New Zealand.

ILLUSTRATIONS: Plate 66; Handbook, Plate 50; Brotherus (1924–5), Fig. 490 (as *Dendrocryphaea*); Wilson (1859), Plate 175.

This is an aquatic moss on stones in forest creeks and is likely to be more widespread than has been recorded, since it is often submerged.

C. ovalifolia (C.Muell.) Jaeg., we have never encountered.

HEDWIGIACEAE

Hedwigia P.Beauv.

Outcrops and boulders of dry acid rock, especially in upland regions, are the usual habitats for both species of this genus in Australia. The distinction between them morphologically is obvious—*ciliata* has conspicuous hyaline hair-points on the mature leaves and *integrifolia* has not—but ecologically the differences are subtle. The pattern of distribution of *integrifolia* seems to involve protection from the direct rays of mid-day sun, but a full investigation is needed since the two are often found intermixed.

The species are common enough and are characterized by broad, very opaque, solid-looking leaves with intensely papillose cells but no nerves, and with almost immersed gymnostomous capsules.

PLATE 67. *Hedwigia ciliata* VIC—Fruiting shoot, dry (left) and moist (right), ×11, cells ×1000

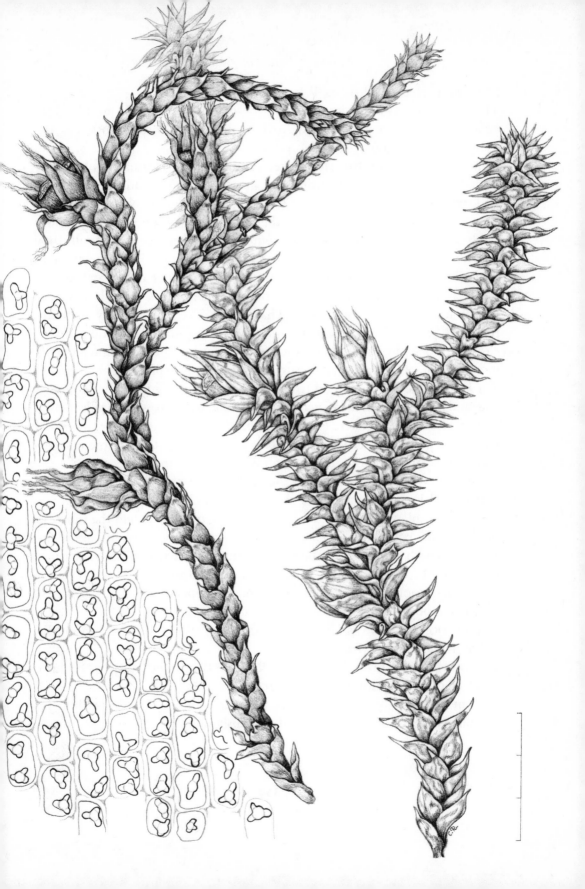

H. ciliata (Hedw.) Ehrh. ex P.Beauv.

The shoots are 4–6 cm long, dark grey-green below, paler and slightly yellow in the top 1 cm of growth. Branching tends to be sympodial, the stems ending in capsule production and being replaced by 1 or 2 side branches from just below. They form loose green cushions or turfs, usually quite hoary with the hyaline leaf tips when dry. The shoots are narrow and wiry, *c* 0·6 mm wide when dry, with very densely imbricate leaves completely obscuring the stem and appressed to it, with only the hyaline hair-points spreading; more rarely the leaves are spreading even when dry. When moistened out, they over-expand and then relax back, as in *Rhacomitrium*. The ovate leaves typically measure 1·4 × 0·8 mm and are shortly decurrent at the basal corners and hyaline at the tip (and sometimes over all the upper ¼ of the leaf) which is contracted to a short, rather broad hair-point. The margins may be gently recurved below. The cells in mid-leaf are in regular longitudinal rows, *c* 10 µm across, squarish, thick-walled, with a single multiple-pointed papilla in the middle of the cell, on each surface, slightly more prominent abaxially. The cells in the leaf base tend to become rectangular with rather rounded ends and bulging sides, thick-walled and porose, with a row of papillae down the centre of the cell. There is no nerve but the cells in the middle of the leaf base at the point of insertion on the stem are yellowish, those at the margins quadrate, scarcely yellow, forming an indistinct group.

There is virtually no seta, and the capsules are completely enclosed by perichaetial leaves which *spread* when *dry* and close in over the gymnostomous capsule when moist, thereby replacing to some extent the function of the missing peristome. The perichaetial bracts are large, hyaline at the tip, and *ciliate* with marginal hyaline multicellular hairs which are sometimes branched. The capsule is cup-shaped, *c* 1·5 mm tall and across, nearly hemispherical, smooth, with a wide red rim. The operculum is low-convex, retained in position as a lid by the enclosing bracts.

DISTRIBUTION: TAS, WA, SA, VIC, NSW, ACT, QLD; also in most regions of the world including New Zealand.

ILLUSTRATIONS: Plate 67; Handbook, Plate 51; Brotherus (1924–5), Fig. 481 (as *H. albicans*); Watson (1955), Fig. 106.

At first glance this could easily be mistaken for a *Grimmia* [11] but the complete absence of nerve readily distinguishes it.

H. integrifolia P.Beauv. (*Hedwigidium* of the Index)
The most immediate difference from the preceding species lies in the absence
of hair-points on the leaves. The shoots of this species are pale green above
and dark brownish-green below, but the green is darker and the brown
blacker than in *ciliata*, the leaves tend to be plicate, and there are no hyaline
points. The stem extends to 6 cm or more, of which the terminal 1 cm is
current growth. The plant branches throughout, as well as from below the
capsules, more commonly than in *ciliata* so that the sympodial growth form
is less striking. Some branches are tapering to form very slender, rhizoidal
shoots with tiny leaves (*flagella*). The nerveless leaves are appressed when dry,
spreading when moist, but do not have the feature of over-expansion when
moistened as in *H. ciliata* and *Rhacomitrium*. The leaves are oblong or ovate–
oblong, 1·4–1·6 × 0·6–0·8 mm, the upper third triangular in outline, tapering
to a short point which, on the flagellar shoots, is prolonged in a multicellular
hyaline hair on young leaves, soon breaking off. The leaves are concave
throughout, with the margins rolled back along almost the whole length
from base to apex, wet and dry, giving the stems, especially when dry, a very
neat clean-cut appearance. The margins are entire, or faintly denticulate near
the apex, slightly decurrent down the stem beyond the rest of the leaf
insertion.

The cells near the apex are mainly shortly rectangular, with thick wavy
walls; those below more elongated; those in the leaf-base (mid) long and
narrow and porose. The outermost cells are roughly isodiametric, right to
the base of the leaf. The cells, in longitudinal rows, are sometimes smooth,
but are commonly densely papillose with many single, low papillae per cell,
becoming uniseriate on the cells of the leaf-base; the alar cells are conspicuous orange, and smooth.

The capsule measures *c* 1·5 mm long × 1·2 mm broad, deeply grooved,
slightly contracted to the mouth which is orange-brown and gymnostomous, far overtopped by the perichaetial leaves, but less completely enclosed
by them than in *ciliata*, with 1 side exposed. The perichaetial leaves are not
ciliate.

DISTRIBUTION: TAS, WA, SA, VIC, NSW, ACT; almost as widespread as *ciliata*.
ILLUSTRATIONS: Plate 68; Handbook, Plate 51.

The species is often separated from the previous one as *Hedwigidium
integrifolium* (P.Beauv.) Dix. on the grounds of grooved capsule, and different
leaf papillae. There are a number of other differences in both capsule and

355

leaves but it is a matter of opinion whether to treat the two genera as separate. We have chosen not to recognize *Hedwigidium* at the generic level, but without any great conviction that this is the right course. The nerveless, opaque papillose leaves will separate the species at once from any others, such as *Rhacomitrium crispulum* [12] or *Orthotrichum rupestre* [53] which it might otherwise resemble. The slender flagelliform shoots are distinctive and always seem to be present.

Rhacocarpus Lindb.

R. purpurascens (Brid.) Par. (=*humboldtii* of the Handbook)
The smooth stiffly pointed shoots of this montane species could easily be mistaken for a small flowering plant of the Epacridaceae. The shoots, to *c* 5 cm, are irregularly pinnately branched, forming loose mats on wet rocks in montane and subalpine regions, the leaves tightly imbricate when dry, little altered when moist, and having a solid texture more reminiscent of a flowering plant than a moss. The leaves are a very opaque greyish yellow, 1·2–1·6 × 0·4–0·6 mm, oblong, almost parallel-sided below and sometimes slightly waisted in the middle, the upper ⅓ triangular or round in outline, with the margins rolled in to form almost a conical tube and the apex prolonged into a hair-point which is sometimes long. Both hair-point and leaf edges tend to be dark and reddish in incident light. The margins are slightly decurrent and recurved below, slightly denticulate above.

The upper cells are roughly rhomboidal, *c* 30–40 × 7–8 μm, those below larger, 45–80 μm or more long × 7 μm wide. All cells have rather thick *porose* walls and the whole surface is finely and intensely wrinkled, sometimes to the extent of obscuring the cell outlines, the ridges in extreme cases joined up to form a reticulum all over the surface. The cells do not seem to be truly papillose although electroscan investigation is needed to be certain. The alar cells are conspicuous, shorter and with very thick porose walls and the lumen often crossed by transverse bars of thickening. All across the base of the leaf the cells are orange-brown and smooth. The marginal cells, for

PLATE 68. *Hedwigia integrifolia* VIC—Fruiting shoot dry (right) and wet (left), detail of fruit (below) × 11, cells × 1000

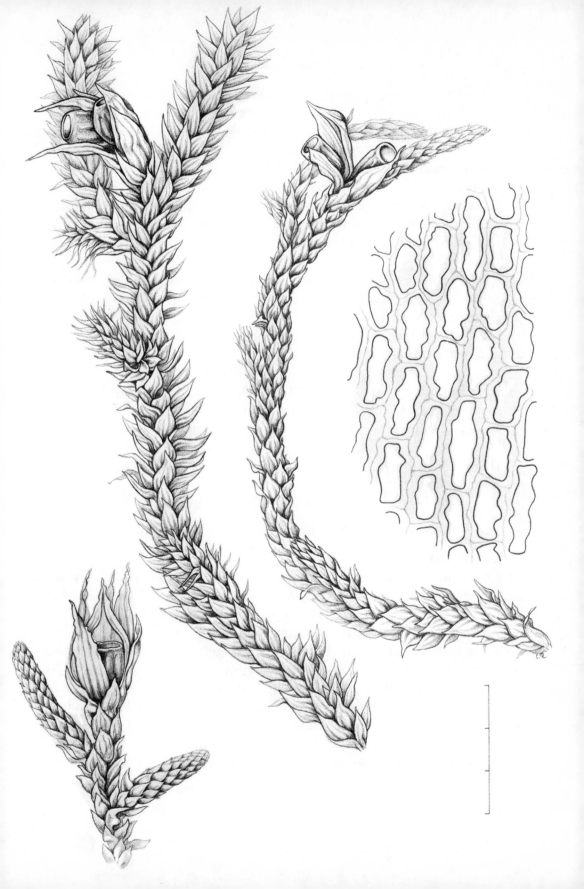

several rows extending to about mid-leaf, are also smooth and extra thick-walled, forming a moderately distinct border.

The capsule is wide, grooved and gymnostomous, exserted on a seta *c* 0·5–1·0 (–2·0) cm long; the operculum with a long oblique beak. The plants are dioicous with short bud-like lateral male branches which are *blunt*, in contrast to the spear-pointed vegetative branches.

DISTRIBUTION: TAS, WA, VIC, NSW, ACT; also in Africa, America, and New Zealand (where it is much commoner than in Australia).

ILLUSTRATIONS: Plate 69; Handbook, Plate 49; Brotherus (1924–5), Fig. 486 (as *australis*).

The name *purpurascens* takes priority over *humboldtii* provided the synonymy suggested in the Index is correct. There is no moss with which this species could easily be confused, except perhaps *Hedwigia* [89] from which it is distinct vegetatively in the bordered leaves with orange alar cells and, when fruiting, in the long seta.

The endemic *R. webbianus* (C.Muell.) Par., a rare dark-green plant of very wet rocks in WA, differs in having neither hair-points nor porose cells; also the alar cells are thin-walled and mostly colourless.

PTYCHOMNIACEAE

Ptychomnion (Hook.f. & Wils.) Mitt.

P. aciculare (Brid.) Mitt.

One of the most striking and robust mosses of fern gullies and rain-forest, this species grows on the ground or rotting logs, sometimes as an epiphyte.

The red, stiff stems are 5–10 cm long, forming shoots *c* 0·5 cm in diameter with a greenish or straw-coloured growing region of 2–3 cm; branched rather sparsely and irregularly to form open, springy cushions or wefts of an unmistakable, harsh woolly texture, caused by the crisp, papery leaves spreading stiffly out from the stem, even at the apex. The leaves are 3·0–3·5 × 1·5–2·0 mm, oval or almost circular, quickly contracted to a toothed, narrow, ribbon-like point which is slightly corkscrewed; the margins

PLATE 69. *Rhacocarpus purpurascens* VIC—Shoot × 11, leaf × 50, cells × 1000

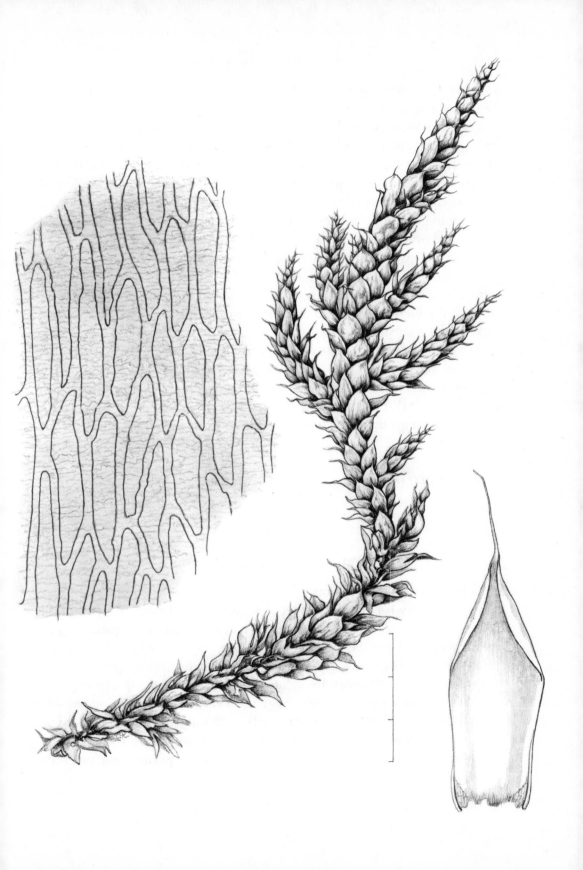

variably toothed in the upper half of the leaf. The nerve is short and double, or almost absent; the leaf yellow and plicate at the insertion. The plications seem to correspond with changes in cell width. Cells, almost throughout, are oblong, very thick-walled and conspicuously porose; those in mid-leaf $30-50 \times c \, 7 \, \mu m$; those above rather smaller, $20-30 \, \mu m$ long; those at the base more elongated, up to $90 \, \mu m$. Marginal cells in the lower half of the leaf are narrower, outside several rows of wider cells, but not differing in colour and not forming a definite border. Cells at the angles are wider, rounder, thick-walled and orange, forming a small group in most cases.

The capsule is deeply 8-grooved when dry, curved downwards, and held horizontally on the end of a blackish, stiff seta c 2–4 cm long; operculum with a long slender beak; calyptra side-split, with a long point. The peristome is double with standard double peristome movement.

CHROMOSOME NUMBER: $n=7$ (NSW, TAS).

DISTRIBUTION: TAS, SA, VIC, NSW, QLD, Lord Howe, apparently not recorded from ACT and WA; also in S. America, New Zealand and Oceania.

ILLUSTRATIONS: Handbook, Plate 53; Brotherus (1924–5), Fig. 514; Allison and Child (1971), Plate 23.

This is perhaps the most unmistakable moss in the entire flora; there is nothing like it.

P. leichhardtii (Hampe) Jaeg. we have never seen.

Glyphothecium Hampe

G. sciuroides (Hook.) Hampe

Tufts of soft, fawn-coloured, rather stringy, scarcely branched stems of this species are not uncommon epiphytes in rain-forest and fern gullies. Commonly 2–3 cm long, they can be much more, especially in a really humid micro-climate. The leaves are very densely imbricate, half-appressed (c 30° to stem) and mixed with short, filamentous, green gemmae, 5–7 cells long (but we have recorded up to 23 cells long) and very narrow, tapering leaf-like paraphyllia 1–5 cells wide at or near the base. Both gemmae and paraphyllia tend to be completely concealed by the leaves which are ovate, $2 \cdot 3–3 \cdot 0 \times 0 \cdot 8–1 \cdot 2$ mm, tapering to a short fine point, variably plicate, rather concave or even channelled and with plane margins often reflexed at the base, denticulate in the upper ½ or less. The nerve is usually virtually absent, or very short and double; cells in mid-leaf are roughly rectangular, 30–50 × 7–10 µm, very thick-walled and porose, with pores especially at the ends of the cells, connecting to diagonally adjacent cells on right and left. This gives a curious and rather characteristic pattern of wall thickenings in the leaf. The cells above are rather shorter, those below longer, a broad band at the basal margin very thick-walled and almost isodiametric, or slightly and obliquely elongated, forming a conspicuous alar group with dark cell contents.

The capsules, 2 × 0·8 mm on a seta *c* 4 mm long, are widest at the base, *c* 0·5 mm across the mouth, oblong–cylindrical and strongly ridged. Our fruiting material is not in good condition but the peristomes on it are double, of hyaline teeth outside and the inner reduced to almost a membrane. The green, papillose spores measure 15–20 µm.

CHROMOSOME NUMBER: n=7 (TAS).

DISTRIBUTION: TAS, VIC, NSW, ACT; also in S.E. Asia, and New Zealand.

ILLUSTRATIONS: Handbook, Plate 54; Hooker, W. J. (1818–20), II, Plate 175 (as *Leskea*).

Although not a conspicuous moss, this species presents little difficulty and, once known, can be recognized easily with the naked eye.

Hampeella C.Muell.

H. pallens (Lac.) Fleisch.

Although not usually seen fruiting, the glossy golden yellow shoots of this epiphytic species have a distinctive growth form which is easily recognizable even vegetatively.

The sparse, spreading prostrate stems form a small open mat on tree bark, bearing broad flattened branches, densely tomentose at the base with dark brown rhizoids, and curving outwards and *upwards* like hooks. The leaves on the undersides of these spread to form a flat frond, apparently complanate from below (a common viewpoint since it grows midway up tree trunks), but with leaves on the upper side curved to point upwards. The leaves on these characteristic branches are densely imbricate, 1·7–1·9 × 0·4–0·7 mm, oblong or lanceolate with a rather broad but sharp point, sometimes asymmetric and curved slightly to one side at the base and apex. The nerve is almost absent or very short and faint, single or double, and the margins are incurved in the upper ½ of the leaf to give an almost tubular point. The prostrate stems have the leaves spreading evenly and much narrower, symmetrical and scarcely tubular. In the axils of the uppermost branch leaves are dense clusters of *green* filamentous gemmae, 9–15 cells long. The cells in mid-leaf are spindle-shaped, typically 90–110×7–9 μm, with moderately thick walls; those in leaf base and apex slightly shorter and wider; those at the angles and extreme base of the leaf quadrate, thick-walled and slightly porose but not conspicuous.

Sporophytes are known to us only from QLD. The Handbook (p. 342) describes the capsules appropriately as "usually suberect, oblong or nearly cylindrical, sharply 8-ribbed".

DISTRIBUTION: VIC, NSW, QLD; also in S.E. Asia and New Zealand.

ILLUSTRATIONS: Handbook, Plate 54; Brotherus (1924–5), Fig. 510.

This predominantly warm-temperate species is rare in the rain-forest gullies of Victoria but seems not uncommon in coastal forest of NSW. The gemmae are almost diagnostic when observed. The difficulty with this species is not so much separating it from similar species, but of identifying it in the first place.

H. alaris (Dix. & Sainsb.) Sainsb., although omitted from the Index, is recorded by Sainsbury (Handbook, p. 345) from TAS. It differs from *pallens* in the hooked leaf tips and conspicuous coloured alar cells, but the relationship between the two species is in need of investigation.

Tetraphidopsis Broth. & Dix.

T. pusilla (Hook.f. & Wils.) Dix.

This charming little plant was thought until recently to be endemic to New Zealand, but it was discovered in the Otway district of Victoria by Prof. D. J.

PLATE 70. *Tetraphidopsis pusilla* VIC—Fruiting female shoot (left) and male shoot with antheridial branches on stem and cluster of gemmae at stem apex ×15, leaf ×50, cells ×1000

Carr, and has since turned up also in Gippsland, Eastern Victoria. The characteristic habitat is on old, dead lianes or small twigs in very humid, dark stream gullies. The stems are very slender, 1–3 cm long, several springing from a single spot in a loose open bunch, each stem ending in a tiny round ball of green gemmae which consist of 4–5 bright green cells in a row; presumably these are shed and grow to produce a new protonema. The leaves are very narrowly triangular, rather widely separated, slightly twisted and shrivelled when dry; the margins entire or faintly crenulate, sometimes partially recurved. There is a single faint nerve, extending to mid-leaf or less. The cells throughout are narrowly rhomboidal, c 35×7 μm, tapering and rounded at both ends, sometimes slightly sigmoid, the walls quite thick; the cells at the angles are short and roughly square with yellowish thick walls, forming a distinct alar group.

Male plants have small (up to 1 mm) bud-like male branches along the main stem. Female plants have larger and more slender reproductive branches, up to 1·5 mm long, in similar positions. The seta is 1·5–2·0 mm long; the capsule narrowly cylindrical, erect and symmetrical, c 1 mm long and strongly ribbed to the point of being almost winged. The double peristome is white; the outer teeth triangular, the inner processes hair-like, inflexed over the capsule opening.

DISTRIBUTION: VIC; also in New Zealand.

ILLUSTRATIONS: Plate 70; Handbook, Plate 52; Wilson (1854), Plate 88 (as *Meteorium*).

Once the gemma-heads are noticed, this species is unmistakable. It is extremely inconspicuous and might well turn up in TAS or NSW.

LEPYRODONTACEAE

Lepyrodon Hampe

L. lagurus (Hook.) Mitt.

This is a rare plant of upland regions, quite large but unobtrusive. The shoots, 1–3 cm long, matted together by conspicuous chocolate-coloured tomentum, are straw-coloured except for the glistening pale green fresh growth of the upper 1 cm. The densely imbricate leaves, very glossy and rather crumpled when dry, are nearly oval, 2–3 mm long × 1 mm wide, concave and abruptly contracted to rather a long fine hair-like point, but the nerve is short and double or almost absent. The cells are long and narrow, slightly sinuose, rather thick-walled and clearly porose, especially near the ends of the cells; those at the leaf base rather shorter and wider.

We have seen no sporophytes on Australian material.

DISTRIBUTION: TAS, VIC; also in New Zealand, Antarctica and S. America.

ILLUSTRATIONS: Brotherus (1924–5), Fig. 515; Allison and Child (1971), p. 106.

There is a resemblance here to *Leptostomum* [69] in size, habit and colouration, but the long thick-walled cells are quite different. The combination of hair-point with no nerve will separate this species from any other of the same appearance. It is very uncommon, on logs or rocks in forest, and is probably mainly a plant of mountainous areas.

CYRTOPODACEAE

Includes: *Cyrtopus setosus* (Hedw.) Hook.f. (TAS, QLD)
Bescherellia brevifolia Hampe (NSW, QLD)

TRACHYPODACEAE

Includes: *Trachypus humilis* Lindb. (QLD)

MYURIACEAE

Includes: *Myurium rufescens* (Reinw. & Hornsch.) Fleisch. (QLD)

PTEROBRYACEAE

Includes: *Pterobryon australiense* Dix. (?NSW)
    P. humile Mitt. (NSW)
    Endotrichella dietrichiae C.Muell. (NSW, QLD)
    E. lepida C.Muell. (NSW, QLD)
    E. subelegans (Broth.) Broth.
    Euptychium mucronatum Hampe (Lord Howe)
    E. robustum Hampe (Lord Howe)
    E. setigerum (Sull.) Broth. (NSW, QLD)
    Garovaglia longicuspis Broth. (QLD)
    G. muelleri (Hampe) Mitt. (QLD)
    Muellerobryum whiteleggei (Broth.) Fleisch. (NSW, QLD)
    Pterobryella praenitens (Hampe) C.Muell. (Lord Howe)
    Pterobrydium australe Broth. & Watts (QLD)
    Pterobryopis filigera Broth. & Watts (QLD)

Trachyloma Brid.

Trachyloma planifolium (Hedw.) Brid.

An uncommon epiphyte, confined to tree trunks in rain-forests, this species has large dendroid, glossy green fronds which are roughly pinnately branched in one plane at the end of a simple, black smooth stem arising from a rhizoidal primary stem. There is usually a completely characteristic silver sheen on old fronds, caused by dead or dying leaves and on many plants there are masses of long chocolate-brown filamentous gemmae partly concealed by the leaves at the stem tips. The leaves are nerveless or faintly double-nerved, complanately flattened in the plane of the frond, and, when old, become curiously split into long narrow strips by lines of dead, brown-walled cells. The margins are strongly toothed above. The cells are long and narrow, *c* 80–100 × 8–10 µm, slightly thickened and porose, especially towards the

PLATE 71. *Trachyloma planifolium* VIC—Fruiting plant (left) with portion of stem removed, shoot tip (inset) showing gemmae at apex × 4, cells × 1000

leaf base; shorter and increasingly thick-walled and porose towards the angles but without distinct alar cells.

Sporophytes are not common in VIC, but are more plentiful in NSW and QLD. The seta is 1–2 cm long; the capsule long and narrow with a long double peristome of white, hair-like teeth.

CHROMOSOME NUMBER: n=8 (NSW).

DISTRIBUTION: TAS, VIC, NSW, extending into QLD; also in New Zealand and possibly much more widespread, extending to Oceania and S.E. Asia if it is conspecific with *T. indicum*, as the Handbook suggests.

ILLUSTRATIONS: Plate 71; Handbook, Plate 55; Brotherus (1924–5), Fig. 530.

This species, with its complanate branches and leaves, could be mistaken in the field only for some forms of *Hypnodendron* [86] and then only when its silvery sheen was lacking. But the strong nerve of *Hypnodendron* will at once separate the two microscopically.

T. diversinerve Hampe in F.Muell. (VIC) is very similar if not conspecific.

We do not know:
T. leptopyxis C.Muell. (NSW)
T. pycnoblastum C.Muell. (QLD)
T. wattsii Broth. (Lord Howe)

Rhabdodontium buftonii (Broth. & Geh.) Broth.

This exceedingly rare species is endemic to Tasmania. It has a long sparsely pinnate stem, with short and few side branches. The closely imbricate leaves are broadly ovate and concave with a short acute apex and entire margins. The cells in mid-leaf are *c* 50–70 × 7 μm but only about 2 × 1 at the leaf apex; those at the leaf base are not conspicuously porose. The capsules are sessile or almost so.

DISTRIBUTION: TAS; endemic.

ILLUSTRATIONS: Brotherus (1924–5), Fig. 538.

This species has not been found for many years and our description is based on material in Herb. BM. A re-discovery is much to be hoped for. The growth form is that of *Trachyloma* and the leaf shape rather like *Lembophyllum* [103].

METEORIACEAE

Includes: *Meteorium baileyi* (Broth.) Broth. (QLD)

> *M. compressum* Mitt. (QLD)
> *M. miquelianum* (C.Muell.) Fleisch. (QLD)
> *Aerobryopsis longissima* (Doz. & Molk.) Fleisch. (QLD)
> *Barbella enervis* (Thwait. & Mitt.) Fleisch. (NSW, QLD, Lord Howe)
> *B. perpinnata* (Broth.) Broth. (QLD)
> *Chrysocladium phaeum* (Mitt.) Fleisch.
> *Floribundaria floribunda* (Doz. & Molk.) Fleisch. (QLD)
> *Meteoriopsis reclinata* (C.Muell.) Fleisch. (QLD)
> *Pilotrichella conferta* Ren. & Card.
> *P. dimorpha* (C.Muell.) Jaeg. (NSW, QLD)
> *P. recurvula* C.Muell.

Papillaria (C.Muell.) C.Muell.

Of all the genera of Meteoriaceae this is by far the largest in Australia with 9 species recorded of which at least 5 extend to the temperate regions of southern Australia. The plants are pendulous, tend to be branched with short branches at right angles, and differ from *Weymouthia* [98] partly in the leaf shape which tapers to a point from a broad, heart-shaped base without alar cells, and in the papillose cells. Capsules are not common, probably because the whole sporophyte tends to drop out of the vaginula when old. Like the calyptra, the vaginula is densely beset with long, conspicuous hairs which persist after the capsule has gone.

Key to species

1. Leaves undulate or longitudinally grooved . . 2
   Leaves, when dry, smoothly curved abaxially, neither plicate nor undulate 4

2. Plants soft and small (texture of *Weymouthia mollis*). Cells clear, scarcely papillose . . . . . . *nitens*
   Plants larger and much stiffer in texture. Cells obscured by dense papillae 3

3. Leaf border narrow or absent; apical points of
leaves never forked . . . . *crocea*
Leaf border of small cells very wide, especially at
the leaf base; apical leaves of the most slender
shoots ending in long fine *forked* points . . . *flavolimbata*

4. Leaves narrow, flattish, tapering to long fine points *flexicaulis*
Leaves wide and concave, oval, with short wide
points . . . . *amblyacis*

P. nitens (Hook.f. & Wils.) Sainsb.

The shoots here are very soft and slender, dull yellowish green, little more
than *c* 0·3 mm diameter *including* leaves, which are scarcely imbricate. The
leaves are erect, appressed, 1·2–2·0 × 0·5–0·6 mm, with a wide base contracted quickly to a very long tapering flexuose point which is channelled
below, as in *flavolimbata*, by the margins bending backwards abaxially even
when moist. The margins are usually denticulate and the nerve is inconspicuous, ending about mid-leaf. The cells are not at all obscured, with only
1–2 minute spiculose papillae per cell. We have not seen capsules.

DISTRIBUTION: TAS, VIC, NSW, QLD; also in New Zealand.

ILLUSTRATIONS: Studies, Plate 10 (as *P. nitidiuscula* and *Meteorium nitens*).

The very soft texture, narrow stems, leaf shape (wide base and long
narrow subula) and sparse papillae amply distinguish this species from the
others. It is inconspicuous and probably overlooked but is certainly rare.
P. fulva (Mitt.) Jaeg. is probably the same thing.

P. crocea (Hampe) Jaeg.

The trailing, yellow-brown, olive-green or dark green shoots of this species
are *c* 0·5–0·6 mm diameter, with densely imbricate opaque leaves, sometimes clearly arranged in 5 longitudinal ranks down the stem, evident even
when the leaves are dry and erect, lying along the stem, but especially conspicuous when they are moist and spreading widely. There is a strong single
nerve reaching nearly to the apex, prominent abaxially and emphasized
by a deep wide fold on either side which gives the leaf a W-shape in section
across the middle; the leaf flattens out only at the very tip. Both this folding
and the slight undulation of the margins persist even when moist. The leaves,
which measure typically 0·9–1·2 × 0·6–0·7 mm, are ovate–triangular with a
short wide point. The upper cells are rounded–rhomboidal, *c* 10 × 6 μm,

densely pluri-papillose to the point of obscuring the cell outlines, with 1–3 marginal rows of cells smooth and thicker-walled, forming a narrow clear border. The cells below, near the nerve, are larger and narrower, becoming smooth and therefore pale in the middle of the leaf base, where the marginal cells remain short and papillose, as above, forming a wide decurrent lobe (auricle).

Female branches are short, *c* 3 × 1 mm, conspicuous, pale golden-green, mainly because of the hairs which extend beyond the bracts forming a dense brush round the base of the seta. The hairs are flexuose, many cells long and up to 3 or so cells wide in the middle, rather like narrow paraphyllia. The seta is 3–4 mm long and the calyptra densely hairy matching the vaginula.
DISTRIBUTION: TAS, VIC, NSW, QLD; also in New Zealand, Oceania and S.E. Asia.
ILLUSTRATIONS: Handbook, Plate 56.

It is rare in VIC (Willis, 1955a, p. 162) but common enough in NSW coastal rain-forest.

P. flavolimbata (C.Muell. & Hampe) Jaeg.
Perhaps the commonest species of the genus, in our area, this is also the most picturesque, with robust bright yellow and brown trailing shoots. These shoots measure about 0·5–1·0 mm across, excluding the spreading leaf tips, but there are also very slender shoots, *c* 0·2 mm diameter, which tend to be greenish. All shoots taper conspicuously. The main leaves, 1·5–2·0 × 1 mm wide, are similar in shape to those of *crocea*. Although the nerve is strong and conspicuous abaxially, the leaf surface is not grooved or ridged as in *crocea* but the margins are undulate and bent back abaxially in mid-leaf (as in *nitens*), sometimes so strongly that they meet, giving a rather similar effect. On the very narrow shoots the leaves end in a slender *forked* point, spiky with papillae, and sometimes multiply branched. The cells are similar to those of *crocea* but with the margin of smooth cells very wide below, extending nearly or fully across the leaf-base to join the large patch of similar cells adjacent to the nerve. The margins are usually toothed near the base.

The hairs on the vaginula are much less prominently displayed than in *crocea*.
CHROMOSOME NUMBER: n = 11 (NSW, TAS).
DISTRIBUTION: TAS, VIC, NSW; also in New Zealand. Commonest in the south.

ILLUSTRATIONS: Handbook, Plate 56; Wilson (1859), Plate 175 (as *Meteorium cerinum*); Allison and Child (1971), Plate 25, p. 109.

The forked tips, which are usually only to be seen on the very narrow shoots, seem to be completely diagnostic. Even without them the strongly swept-back margins are sufficient to identify the species.

P. flexicaulis (Wils.) Jaeg.

The stems are about as slender as in *nitens*, dull green but *stiff*, not soft as in *nitens*. The leaves are triangular–ovate, 1·0–1·5 × 0·6 mm, with auricles at the base, *very widely* spreading when wet, sometimes at right angles to the stem, very tightly wrapped round the stem when dry, sheathing it. The nerve, which reaches to ¾ of the leaf, is not at all conspicuous abaxially. The cells above are oval–rhomboidal and densely papillose; those in the middle of the leaf base only slightly longer, still rhomboidal but scarcely papillose, the marginal cells more or less papillose throughout (although the Handbook p. 356 does speak of a pale border). The margins are not toothed but are very finely crenulate with papillae.

CHROMOSOME NUMBER: n = 11 (NSW).

DISTRIBUTION: TAS, NSW, QLD.

ILLUSTRATIONS: Wilson (1859), Plate 175 (as *Meteorium filipendulum*).

Material of this species in Herb. MEL has been misidentified by both Carl Müller and Bartram as *Cryphaea tenella* [88], which superficially resembles it but has smooth cells.

P. amblyacis (C.Muell.) Jaeg.

The robust bright yellowish and brownish green shoots, *c* 6 mm diameter, and *concave* leaves give this species quite a different appearance. The leaves, *c* 1·2 × 0·7 mm, are oblong with almost parallel sides, and a short wide point, densely imbricate, tightly appressed to the stem when dry, and not very different when moist, with only the short points spreading. There is a nerve extending to about ¾, and rather wide and conspicuous auricles, with a toothed margin. The upper cells are rhomboidal (the walls having very strong and irregular thickenings) with dense low papillae but not as obscure as in *crocea*. There is a patch of smooth or almost smooth elongated cells in middle leaf-base and 1–2 rows of cells form a smooth border just above the auricles. The cells in the auricles themselves are only sparsely papillose, and not as different from those of *flavolimbata* as published accounts might suggest. We have seen no capsules.

CHROMOSOME NUMBER: n=11,22 (NSW).

DISTRIBUTION: NSW, QLD; also in New Zealand and Oceania.

ILLUSTRATIONS: Brotherus (1924–5), Fig. 555.

This is a rare species in our area.

Other species recorded by the Index are:

P. eavesiana (Hampe) Jaeg. (QLD, endemic).

P. reginae (Hampe) Jaeg. (QLD, endemic).

P. squamata (Hampe) Hampe (endemic).

Weymouthia Broth.

Together with *Papillaria*, this genus forms a temperate zone outlier of the tropical family Meteoriaceae, characteristically rain-forest plants which hang in festoons from branches and twigs. The pendulous stems of both genera are pinnately branched with branches almost at right angles to the main stem and often tapering out to slender ("flagellate") tips. *Weymouthia* differs from *Papillaria* in having smooth instead of papillose cells, and very rounded, obtuse, concave leaves instead of pointed. There are only 3 species in the genus. The two Australian species are dioicous and capsules are uncommon.

They can be distinguished very readily by appearance, and by the following characters:

Plant firm. Leaves almost circular in outline, or wider
than long, very concave, wrinkled when dry . .   *cochlearifolia*

Plant very soft. Leaves oblong, longer than wide,
predominantly U-channelled rather than concave-
hemispherical, not wrinkled . . . . . .   *mollis*

W. cochlearifolia (Schwaegr.) Dix.

This rather coarse plant tends to be found on trunks and branches rather than twigs, except in extremely humid forest. It may also persist for many years on the ground after falling and can then be difficult to separate from other ground mosses.

The stems are rather irregularly branched with very concave, almost hemispherical leaves, tightly imbricate and usually wrinkled but rather spreading at the stem tips and not forming a spear point. At the branch tips, the leaves may be much smaller giving tapering, flagellate branches. The branch leaves tend to be narrower than those on the main stem, but all are very obtuse and very wide, 1·0–1·2 mm long and 0·6–2·0 mm wide, so deeply concave as to be hooded at the tip, with margins entire or at most slightly crenulate at the apex. The nerve is faint and double or absent. The cells in mid-leaf are long, tapering at both ends and usually with a slight sigmoid twist (c 40–80 × 6 µm) with thick walls which tend to thin out and become porose particularly near the cell ends. Near the leaf apex and in the leaf base, cells are shorter and usually distinctly porose. The alar cells are short, squarish, or oblong, very thick-walled and porose with dark granular contents, forming a distinct alar group.

The female branches are small, pointed, with tightly rolled up bracts, and project c 1–2 mm out from the stem. The capsule is short and wide, c 1·5 mm long, horizontally held on a short red seta (1·0–1·5 cm) and with a blunt conical operculum. The inner peristome has the teeth united into a basal membrane for about one third of the length. There is the usual double peristome dispersal mechanism but when the capsule is extremely dry (or possibly very old) the outer teeth, digging deeper into the capsule mouth, pull back the membrane to open the capsule completely and release the last of the spores. The exact conditions for this occurrence still need to be established. The spores, in what Australian material we have seen, measure 15–20 µm, unlike the 20–30 given in the Handbook. The significance of this difference is not known.

DISTRIBUTION: TAS, VIC; also in New Zealand and S. America.

ILLUSTRATIONS: Handbook, Plate 56; Brotherus (1924–5), Fig. 581 (as *Lembophyllum*).

This species can be very hard to separate from certain species in other genera which also have very concave leaves, namely *Acrocladium chlamydophyllum* [117] and especially *Lembophyllum clandestinum* [103]. The former differs in having pointed shoots, leaves rather flat instead of hemispherical, and big, inflated, empty hyaline cells in the auricles. *Lembophyllum* is more troublesome but differs in having the cells predominantly rhomboidal and shorter (*c* 20 × 6 μm typically) and a distinctly crenulate-denticulate apex. *Camptochaete* species [104] may also give some trouble to a beginner.

W. mollis (Hedw.) Broth.

This species is much softer than the foregoing, and forms long, pendulous, fawn-coloured tangled skeins on the twigs of trees in very humid fern gullies and rain-forest. The shoots commonly reach 17 cm and sometimes much more, with abundant short branches, 0·5–2·0 cm long, spreading horizontally. The leaves, loosely imbricated, are rather oval in shape, 1·5–2·0 mm long × 0·5–0·7 mm wide, channelled from side to side, but only slightly arched from base to apex, usually more strongly channelled just above the widened base, to give a narrow waist. Below this, the base is strongly contracted to the insertion on the stem. The apex is obtuse and the margins entire, or very slightly denticulate at the apex. The alar cells are dark, thick-walled, conspicuous, often yellowish and sometimes slightly porose. The rest of the cells are similar to those of *cochlearifolia*, long and thick-walled, 50–80 × 3–5 μm; shorter and wider at the base and apex, but not porose. Capsules seem to be very uncommon.

DISTRIBUTION: TAS, VIC, Lord Howe; also in New Zealand and S. America.
ILLUSTRATIONS: Brotherus (1924–5), Fig. 552; Allison and Child (1971), Plate 24.

Once recognized it is unlikely to be mistaken for anything else. It is not usually found except hanging from twigs, but fallen plants might be mistaken for *Camptochaete gracilis* [104] which, however, has much shorter cells (usually 20–30 μm) which usually project from the abaxial surface of the leaf as papillae.

NECKERACEAE

Includes: *Calyptothecium acutum* (Mitt.) Broth. in Par. (QLD)
C. australinum (Mitt.) Par. (QLD)
C. caudatum Bartr.
C. recurvulum (Broth.) Broth. in Thér. (QLD)
C. subecostatum Dix. (QLD)

Himantocladium loriforme (Bosch & Lac.) Fleisch. (QLD)
H. plumula (Nees) Fleisch. (QLD)
Homalia falcifolia (Hook.f. & Wils.) Wils. in Hook. (TAS)
H. punctata (Hook.f. & Wils.) Wijk & Marg.
Homaliodendron exiguum (Bosch & Lac.) Fleisch. (QLD)
H. flabellatum (Sm.) Fleisch. (QLD)
Pinnatella intralimbata Fleisch. (QLD)

Neckera Hedw.

N. pennata Hedw.

It is a pity that this moss of rain-forest and wet forested gullies is uncommon for it is one of the most charming. The very strongly undulate, even corrugated, leaves give it an unmistakable texture not matched by any but its closest relatives. The stems are 5 cm or so long, sparsely pinnately branched and with the leaf arrangement strongly complanate in the same plane as the branches, forming flat pale golden green to golden brown fronds, varying from erect to pendent. The smooth rhizoids arise in bunches from stems and branches. The leaves are narrowly ovate or oblong–lanceolate, tapering to a point, $1\cdot5$–$2\cdot8 \times 1\cdot0$–$1\cdot3$ mm, asymmetrical with one side of the leaf almost straight, strongly transversely undulate. The nerve is almost missing, faint, short and forked, although occasionally with one branch of it reaching ¼ way up the leaf. The margins are crenulate below and rather more strongly denticulate in the upper ½ of the leaf. The cells are long and narrow, thick-walled and porose, 50-90 × 6 µm in mid-leaf, rather shorter near the apex; those at the angles roughly isodiametric, forming a distinct alar group.

Sporophytes are easily overlooked as the seta is very short and the capsule virtually immersed, concealed by the leaves.

DISTRIBUTION: TAS, VIC, NSW, QLD; widely distributed in all continents.
ILLUSTRATIONS: Plate 72; Handbook, Plate 57; Allison and Child (1971), Plate 26.

Most commonly, this is an epiphyte on twigs and branches of trees in

PLATE 72. (upper) *Neckera pennata* TAS—Shoot × 7, cells × 1000; (lower) *Catagonium politum* VIC—Shoot × 7, cells × 1000

humid forest. Sainsbury (1955f, p. 51) considers the Tasmanian plants to be this species and we know of no substantial differences between it and *N. aurescens* Hampe.

N. leichhardtii Hampe (NSW) is the only other Australian species in the Index but *Neckeropsis sparvelliae* Dix., endemic to QLD, *N. lepineana* (Mont.) Fleisch. and *N. nano-disticha* (Geh.) Fleisch. are also recorded.

Homalia, a related genus with very complanate shoots but smooth flat leaves, has been recorded from TAS but Sainsbury (1955f, p. 51) doubts the record.

Leptodon Mohr

L. smithii (Hedw.) Web. & Mohr
When dry, this uncommon moss is easily recognized by the densely branched flat fronds, *rolled up* and even coiled at the tips. The bipinnate fronds from a creeping stem are up to 4 cm long by 1 (–2) cm broad, with the leaves oval, very obtuse, slightly concave and plicate, appressed and somewhat complanate when dry, spreading when moist. The nerve is very strong below, fading out in mid-leaf. The cells are diamond-shaped, oval or squarish, *c* 13–15 μm long. The branches bear paraphyllia. We have not seen capsules.
DISTRIBUTION: VIC, NSW, ACT, QLD; also in Europe, Asia, Africa, America and New Zealand.
ILLUSTRATIONS: Dixon (1924a), Plate 50 (K); Brotherus (1924–5), Fig. 566.
   It is only recently that the species has been discovered in Victoria. It grows on rocks (especially limestone) and trees in rather dry sclerophyll forest.

Thamnobryum Nieuwl.
(= *Thamnium* of the Handbook)

Thamnium has recently been shown to be illegitimate and *Thamnobryum* apparently the proper replacement for it (Crundwell, 1971). It is a genus of sombre-coloured, often dendroid, mosses of wet forests, sometimes more or less aquatic, and often complanately branched with single nerves and short rhomboidal cells.

T. pumilum (Hook.f. & Wils.) Nieuwl. is not an uncommon plant of rocks and earth, especially under overhangs in fern gullies and similar places where it forms dull green mats of rather loose, straggling, pinnately branched more or less dendroid shoots *c* 2–4 cm long, emerging from creeping, leafy, unbranched stems, dark and sombre in the field but bright deep green in a good light. The stems are green but very fine and wiry, 120–160 μm diameter, and not much thicker at the base than at the apex. The leaves are short, *c* 0·6–0·7 × 0·2–0·4 mm, elliptical in outline below and narrowed to a sharply angled or rounded usually asymmetrical point. They are spirally inserted but complanately flattened in the same plane as the branches and have strongly denticulate margins in the upper half, especially near the apex, and a single strong nerve to *c* ¾, occasionally forked at the tip. The cells are hexagonal, those above nearly isodiametric, those below mid-leaf much narrower and *c* 22 × 7 μm. Cell walls are not thickened.

Capsules are not common. The male branches, not seen by Sainsbury, replace other, vegetative, branches in the branching pattern; they are sessile, very short and bud-like, enclosing a cluster of antheridia, the whole branch little bigger than the adjacent leaves.

DISTRIBUTION: TAS, VIC, NSW, ACT, Lord Howe; also in New Zealand and S. America.

ILLUSTRATIONS: Handbook, Plate 59; Wilson (1859), Plate 175 (as *Isothecium*)

This species is not likely to be confused with anything else; the strong single nerve and short rhomboidal cells above mid-leaf will separate the species from all but a close relative.

T. pandum (Hook.f. & Wils.) Stone & Scott is much larger and more dendroid and differs in the nerve which reaches almost to the apex and in the slightly plicate, coarsely serrate leaves.

CHROMOSOME NUMBER: n = 11 (NSW).

DISTRIBUTION: NSW; also S.E. Asia and New Zealand.

ILLUSTRATIONS: Brotherus (1924–5), Fig. 578; Wilson (1854), Plate 89 (as *Isothecium*).

Thamnium eflagellare Aongstr., which Sainsbury considers the same as *pandum*, is said by Burges (1935, p. 87) to be "frequent throughout the coastal rain forest; often mistaken for *T. pumilum*". We have not seen it in Victoria unless we have fallen into this same trap. Provisionally we must follow Sainsbury's treatment.

Thamnium novae-walesiae Kindb. is unknown to us.

ECHINODIACEAE

Echinodium Jur.

This small genus is a famous example of an almost antipodal disjunction in distribution, occurring in Madeira and the Azores as well as in Australia, Fiji, New Zealand and some neighbouring islands.

E. hispidum (Hook.f. & Wils.) Reichdt.

Dark dull green is the prevailing colour of the trailing shoots of this rain-forest moss, although sometimes it can have a yellowish tint. The stems are long and quite robust, rather distantly branched, commonly to 6 cm long but sometimes very much longer. The leaves are very densely imbricate, slightly turned to one side, *c* 3 mm long of which slightly more than ⅓ is taken up by a wide, triangular, concave base, 0·6–0·7 mm wide, and the rest by a very long bristle-like subula mostly filled by the nerve, and producing the characteristically harsh and hairy texture of the shoots. The margins are entire, somewhat recurved in the leaf base. The cells are very irregular and variable in shape but are more or less isodiametric, not usually exceeding 2 × 1, *c* 10 μm in average diameter, slightly elongated near the nerve in the leaf base.

DISTRIBUTION: TAS, VIC, NSW, Lord Howe; also in New Zealand and Oceania.
ILLUSTRATIONS: Plate 73; Handbook, Plate 60; Brotherus (1924–5), Fig. 587; Wilson and Hooker (1845), Plate 61 (as *Hypnum*); Allison and Child (1971), Plate 26.

PLATE 73. *Echinodium hispidum* VIC—General view of plant (top left) × 1·5, detail of shoot × 7, leaf × 50, cells × 1000

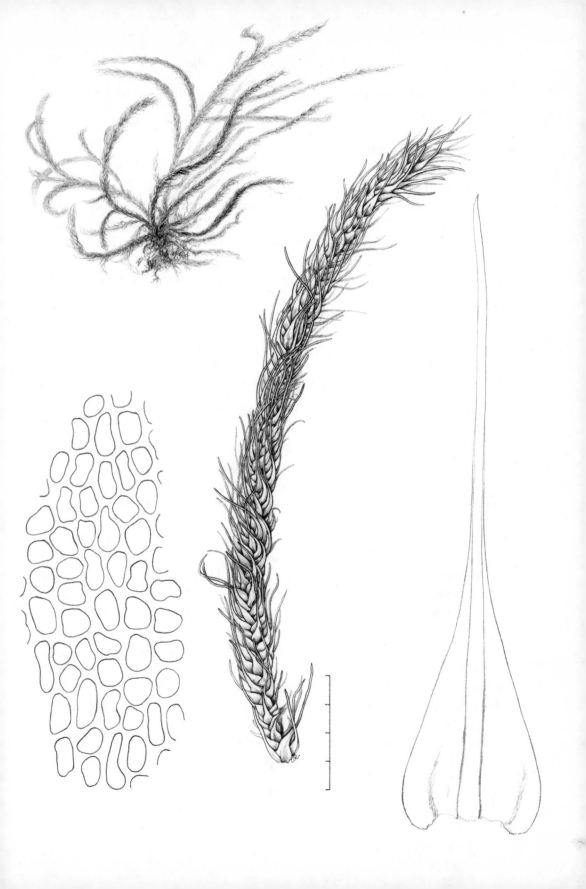

It is reasonably common in rain-forest and wet gullies, mostly on the bases of trees or rocks.

E. arboreum Broth. from NSW has much shorter leaves.

E. parvulum Broth. & Watts is endemic to Lord Howe Island.

Both of these species need to be evaluated again. The only species with which *hispidum* is likely to be confused in the field is *Dicranoloma menziesii* [25] which can have similar colouration and leaf shape but the long narrow cells in the leaf base will distinguish that species microscopically and the much shorter stems and white tomentum will often distinguish it even to the naked eye.

LEMBOPHYLLACEAE

Lembophyllum Lindb.

L. divulsum (Hook.f. & Wils.) Par. (=*clandestinum* of the Handbook)
The rather stiff short branches of this common species have a characteristic neat appearance produced by the even-sized leaves throughout. The stems are roughly pinnately branched, but the branches tend to lie along the main stem, not diverging widely, forming rather tight mats on rocks and soil, sometimes on trees, in a wide range of habitats from wet or dry sclerophyll forest to rocky grassland.

The leaves are oblong or circular, or wider than long, c 1·2 mm long × 0·8 mm wide to 1·5 × 1·5, or 0·6 × 0·8 mm wide, closely and densely set on the axis to give a shoot 0·5–1·0 mm wide. They are very deeply concave above, often so much that the apex of old leaves is actually inflexed to make a pouch-shaped tip, but contracted just above the insertion to form a waist. The nerve is very variably developed: sometimes absent but more commonly wide at the base, forked and fairly pronounced even though ill-defined, reaching near mid-leaf; sometimes almost single, very wide to beyond mid-leaf. The cells are predominantly rhomboidal, variable in size, shape and pattern of distribution. Generally the cells are short and rhomboidal above and at the margins, oblong or spindle-shaped below, especially in the neighbourhood of the nerve, (c 10 × 25 × 2–8 μm in mid-leaf) squarish or rounded in the distinct alar group, porose only in the leaf-base and sometimes near the apex. The margin is usually crenulate and sometimes hyaline too, at the apex.

The short fruiting branches poke out from between the stem leaves; the innermost perichaetial bracts long and sheathing, the outermost short and squarrose. The seta is rather short, 1·0–1·5 cm or sometimes longer, and orange-yellow with an ovoid–cylindrical capsule of *c* 1·5 mm and a conical operculum. The pale peristome teeth are densely transversely striate on the lower half of the outer face, papillose on the upper half. The inner peristome is similar in height, the basal half, or slightly less, joined to form a membranous tube, and there are 1 or 2 nodulose cilia. The papillose spores measure 13–20 μm.

CHROMOSOME NUMBER: n = 10 (TAS).

DISTRIBUTION: TAS, SA, VIC, NSW, ACT, QLD; also in New Zealand.

ILLUSTRATIONS: Handbook, Plate 60; Wilson (1854), Plate 90 (as *Hypnum*); Allison and Child (1971), Plate 27.

This species is not unlike *Pseudoleskea imbricata* [115] which differs in the strong single nerve and in the triangular, rather than circular, leaf outline. More difficulty is likely in separating it from forms of *Weymouthia cochlearifolia* [98] and from *Camptochaete arbuscula* [104] where, however, the shoots are dorso-laterally branched and the leaves narrower and more pointed.

There are two forms of this species: one, usually infertile, with very small, regular leaves, *c* 0·6 mm wide, and the other much less regular in size, commonly fruiting, and with leaves up to twice as large. The former tends to have a strong single nerve and small thick-walled cells, var. *divulsum*. The latter, which tends to have a weak or absent nerve and larger thinner-walled cells, is var. *clandestinum*; this is the form which most closely resembles *Weymouthia*. Sainsbury, after prolonged study, came to the conclusion that the two kinds of plants intergraded and were conspecific, but Australian specimens seem to show more distinction between the two taxa and further study might reveal two species.

Camptochaete Reichdt.

This genus of rather big, often dendroid, forest mosses is characterized also by the concave leaves with cells narrow and sinuose below but short and rhomboidal above, distinct alar cells, but no nerve or only a faint and forked trace of one. Branching is irregularly pinnate; in all species examined the branches emerge consistently from the *upper* side of the stem (as seen in section), i.e. dorso-lateral, so that the underside of the stem presents an unbroken pattern of leaves, while the upper side is interrupted by lateral branches, including fruiting ones. The leaves tend to be dimorphous: the main stem leaves very much larger than the ultimate branch leaves.

Key to species

1. Plants creeping, not dendroid . . . . . .   *gracilis*
    Plants with erect, dendroid fronds   2

2. Fronds large, untidy; main shoot in mid-frond
    c 1·5–3·0 mm wide, including leaves . . .   *arbuscula*
    Fronds smaller, neater; main shoot *c* 0·5–1·5 mm
    wide, including leaves . . . .   *vaga*

C. gracilis (Hook.f. & Wils.) Par.
This is probably quite a common species of damp rocks in forest gullies but seems to be seldom collected, because of its nondescript appearance. It is irregularly, somewhat pinnately branched with prostrate and erect stems but not dendroid, forming loose or tight mats. The leaves are oval, ¼–½ mm long, somewhat secund, pointing towards one side of the plant but not at all falcate, very blunt or quite rounded at the apex where they are slightly toothed. They are concave, but not deeply so, at the apex, and therefore are scarcely cucullate. The cells are rhomboidal to narrowly rhomboidal, *c* 18–35 × 6 μm, with 1–2 rows of marginal cells at the base squarish, thicker-walled, but not porose. Typically the cells project abaxially from the leaf to give a rough surface on the back.
DISTRIBUTION: TAS, VIC, NSW, QLD, Lord Howe; also in New Zealand and South America.
ILLUSTRATIONS: Handbook, Plate 61; Wilson & Hooker (1845), Plate 61 (as *Hypnum*).

C. arbuscula (Sm.) Reichdt. (including *C. ramulosa* (Mitt.) Jaeg.)
This is a big, untidy-looking semi-dendroid moss, usually pale khaki-olive in colour, very variable in growth form although the kind of variation found is itself a diagnostic feature. The stems are prostrate, creeping and anchored

to the substratum at intervals by rhizoids, producing erect, dendroid branches of which the main stem may again extend, drop down to the substratum and "root" at the tip or trail in a long festoon. The erect stems are rather irregularly pinnately branched, sometimes confined to one plane, sometimes not. Often the fronds taper out into long, very slender, small-leaved flagelliform branches. The unbranched stems at the base of dendroid shoots have imbricate leaves in the young state, which split at the tips and eventually fall off to leave an almost bare stem. These stem leaves are widely ovate, concave and cucullate (hooded) at the tip before they split. The branch leaves are sometimes complanate or even apparently distichous, pointed or very blunt, usually very concave, somewhat wrinkled when dry. Leaves which are completely obtuse or virtually circular or quite strongly pointed, can be found even on the same branch. They are extremely mixed in size, but 1·5 × 1·5 mm is perhaps average. The nerve is very faint, short and double or forked. The cells are very narrowly rhomboidal, usually with a sigmoid twist, throughout much of the leaf but shortly rhomboidal above and square, dark and often porose in the alar group.

There is a short reddish seta with an inclined ovoid theca at the tip. The outer peristome is of 16 greenish yellow teeth which are very densely striated transversely, papillose at the tips, and with hyaline membranous margins. The inner peristome has a tall basal membrane half its height and appendiculate cilia in groups of three. Spore discharge is by the outer peristome moving in and out between the inner teeth, plucking and jerking on the projections. The operculum is conical.

CHROMOSOME NUMBER: n = 10 (VIC), 11 (NSW, QLD), 22 (NSW).

DISTRIBUTION: TAS, VIC. NSW, ACT, QLD, Lord Howe; also in New Zealand and Chatham Islands.

ILLUSTRATIONS: Handbook, Plate 61; Brotherus (1924–5), Fig. 579.

This is a perplexingly variable plant. The distinction between *arbuscula* and *ramulosa* given by the Handbook does not seem to hold consistently and we are provisionally treating the two species as conspecific. Sainsbury says of *ramulosa* (Handbook, p. 380) "This species is very near *C. arbuscula*, and its characters are not of great structural importance, but it is usually recognizable by the rigidly imbricated smooth leaves, with longer points, and by the longer seta", for which he gives measurements of 1·0–1·5 cm as against 0·25–0·5 cm, rarely to 1 cm. But both kinds of leaves, smooth and undulate, obtuse and pointed, can occur on the same plant and the difference in seta length is too feeble a specific criterion. Some other features are needed to give a convincing separation. The name *arbuscula* (1808) has priority over *ramulosa* (1856).

The likeliest confusion is with *Lembophyllum divulsum* [103] and *Weymouthia cochlearifolia* [98] from which it differs especially in the dorso-lateral mode of branching, and probably in the less porose or non-porose cells.

C. vaga (C.Muell.) Broth. is more sub-tropical in distribution than the previous species, deep bright green in colour and very much neater in growth form which is tidily and consistently dendroid with rather elegant fronds. Despite Brotherus' statement to the contrary, the branches are sometimes flagelliform, as in the preceding species.

The stem leaves are wider than long, almost distichously inserted, flat or nearly flat, and slightly plicate, 2 mm long × 2–3 mm wide, abruptly pointed, with narrowly oval cells, 15–25 × 11 μm, thick-walled and not porose; those at the base are more or less isodiametric. The branch leaves are almost circular in outline, 0·6 × 0·8 mm diameter or slightly wider than long, with a short abrupt mucro, the cells long and narrow, 20–45 × 5–6 μm, those above shorter and rhomboidal. The alar cells are dense and isodiametric but not at all sharply defined and they are hard to see because of the concavity of the leaf; unlike the other cells in the leaf, they are porose. The leaves are finely crenulate with the projecting cell ends.

The sporophyte, from the upper side of the frond, has a *red* seta *c* 1 cm long, and a capsule *c* 2 × 1·0–1·5 mm, with a domed operculum abruptly contracted to a conical beak, up to 1 mm long.

CHROMOSOME NUMBER: n=11 (NSW, QLD).

DISTRIBUTION: NSW, QLD, Lord Howe; also recorded for New Zealand, but the record is doubted by Sainsbury (Handbook, p. 383).

ILLUSTRATIONS: ?

At first glance, this could be easily mistaken for *arbuscula* but is distinguished, apart from the much more regular growth form, by the almost hemispherical and smaller leaves.

Other species given by the Index are:
C. excavata (Tayl.) Jaeg. (NSW).
C. leichhardtii (Hampe) Broth. (NSW).

C. pulvinata (Hook.f. & Wils.) Jaeg.
C. schlosseri (Sendtn.) Broth.ex Par. (NSW).

EPHEMEROPSIDACEAE (Schultze-Motel, 1970b)

Ephemeropsis Goeb.

E. trentepohlioides (Renn.) Sainsb., perhaps the most extraordinary moss in our flora, has already been described by Willis (1953a). The capsules are not unlike those of *Daltonia splachnoides* [107] in size and structure, but the gametophyte is reduced to a soft velvety pad of chocolate-coloured tomentum growing on fine twigs in very shaded humid forest. It has been found only once but could very easily be passed over for the alga *Trentepohlia* which is almost identical in appearance although perhaps slightly less brown. When fruiting there is no confusion but the capsule and seta are only 2–3 mm long and could easily be missed. Microscopically, the oblique cross walls of the *Ephemeropsis* gametophyte are conclusive.

DISTRIBUTION: TAS; also in New Zealand.

ILLUSTRATIONS: Handbook, Plate 62; Willis (1953a).

PILOTRICHACEAE

Includes: *Pilotrichum sieberi* Hampe

HOOKERIACEAE

Includes: *Callicostella baileyi* (Broth.) Kindb. (QLD)
    C. kaernbachii Broth. ex Fleisch. (QLD)
    C. rugiseta Dix. (QLD)
    Chaetomitrium entodontoides Broth. & Watts. (QLD)
    C. tahitense (Sull.) Mitt. in Seem. (QLD)
    Cyclodictyon lepidum (Mitt.) Broth. & Watts. (NSW, QLD).

Daltonia Hook. & Tayl.

D. splachnoides (Sm.) Hook. & Tayl., which is rather similar in size and appearance to *Sauloma tenella*, is not uncommon in wet forest in Victoria, forming small tufts only 5 mm high or so, usually on fine twigs in very humid micro-habitats. The leaves are very narrow, *c* 2 mm long, pale and glossy, erect and almost straight even when dry. The cells are roughly rhomboidal but with the marginal rows narrowed to form a distinct border.

Capsules are common, dark brown and elegant. There is a long seta, *c* 7 mm tall, with a tiny oval capsule and a flesh-coloured double peristome, the outer teeth papillose and tapering to long hair-like points, the inner processes almost hair-like throughout. The calyptra is fringed with long hyaline spiny hairs at the base.

DISTRIBUTION: TAS, VIC; also in New Zealand, Europe, Central Africa and Central America.

ILLUSTRATIONS: Handbook, Plate 63 (as *D. angustifolia*).

This charming little moss is not particularly rare in suitably wet forest. It is likeliest to be confused with *Sauloma tenella* [110, q.v.].

D. pusilla Hook.f. & Wils. from TAS, is considered conspecific by Sainsbury (1956a). It is illustrated in Wilson (1859), Plate 177.

Distichophyllum Doz. & Molk.

There is a strong family resemblance between this genus, *Pterygophyllum* [109] and *Eriopus* [108], since all three share the same wide, round leaf shape and complanate arrangement. There is, however, a single unbranched nerve to beyond mid-leaf (which separates it from *Eriopus*) and a border, in all but *D. microcarpum*, to separate it from *Pterygophyllum*. The calyptra is split at the base into a spreading frill of fine segments (in *Pterygophyllum* it is perhaps more coarsely lobed). The outer peristome has the 16 papillose teeth densely transversely striated and with strong bars which project widely on either side and may be joined by a marginal web of hyaline material; the teeth are split vertically in the lower two thirds, the two halves separated but connected by a network of cross strands.

Key to species

1. Leaves unbordered . . . . *microcarpum*
   Leaves bordered 2

2. Leaves completely obtuse and rounded, or sometimes with a very short stubby point . . . *pulchellum*
   Leaves wide but with a distinct sharp tapering
   point on top . . . . *crispulum*

D. microcarpum (Hedw.) Mitt.

This uncommon species forms open colonies on rocks or logs in wet forest. The shoots, *c* 2–4 cm × 3–4 mm, are dark bright green and strongly complanate. The leaves are clearly obovate, 2·3–2·7 × 1·6–1·7 mm, tapering gradually to a narrow insertion and *decurrent* there, single-nerved to about ¾ and completely rounded at the apex; unbordered and entire, although the margins may be crenulate with projecting cells. The cells are wholly characteristic: those above mid-leaf, small and roughly hexagonal, *c* 10–18 μm, those at the margins more nearly square; but there is a big patch of enlarged lax and hyaline cells in the centre of the leaf, extending to the leaf base, forming a different area which is especially conspicuous under low magnification.

The erect to pendulous capsule is *c* 1 mm long (with an erect-beaked operculum almost as long again) on a dark purple-black seta. The calyptra, flared out at the base in a whitish frill, is rather thick and fleshy and warted with single papillae, especially near the apex. The inner peristome is similar in length to the outer, hyaline and finely papillose throughout, with a basal membrane to *c* ⅓, the processes finely tapering and mostly made up of a single chain of cell-like segments with coloured thickenings between the cells and down one side of the process.

DISTRIBUTION: TAS, SA, VIC, NSW, ACT, QLD(?); also in New Zealand.

ILLUSTRATIONS: ?

The lack of border at once suggests *Pterygophyllum* but the patch of enlarged cells which occupies most of the centre of the leaf is quite diagnostic of this species.

D. pulchellum (Hampe) Mitt.

There is usually a yellowish colour here which is different from other Australian species of the genus, but it is not always present. At its most distinctive, this species has the shoots 1 or 2 (–5) cm long × 2·0–3·5 mm wide, strongly flattened and pressed together to form quite dense little prostrate mats. The leaves are not obviously in ranks, and tend to be undulate and crisped when dry, 0·5–2·3 × 0·6–1·2 mm, obovate or oval, rounded and completely obtuse at the apex, or with a small projecting apiculus, nerved to *c* ¾ or sometimes more and with the nerve rarely slightly forked at the tip. The cells are hexagonal, wide, isodiametric or slightly elongate above and round the leaf edges, 12–18 µm; those below oblong-hexagonal, *c* 40 × 20 to 80 × 30 µm, 2–3 rows of marginal cells very long and narrow but not thick-walled, forming a distinct entire border. Capsules are small, *c* 1 mm long, with an evident tapering neck held almost horizontally on a very slender wine-red seta, 1–2 cm long and slightly twisted (left hand).

DISTRIBUTION: TAS, SA, VIC, NSW; other states? Also in New Zealand.

ILLUSTRATIONS: Handbook, Plate 65; Wilson (1854), Plate 93 (as *Hookeria amblyophylla*); Wilson (1859), Plate 177 (as *H. sinuosa*).

   This is the commonest of the species, on stones, logs, trees etc. in damp forest. While it is unmistakably a *Distichophyllum* because of the very complanate leaf arrangement, single nerve and border, it is not always possible to separate it satisfactorily from *crispulum*, which mainly differs in the much more conspicuous leaf-point and in the more undulate leaves.

D. crispulum (Hook.f. & Wils.) Mitt. differs from the preceding species mainly in the very strong crisping of the leaves when dry and in their shape which is oblanceolate or oblong–oval, *c* 1·4–1·7 mm × 0·5–0·7 mm, tapered below to a very narrow insertion, not decurrent, and *tapered above* to the apex where the leaf borders join to form a rather long spike-like point. It is not uncommon, especially on stones on the forest floor in wet sclerophyll or rain-forest.

DISTRIBUTION: TAS, VIC, NSW; also in New Zealand.

ILLUSTRATIONS: Wilson (1854), Plate 93 (as *Hookeria*).

D. rotundifolium (Hook.f. & Wils.) C.Muell. & Broth. differs from *crispulum* in the denticulate leaf margin and short, thick leaf-point. We have not seen Australian material. In the Handbook, it is recognized that "more or less intermediate forms occur".

DISTRIBUTION: TAS, NSW; also in New Zealand.

ILLUSTRATIONS: Brotherus (1924–5), Fig. 596; Wilson (1854), Plate 93 (as *Hookeria*).

The other species given by the Index are all endemic:

D. baileyanum C.Muell. (NSW, QLD).

D. beccarii (Hampe & Geh.) Par. (TAS).

D. complanatum (Hampe) Mitt. (VIC).

D. levieri (Geh.) Broth. (TAS).

D. longicuspis Broth. (Lord Howe).

D. minutifolium C.Muell. (NSW).

D. subminutifolium (Broth. & Geh.) Fleisch. (NSW).

D. whiteleggeanum C.Muell. (NSW).

Eriopus C.Muell.

The flattened shoots of this genus, with rounded very complanate leaves, are typical of several related genera from which this can be separated by the strong leaf border, very short forked nerve, and hairy or papillose seta.

E. apiculatus (Hook.f. & Wils.) Mitt.

This is an uncommon or overlooked species which we have found in a water-trickle down a sandy sea cliff, but is otherwise more typical of damp stones and earth in shade, or epiphytic in swamps.

The yellowish green shoots are 2–3 cm long (of which *c* 1 cm is current growth) × 3 mm broad when moist. The leaves are complanate but very undulate and crisped when dry, obscuring the flattened arrangement. There are 6 rows of leaves: 2 rows of large lateral leaves, 2 of dorsal and 2 ventral. The lateral leaves are *c* 2·0 × 1·2 mm, the others slightly smaller; oval to oblong or, at times, almost circular, abruptly ending in a big wide point. In mid-leaf the cells are wide, short, hexagonal and thin-walled, *c* 50–70 × 30 μm, becoming smaller near the apex, *c* 45 × 24 μm, all strikingly contrasted with the broad conspicuous border of 4–7 rows of long narrow thick-walled porose cells. This border is denticulate near the apex and slightly crenulate below. The nerve is not always missing, as the Handbook gives it, but is never more than short, forked and faint.

We have not seen sporophytes, but the Handbook describes and illustrates an erect straight seta, papillose with low rounded papillae, the calyptra smooth above and with a fringe of hairs below, and an erect or slightly tilted capsule.

DISTRIBUTION: TAS, SA, VIC, NSW: also in S. America and New Zealand.
ILLUSTRATIONS: Handbook, Plate 64; Wilson and Hooker (1847), Plate 155 (as *Hookeria*).

This uncommon plant is superficially like the much commoner *Distichophyllum pulchellum* [107] but the short or missing nerve will immediately distinguish it.

E. cristatus (Hedw.) Brid. differs in the much weaker border above, 4 rows wide, conspicuous forked nerve, and much larger cells—the Handbook gives 70–100 × 50–60 µm. When fruiting it has a seta curved at the tip and densely hairy, with unicellular hairs forming a crest at the top of the arch. It is stated to be endemic to New Zealand by Sainsbury, but is recorded in the Index from S. America, Madagascar, Oceania and Australia as well as New Zealand.
ILLUSTRATIONS: Handbook, Plate 64; Brotherus (1924–5), Fig. 599.

E. flexicollis (Mitt.) Jaeg., which Sainsbury (1956a, p. 36) records from TAS and we have found in VIC, is not given for Australia by the Index. It is similar to *cristatus* but with much smaller cells, 24–40 × 18–24 (up to 50 µm long × 30 µm in our material), a rather long leaf-point, strong, toothed leaf border, and a few erect hairs at the crest of the seta two or more cells long.

E. tasmanicus Broth. may be a synonym of *flexicollis* according to Sainsbury (1956a, p. 36).

E. brownii Dix. has been recorded from TAS by Willis (1957b, p. 102) and is distinct in the very narrow entire border of 1–2 cell rows, spathulate nerveless leaves, and papillose, not hairy, seta. We have recently found it in VIC.

E. brassii Bartr. is from QLD.

Pterygophyllum Brid.†

The generic name *Pterygophyllum* is illegitimate according to Crosby (1972), despite the arguments of Miller (1971), and has been replaced by *Achro-*

phyllum (Vitt and Crosby, 1972). It has, however, been proposed for conservation (Margadant *et al.*, 1972) which would be a blessing to southern
hemisphere bryologists.

P. dentatum (Hook.f. & Wils.) Dix.

Wet sclerophyll forest, rain-forest and fern gullies are the typical habitats of
this species. It is quite common, mainly on earth and small stones on the
forest floor and beside streams etc., but is hard to distinguish in the field from
the closely related *Distichophyllum* [107] and *Eriopus* [108] which have the
same extremely complanate leaf arrangement.

The shoots are 1–3 (–6) cm tall × 4–5 mm broad, dark but translucent and
almost unbranched, forming sparse tufts. The leaves, typically 2–3 × 1·5 mm,
are obovate or oblong, asymmetric especially at the base, flat and strongly
compressed into a flat, complanate frond. When dry, the leaves become
opaque, black and crisped, rarely remaining quite pale green. The margins
are plane, coarsely toothed at the apex (hence the epithet *dentatum*) and part
way down one side. There is a clear nerve to about ½–⅔ of the way up the
leaf, forked at the apex. The cells are rounded–hexagonal with triangular
corner thickenings, and measure *c* 30–50 µm in mid-leaf, smaller towards
the margins but not forming a distinct border. Not uncommonly there are
abundant bright pale green gemmae on the leaves, concentrated in a band
just inside the margins. These consist of one long filament, *c* 6 cells long or
more, with 1–3 short branches each 1–3 cells long, emerging at right angles
from the base of the filament. These are the "L-shaped" gemmae of Sainsbury
(1935, p. 101) and may represent prostrate and erect filaments of protonema.
The leaves near the apex may also be fragile, falling off easily to leave a bare
stem, and no doubt acting as propagules.

The male branches are small and bud-like on the stems, smaller than and
concealed by the vegetative leaves. The female branches are slightly larger
than the leaves and therefore project more conspicuously. There is a twisted
seta (left hand) of *c* 1·5 cm, curved at the top to a half-pendent capsule which
is *c* 1·5 mm long × 1·0–1·3 mm wide, obovoid, widest below the mouth and
with tubercles on the lower half or apophysis. There is a long-beaked
tapering operculum almost as long as the rest of the capsule. The calyptra is
membranous, black at the tip and white at the base where it flares out
slightly to form a hyaline, split frill. The peristome teeth are much as in

Distichophyllum but split into 2 strips from near the base almost to the apex, transversely striolate and barred with very widely projecting bars.

CHROMOSOME NUMBER: n=20 (VIC, NSW).

DISTRIBUTION: TAS, SA, VIC, NSW, ACT, Lord Howe; also in New Zealand and S. America.

ILLUSTRATIONS: Handbook, Plate 66; Allison and Child (1971), Plate 27; Wilson (1854), Plate 93 (as *Hookeria*).

In the field, when moist, this can be quite difficult to separate from *Eriopus* and *Distichophyllum* (q.v.) but the lack of evident borders will distinguish *Pterygophyllum* except from *D. microcarpum* which has no border but is characterized by the entire leaf margins and an area of enlarged rather empty cells in mid-leaf. The other Pterygophyllums recorded are:

P. bryoides Broth. (VIC).

P. flaccidissimum Broth., (TAS; not mentioned by Sainsbury).

P. subrotundum (Hampe) Jaeg. (VIC), and

P. wattsii Broth. (NSW) which are all supposedly endemic, and

P. obscurum Mitt. (TAS), which is also recorded from S. America.

Sauloma (Hook.f. & Wils.) Mitt.

S. tenella (Hook.f. & Wils.) Mitt.

This uncommon plant of rocks and logs in wet forest and fern gullies might easily be mistaken in the field for a small cushion of *Leucobryum* [28]. Although the leaves are glossy, the plant has the same pallid tint. The individual stems are *c* 1 cm long, very densely leafy and usually slightly bent to one side at the stem apex, and are sometimes silvery at the tips by death of the cells. The leaves, *c* 2·0×0·4-0·5 mm, are narrowly triangular, or ovate–triangular, and are very glossy. The nerve is absent or very short, double and faint, but commonly there is a strong single fold (plica) up the middle, which can be surprisingly hard to tell from a nerve without sectioning. As a result of this, the plant is liable to be misidentified on first acquaintance. The leaf margins are slightly recurved, especially above. Characteristically there are tufts of

PLATE 74. *Sauloma tenella* VIC—Fruiting plant × 11, leaf × 50, cells × 1000

whitish richly branched rhizoids in the leaf axils, springing from brown rhizoidal strands. The cells are spindle-shaped, 80–120 × 10 μm, unaltered at the margins but usually with a row of short, almost isodiametric cells at the very base of the leaf and often continuing into a short decurrent part of 1 or 2 larger inflated cells.

Sporophytes seem to be quite common, with a dark seta, deep red–brown above and blackish crimson below, springing from the base of the stem and usually abruptly bent at right angles at the very tip, although sometimes erect. The capsule, *c* 1 mm long, has a relatively long-beaked operculum, *c* 0·5 mm long, tightly clasped by the smooth membranous calyptra which is conical above and spreads out into a split brim below (mitriform). The calyptra is beautifully coloured, black at the tip, white at the base, and golden brown in between.

DISTRIBUTION: TAS, VIC, NSW; also in New Zealand and S. America.

ILLUSTRATIONS: Plate 74; Handbook, Plate 63; Brotherus (1924–5), Fig. 598; Matteri (1972), Fig. 8.

Once the leaf structure is seen, no confusion with *Leucobryum* is possible.

S. zetterstedtii (C.Muell.) Jaeg. is recorded from WA (Bartram, 1951). It is doubtfully distinct from *tenella*.

Daltonia splachnoides [106], which is rather similar in size and appearance, has the marginal cells narrowed to form a distinct border and the calyptra fringed with long hyaline spiny hairs at the base. The micro-habitat is rather different, with a preference for twigs and lianes rather than rocks or logs.

LEUCOMIACEAE

Includes: *Leucomium hillianum* (Hampe) Jaeg. (VIC)

HYPOPTERYGIACEAE

Hypopterygium Brid.

H. rotulatum (Hedw.) Brid.

This charming little umbrella-moss is quite common in rain-forest and fern gullies. The erect stem, *c* 0·5–1·0 cm tall, is black either throughout or just below, and can be smooth or densely matted with tomentum. The cluster of branches forming the umbrella at the top of the stem commonly measures only 1·0–1·5 cm across the diameter, except in very moist and shady situations; this is apparently smaller than in New Zealand forests.

As in many other members of this family, the leaf arrangement is one more typical of leafy liverworts than mosses. The branches have leaves in 3 ranks with asymmetrically oval lateral leaves and a row of smaller almost circular underleaves (sometimes unnecessarily termed *amphigastria*). The nerve of the lateral leaves varies in length, extending anywhere from mid-leaf almost to the apex. In the underleaves the nerve is even more variable from shortly excurrent to virtually absent. All leaves are clearly bordered by narrow cells, the border being either strong or weak and almost entire or strongly toothed. The leaves are usually downwardly curved when dry, rolled under the branch to form almost a tube. The lamina cells are shortly rhomboidal, pointed at both ends, and have an excessively finely wrinkled cuticle, almost smooth in dull green plants, but quite conspicuously roughened in glaucous plants; such a cuticle is a common cause of glaucousness in bryophytes.

The sporophytes are carried at the tip of the stem in a cluster of one to several very short fertile branches. The short seta, c 5–6 mm or more, is reddish brown and rather robust, curved over at the tip to the small, fat, horizontal or pendulous beaked capsule which is c 2 mm long with a fleshy bluntly pointed calyptra just covering the long operculum. The spores are c 15 µm, smooth.

CHROMOSOME NUMBER: n=9, 18, c 27, 36 (NSW). See Ramsay (1967a, 1974).

DISTRIBUTION: TAS, SA, VIC, NSW, ACT, QLD; also in New Zealand, S. America and the Pacific Islands.

ILLUSTRATIONS: Handbook, Plate 67 (as *novae-seelandiae*); Matteri (1973), Plate 3 (as *didictyon*); Allison and Child (1971), Plate 29 (as *novae-seelandiae*).

There seems to be no satisfactory distinction between this species and *H. novae-seelandiae* since the features given in the Handbook as distinguishing them are often not correlated. Matteri (1973) has recently reduced *novae-seelandiae* to a synonym of the South American *didictyon* C.Muell. Not uncommonly, where the plants are growing on a steep bank, the growth form is more of a flattened frond than an umbrella. It is then rather like *Lopidium concinnum* [113] from which it can be separated by the more pinnate branching, excurrent nerves of the underleaves and the isodiametric leaf cells of that species.

H. commutatum C.Muell. (=*setigerum* according to the Index) is given by Sainsbury (Handbook, p. 410) for Australia and Tasmania. It is distinguished by the transformation of alternate underleaves into fine bristles, and is recognized in the field by its dull colour. We can find no record of its occurrence in Australia and Tasmania and it is not given from there by the Index.

H. muelleri Hampe (NSW, QLD, Lord Howe) including *H. scottiae* C.Muell. is very like a large *H. rotulatum*, vegetatively.

H. planatum Hampe and *H. rigidulum* Mitt. ssp. *balantii* Kindb. have also been recorded for Australia, but we do not know them.

Catharomnion ciliatum (Hedw.) Wils. is dubiously recorded for TAS (Sainsbury, 1956a).

Cyathophorum P.Beauv.

C. bulbosum (Hedw.) C.Muell.
This is probably the most individualistic bryophyte in Australia, if not, indeed, in the world. Even in its most bizarre habitat modifications, it can be confused with no other species. Typically the fronds are single and unbranched, *c* 3–13 cm long (but fronds well over 20 cm can occur), black stemmed, springing from a prostrate underground rhizoid-covered stem and standing out from the substrate whatever the slope. Branched forms are very uncommon and probably always the result of injury. The leaves are strictly in 3 ranks; 2 rows of fairly narrowly oblong lateral leaves, commonly *c* 8 mm long × 1·5 mm wide, but sometimes much larger or smaller, tapering to a short point and expanded and rounded at the base, with the leaf bases imbricated and overlapping. They spread distichously except when dry

PLATE 75. *Cyathophorum bulbosum* VIC—Under-view of fruiting plant (moist) with portion of stem removed (right) and upper side of dry plant (left) × 4, cells × 1000

when they tend to curve downwards. The underleaves are almost circular with an abrupt little point, and arranged in a single row, always along the underside of the stem. There is a short forked nerve, sometimes almost undetectable, sometimes with one side suppressed to give a short single nerve. The cells are rhomboidal, pointed at both ends, c 100–120 × 18–30 μm, the marginal cells somewhat narrower but not forming a border.

The tiny male and female branches (on separate plants) are in the axils of the lateral leaves, exposed only to the underside of the stem and partly concealed by the underleaves. The male buds are swollen and orange with the antheridia, while the female buds are smaller and with more pointed bracts. Fruit is not uncommon, with a double and usually incomplete row of capsules along the underside of the stem. The capsules are c 2·0 × 1·3 mm, short and rather globose, on a seta not much longer than the capsule. The mouth is rather narrow, c 0·5 mm, with a double peristome, a straight beak on the operculum and a short conical fleshy calyptra, black at maturity, and then covering only the beak of the operculum. The fusion of cilia in the inner peristome, mentioned in the Handbook, may be correlated with the retention of spores in the pendulous capsules.

Short green filamentous gemmae sometimes occur in the axils of the underleaves.

CHROMOSOME NUMBER: $n = 5$ (VIC).

DISTRIBUTION: TAS, VIC, NSW, QLD, Lord Howe; also in New Zealand.

ILLUSTRATIONS: Plate. 75; Labillardière (1806–7), Fig. 253; Troughton and Sampson (1973), Plate 53; Brotherus (1924–5), Fig. 630; Allison and Child (1971), Plate 28.

This is an abundant species on rocks, roots and trees in rain-forests and fern gullies, the size depending on the moistness and shadiness of the habitat. In wet sclerophyll forest it can just hold its own but tends to be small and stunted. At its most luxuriant it is a splendid plant, not unlike a small *Blechnum* or *Asplenium*, and it has often been mistaken for a fern by beginners.

Lopidium Hook.f. & Wils.

L. concinnum (Hook.) Wils.

A close relative of *Hypopterygium*, in which it was formerly placed, this too is a reasonably common plant of wet forest where it may be terrestrial but is perhaps commoner as an epiphyte low down on tree trunks, growing out sideways or even pendulous rather than erect.

The soft, pallid fronds are pinnately branched in one plane, the branches being quite distant and not in a close cluster as in *Hypopterygium*. The leaves have strong nerves, which are far excurrent in the underleaves, and the cells differ from those of *Hypopterygium* by being isodiametric and more or less round.

CHROMOSOME NUMBER: n=12 (NSW).

DISTRIBUTION: TAS, VIC, NSW; also in New Zealand and S. America.

ILLUSTRATIONS: Handbook, Plate 62; Allison and Child (1971), Plate 28; Matteri (1973), Plates 1, 2.

L. nematosum (C.Muell.) Fleisch. (NSW), *L. pinnatum* Hampe (QLD) and *L. plumarium* (Mitt.) Hampe are the Australian species given in the Index. The last of these is considered by Thériot to be conspecific with *L. concinnum*.

L. daymannianum (Broth. & Geh.) Fleisch. is recorded by Dixon (1942) for QLD, but omitted from the Index.

THUIDIACEAE

Includes: *Claopodium assurgens* (Sull. & Lesq.) Card.
Haplohymenium pseudotriste (C.Muell.) Broth. (NSW)
Herpetineuron toccoae (Sull. & Lesq.) Card. (QLD)

Thuidium B.S.G.

The paraphyllia and bipinnate fronds make this genus easy to recognize, but the distinctions between species are far from straightforward; in Australia, as in other parts of the world, *Thuidium* is a taxonomically critical genus. The leaves are of two sorts: broad stem leaves contracting to rather fine points and branch leaves narrower, much smaller and usually much less pointed. There is a strong single nerve ending below the leaf apex. The terminology of the branching system is complex. In the pinnately branched frond there is a main *stem* bearing, as well as a dense mat of paraphyllia, *stem leaves* and side *branches* (pinnae); these branches in turn are clothed more sparsely in paraphyllia and bear *branch leaves* (rameal leaves of the Handbook) and *branchlets* (second order branches; pinnules). These branchlets, which have *branchlet leaves* (ramuline leaves of the Handbook) but commonly only rudimentary paraphyllia, are sometimes also provided with *third order branches*. Paraphyllia are a mixture of small leaf-like structures, filaments and some intermediates which are a few cells wide.

The two species dealt with here are dioicous, with short bud-like male or female branches replacing side branches, i.e. carried directly on the main stem. The capsule has a double peristome with the standard dispersal mechanism, a pointed operculum, and a side-split membranous calyptra. Because of the difficulty of separating the two species, no key is given here.

T. furfurosum (Hook.f. & Wils.) Reichdt. (= *Thuidiopsis furfurosa* (Hook.f. & Wils.) Fleisch. of the Index)

At its most conspicuous, this very common plant of grassland is almost orange in colour when exposed to sunlight, yellowish green when more heavily shaded. It is basically 1-pinnately branched, sometimes regularly and sometimes irregularly, with some branches themselves pinnate giving a partly bipinnate frond. The main stem tends to develop branches very close behind the apex, and to continue growth without interruption, forming long straight pinnate fronds. The branch and branchlet leaves, when dry, are strongly arched into an almost semicircular shape with the apex pointing straight in to the stem giving a chain-like (catenulate) effect when the branchlets are seen in profile. In general, the stems are orange in colour and have an outer rind of some 3 layers of thick-walled, pigmented cells, clearly distinct from the thin-walled central stem cells. Most cells, especially on the back of the leaves, have a single, tall, often slightly curved, spike-like papilla. Sporophytes seem to be very rare.

CHROMOSOME NUMBER: n=11 (NSW, QLD); 22 (NSW).

DISTRIBUTION: TAS, WA, SA, VIC, NSW, ACT, QLD, Lord Howe; widespread in the southern hemisphere.

ILLUSTRATIONS: Plate 76; Mueller (1864), Plate 12 (as *Hypnum suberectum*).

PLATE 76. *Thuidium furfurosum* VIC—Plan of shoot (below) ×1·5, detail (above) ×30, cells ×1000

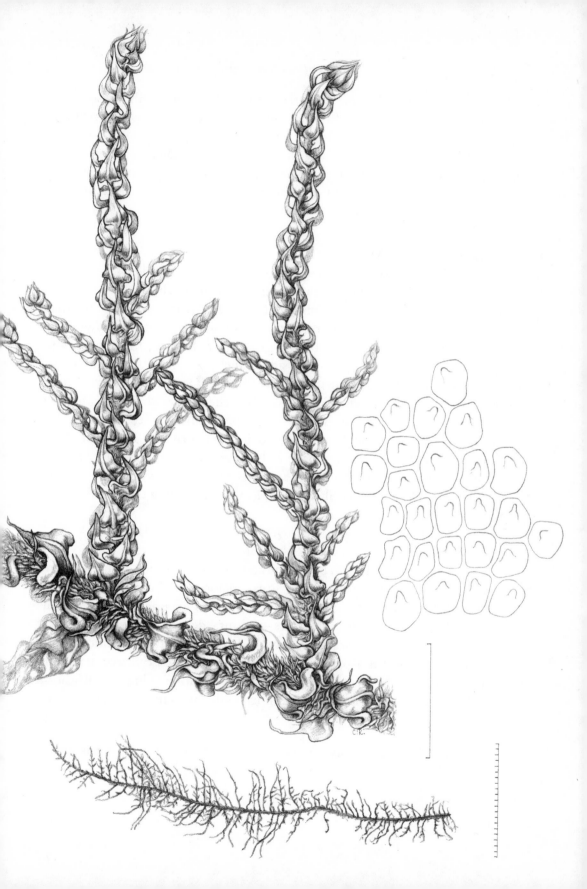

The genus *Thuidiopsis*, separated from *Thuidium* on the basis of leaf arrangement and degree of branching, is too weak to be worth retaining.

T. stuartii (C.Muell.) Jaeg. (= *Thuidiopsis stuartii* (C.Muell.) Broth.) is the same as *furfurosum* according to Sainsbury (1956a, p. 38). The other species in the same group, given by the Index, are: *T. hastatum* (Mitt.) Reichdt. (SA, NSW) and *T. liliputanum* Broth. (NSW, QLD).

T. laeviusculum (Mitt.) Jaeg.

This beautiful, feathery, dark green plant of damp shaded forest soils is primarily bipinnate, with some of the branchlets themselves pinnately branched to give a partially tripinnate frond. The year's growth of the main stem tends to end with a length, bare of side branches, which grows down to the soil and anchors there by the growth of rhizoids; the next season's growth takes place in replacement side branches, giving a system of arched fronds with interrupted branching. The branchlet leaves, when dry, tend to lie more closely to the axis, not so strongly arched and with the apex tending to lie parallel to the axis instead of perpendicular to it; the chain-like appearance of the branchlets is consequently less striking. The stems, typically, are green in colour and have rather thin-walled outer cells. The cells of the branch and branchlet leaves are densely papillose with many (usually 3–5) low papillae on each cell. The whole plant, including the capsule, seems to be smaller than in New Zealand.

DISTRIBUTION: TAS, VIC, NSW, ACT, QLD; also in New Zealand.

ILLUSTRATIONS: Plate 77; Handbook, Plate 68.

On paper, the differences between these two species look impressive and very many specimens, perhaps the great majority, fall clearly into one or other. The distinction has even been considered a generic one. The trouble is that the macroscopic characters—colour, degree of branching, leaf arrangement—are not invariably matched by the appropriate microscopic character, uni- or multi-papillose cells. Without microscopic confirmation, deviant growth forms cannot be certainly identified, but identification based on cell characteristics may conflict with the general appearance and

PLATE 77. *Thuidium laeviusculum* VIC—Plan of shoot (below) × 1·5, detail (above) × 30, cells × 1000

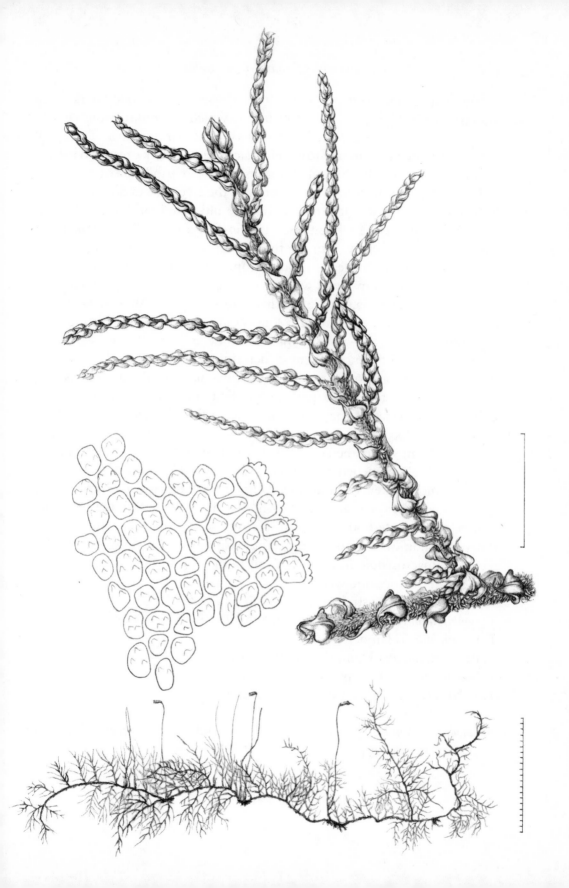

morphology of the plants. *T. furfurosum* var. *sparsum* is a slender, dark green variety usually from tree trunks in forest, which has multi-papillose cells and seems to us better treated as a variety of *T. laeviusculum*. There still remain a few plants of apparently typical *laeviusculum*, dark green and irregularly tripinnate, which have only a single spicule-like papilla on each leaf cell.

The other published differences between the two species do not seem to hold good. Rodway (1914) claims that the ultimate branchlets of *laeviusculum* have paraphyllia while those of *furfurosum* do not. This is sometimes true but not consistently. Nor are the differences in cell size, abaxial nerve crest, ciliate perichaetial leaves, and sporophyte size as helpful as the Handbook suggests. The capsule of what seems to be, vegetatively, unquestioned *laeviusculum* in Victoria, can be less than 2 mm long on a short slender seta; the Handbook gives 3–4 mm on a long stout seta. In any case, capsules are not common enough—and in *furfurosum* they are actually rare—to make any sporophyte differences useful as identifying characters.

The kind of papillosity on the leaves has undeniable attractions as a taxonomic character and is perhaps the best single discriminator to use between the species. If they are separated on this character, the resulting allocation of specimens makes more sense than using any other single character. It must then be admitted that *furfurosum* can have interruptedly pinnate, densely branched dark fronds, very much like those of *laeviusculum* and that the differences in arrangement of the branch leaves in the two species is not always clear cut. The only satisfactory way, in the long run, will be to cultivate both species to discover which characters are stable in different conditions. Until that time, the taxonomist has to use either microscopic or macroscopic features, but not both. The former would seem to have greater taxonomic weight.

The remaining Australian species in the Index are:

T. cymbifolium (Doz. & Molk.) Doz. & Molk. (NSW, QLD, Lord Howe).

T. meyenianum (Hampe) Doz. & Molk. (QLD).

T. nano-delicatulum (Hampe) Jaeg. (NSW, QLD).

T. plumulosiforme (Hampe) Jaeg. (NSW, QLD).

T. protensulum C.Muell. (NSW, QLD, Lord Howe).

Anomodon tasmanicus, possibly also in the Thuidiaceae, has been dealt with under *Triquetrella* [49].

LESKEACEAE
Pseudoleskea B.S.G.

P. imbricata (Hook.f. & Wils.) Broth.

The wiry little shoots of this moss are usually of a characteristic dull bronze green, except in shade where they tend to be more green than bronze. Basically, the shoots are irregularly pinnately branched, but the branches are prostrate and club-shaped, wider at the apex than at the base, but so densely overlapping that this shape is obscured. The branches measure roughly 1 cm long × 1·4 mm diameter and are, themselves, parallel and almost unbranched so that mats of shoots have an elegant, combed-out neatness which they share with almost no other moss. This effect is enhanced when dry by the leaves which are tightly overlapping and closely pressed to the stem, giving a very smooth julaceous shoot. The leaves are concave and not far from circular, c 0·6 mm long × 0·5 mm wide, tapering above to a broad point and slightly spreading when moist. There is a broad strong nerve which fails c ¾ way or more up the leaf. The cells in mid-leaf are mostly rather longer than wide, c 12–15 × 8–9 μm, oval to diamond shaped, the marginal and basal cells slightly smaller and squarish. The lower leaves, hidden under the overhanging pad of branches, tend to be eroded down to persistent nerves only.

The female and male branches are much alike, small, bud-like, almost the size of a full-sized leaf and formed at the base of the current year's growth, hidden under the overlying stems. The seta is dark red-brown, c 1 cm long, surrounded at the base by whitish bracts which are much elongated after fertilization. The capsule, c 1·5 mm long, is cylindrical and curved, with inner and outer peristomes equal in height.

DISTRIBUTION: SA, VIC, NSW, ACT; also in New Zealand. The records from TAS seem in doubt (Sainsbury gives conflicting opinions in the Handbook and in 1956a, p. 39).

ILLUSTRATIONS: Plate 78; Handbook, Plate 68; Brotherus (1924–5), Fig. 655; Wilson (1859), Plate 175 (as *Leskea*).

This is a fairly common species on exposed dry rocks, apparently favouring calcareous ones. It might perhaps be confused at first with the much larger *Hedwigia integrifolia* [89] which, however, lacks any nerve but, once known, it is almost unmistakable. There is a surprising variability in the size and shape and arrangement of the cells, but there seems to be only one species involved.

We do not know of adequate grounds for transferring it to the genus *Pseudoleskeopsis*.

AMBLYSTEGIACEAE

Includes: *Calliergidium pseudostramineum* (C.Muell.) Grout
Sciaromium elimbatum Broth. & Watts (NSW)
S. forsythii Broth. (NSW)

Amblystegium B.S.G.

A. serpens (Hedw.) B.S.G.

Of all the pleurocarpous mosses in Australia, this may well be the smallest, and is easily distinguished from other small species by the short cells. It appears to be very rare, but is also likely to be overlooked. The stems are irregularly or roughly pinnately branched, with prostrate main stems which are larger and have larger leaves, bearing very slender erect branches only *c* 0·2 mm in diameter, forming straggling tangled dull green mats. The leaves are either tightly appressed or somewhat spreading when dry, but spreading quite widely when moist, at *c* 45° or more; in shape, ovate below and prolonged above into a long fine tapering point, and measuring *c* 0·6–0·7 mm long overall × 0·3 mm wide, for branch leaves, and 0·5 mm wide for stem leaves. There is a single nerve, not strong, failing ¾ way up the leaf. The cells are rather narrowly rhomboidal, typically 23 × 7 µm in mid-leaf, shorter and more rectangular below and towards the margins. Those at the angles are almost square but not inflated.

The capsules are relatively conspicuous, held horizontally at the tops of red setas, *c* 1·5 cm tall; narrowly cylindrical, *c* 2 mm long, and contracted just below the rather wide mouth. There is a double peristome and a blunt

PLATE 78. *Pseudoleskea imbricata* SA—Mat of shoots (above) × 4, detail of shoot (below) × 15, leaf × 50, cells × 1000

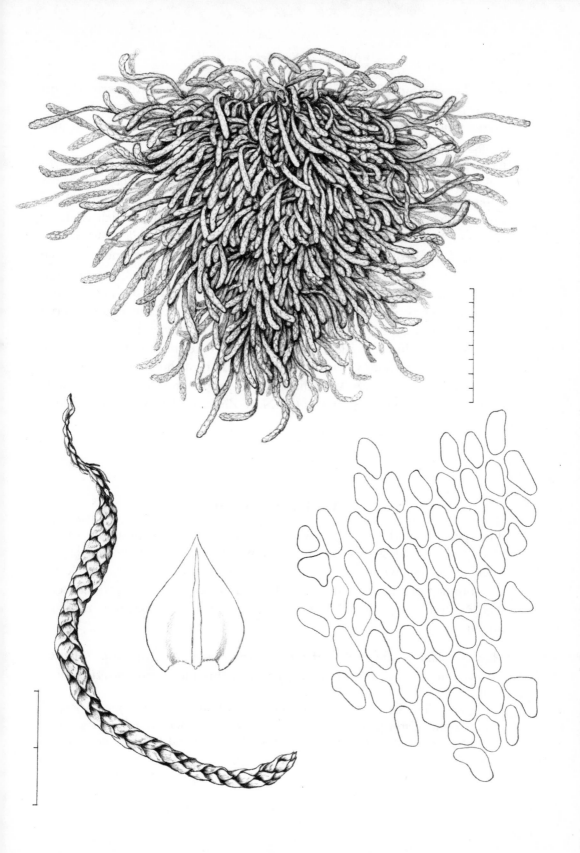

conical operculum. The spores are green, *c* 12–15 µm, faintly wrinkled on the surface.

DISTRIBUTION: TAS; and probably elsewhere; also in New Zealand and temperate regions of both hemispheres.

ILLUSTRATIONS: Plate 79; Handbook, Plate 69; Brotherus (1924–5), Fig. 676.

The only plant which normally rivals this one for smallness of size is *Eurhynchium muriculatum* [123] but that has much longer cells and a long beak on the operculum. The specific identity of this plant, which we have collected on damp shady rock, is doubtful. *A. austro-hygrophilum* Broth. and *novae-valesiae* Broth. & Watts have been recorded from NSW, and *A. varium* (Hedw.) Lindb. is also given for Australia in the Index, but our material seems to match the New Zealand plant which Sainsbury considered either *serpens* or *juratzkanum*. It is not without significance that Tasmanian material has been described as *A. austro-serpens* Broth. in Rodway, but that name is rejected by the Index as illegitimate. It would be unprofitable to speculate on the matter without a very extensive revision of the genus and a much wider knowledge of it in Australia than we can pretend to. It seems sensible to leave it in the same species as the New Zealand plant in the meantime.

Acrocladium Mitt.

The spear-pointed shoot tips which give this genus its name are highly characteristic and shared by only a few genera, such as *Wijkia* [132= *Acanthocladium*], which are easily distinguished in other ways. The leaves are rounded at the apex and have large and very conspicuous alar cells. The Australian species are both mosses of damp places but not normally aquatic.

A. cuspidatum is often split off into *Calliergonella* as *C. cuspidata* (Hedw.) Loeske but we have retained it as part of the broader genus *Acrocladium*.

PLATE 79. (below left) *Amblystegium serpens* TAS—Fruiting shoot ×7, cells × 1000; (top right) *Eurhynchium cucullatum* VIC—Whole shoot (above) ×1·5, detail (centre) ×7, leaf ×50, cells from mid-leaf (above) and leaf-base near margin × 1000

Key to species

1. Leaves *c* 2 × 1, the insertion almost as wide as the
   leaf . . . . *cuspidatum*
   Leaves *c* 1 × 1, much contracted to the narrow
   insertion . . . . *chlamydophyllum*

A. cuspidatum (Hedw.) Lindb.

The irregularly pinnate stems, 5–7 cm long, olive green above and yellowish green below, form loose wefts in damp grassland in parts of Victoria and probably elsewhere. The main shoots and principal branches are spear-pointed, the tightly imbricate leaves forming a shiny, firm point 2–5 mm long. The leaves, 1·5–2·0 × 0·8–1·0 mm wide, are roughly triangular in outline, but obtuse at the apex, sometimes with a small abrupt apical point, and generally rather flat or forming a shallow U-shaped channel; slightly undulate and/or plicate when dry. The cells are thin-walled, linear–rhomboid, 50–60 × 5 μm, slightly shorter near the apex, not at all different at the margins but with a very conspicuous alar group of inflated hyaline cells with a few small dense cells below, forming little strips, decurrent down the stem. The nerve is faint and double and the insertion on the stem takes up most of the width of the leaf base.

DISTRIBUTION: VIC; widely distributed in the northern hemisphere and also found in S. America and New Zealand.

ILLUSTRATIONS: Handbook, Plate 71; Watson (1955), Fig. 127.

For the distinctions between this and *A. chlamydophyllum* see under the latter.

A. chlamydophyllum (Hook.f. & Wils.) C.Muell. & Broth. (= *auriculatum* of the Handbook).

This is a much commoner species than the foregoing, in rather similar habitats but widespread in wet forest communities. The shoots, 2–4 cm long, are only sparsely branched, with unbranched terminal shoots 1–2 cm long, and spear points of *c* 0·5 cm. The leaves are rounded–triangular in outline, almost as wide as long (1–2 mm long × 0·8–1·0 mm wide), very concave and obtuse, single nerved to about ½ way up the leaf or else with a faint double nerve. The cells are slightly longer and narrower than in *cuspidatum*, *c* 60–80 × 4 μm, moderately thick-walled, shorter and narrower at the leaf tips, and shorter, wider, thick-walled and porose in the leaf base.

The leaves are very strongly contracted to the narrow insertion which only takes up a relatively small part of the wide leaf base, so that the alar cells are positioned well in towards the middle of the base of the triangular leaf outline. The alar cells are inflated, thin-walled and hyaline, very conspicuous and sometimes with decurrent strips.

Capsules, on long reddish setas, are more or less horizontal and have a normal double peristome.

CHROMOSOME NUMBER: n=11 (VIC).

DISTRIBUTION: TAS, SA, VIC, NSW; also in New Zealand. If Karczmarz (1966) is correct in separating this species from *A. auriculatum* of S. America, the distribution is limited to Australia and New Zealand.

ILLUSTRATIONS: Handbook, Plate 71; Brotherus (1924–5), Fig. 582; Allison and Child (1971), Plate 31.

Apart from the general appearance, which is quite different, the species is distinguished from *A. cuspidatum* by the very concave leaves, relatively much more nearly isodiametric, and by the shape of the leaf base. The inflated alar cells also prevent confusion with other concave-leaved mosses like *Weymouthia* [98] and *Lembophyllum* [103].

Campylium (Sull.) Mitt.

C. polygamum (B.S.G.) C.Jens, is hard to separate from *Leptodictyum riparium* [121], differing principally in being much more densely leafy, in having the leaf point channelled instead of flat, and the alar cells bulging out from the leaf outline to form *definite auricles*. The seta, too, is long by comparison, *c* 3–4 cm.

DISTRIBUTION: VIC, ACT; very widely distributed in Europe, Asia, America, New Zealand and Oceania.

ILLUSTRATIONS: Dixon (1924a), Plate 57 (as *Hypnum*).

We do not know the source of the Australian record given in the Index; the only material we have seen was collected recently in marshy sand not far from Wilson's Promontory, VIC, by Mrs E. Leech, and the identification was kindly confirmed by K. W. Allison.

C. molle Broth. in Rodway (1914), the only other species recorded from Australia, is probably *Cratoneuropsis relaxa*, according to Sainsbury (1956b, p. 41).

Cratoneuropsis (Broth.) Fleisch.

C. relaxa (Hook.f. & Wils.) Fleisch. in Broth.

Two very different-looking plants are encompassed by the range of variation of this species. In its most perfect development, it has long shoots to 20 cm (the Handbook gives an extreme of 15 cm in New Zealand), densely pinnately branched to give a feather-like frond, soft and woolly with very closely imbricated leaves which are widely spreading, with the leaf tips tightly hooked backwards like a grapnel (Plate 80). The colour is vivid green in the young parts with a bright yellow-brown tinge in older parts; it is very beautiful in shape and colour and is seen to perfection in seepages from cliffs where it can develop undamaged. On river banks and submerged boulders or in marshes, where it is quite a common species, the stems are often badly water-worn, the leaves scarcely reflexed and the colour dull green. Such forms can be hard to identify and are nondescript and unattractive. Almost always a few leaves will have the wide base and strongly reflexed point to reveal the species.

The stems have hair-like or narrowly tapering paraphyllia, sometimes very abundant but sometimes, in water-worn specimens, rather hard to find. The leaves are wide at the base, tapering to a short, wide point and measure c 0·8–1·5 × 0·4–0·7 mm, usually with a strong nerve failing well below the apex, but sometimes excurrent. The margins, which are normally serrulate throughout and especially strongly near the apex, are still almost parallel to the nerve at the point of insertion on the stem, not much curved in at the base. The cells are very variable, 20–50 × 6 µm, rhomboidal near the apex becoming rectangular below and square in the alar region where they form a rather ill-defined group.

PLATE 80. *Cratoneuropsis relaxa* VIC—Whole frond × 4 (right) with detail × 22, leaf × 50, cells × 1000

Capsules appear to be rare but the conical operculum, when present, is a useful identifying feature.

DISTRIBUTION: TAS, SA, VIC, NSW, ACT; also in New Zealand and S. Africa.
ILLUSTRATIONS: Plate 80; Handbook, Plate 71; Brotherus (1924–5), Fig. 672 (as *decussata*); Allison and Child (1971), Plate 30.

The Handbook (p. 431) comments that this species is "very variable in size and branching and in the set of the leaves" and that "the range of variation in the size of the plants is almost incredible". Despite this, the only other aquatic moss with which it is likely to be confused is *Eurhynchium austrinum* [123] which, however, has longer cells and a weaker nerve. When fruiting, the papillose seta and long-beaked operculum of *E. austrinum* are distinctive.

C. decussata (Hook.f. & Wils.) Fleisch. (VIC) is conspecific, according to the Handbook.

C. subrelaxa (Broth.) Broth. we have not seen.

Drepanocladus (C.Muell.) Roth

The strongly falcate leaves and single nerves of this genus are sufficient to distinguish it from almost all other pleurocarpous mosses except *Brachythecium paradoxum* [122], which is distinct in its fruiting characters. Apart from clearly defined alars, the cells throughout the leaf are long and very narrow. The capsule is rather short and curved, and the seta sheathed by very long perichaetial bracts at the base. The peristome is double and complete. Predominantly, it is an aquatic genus.

Key to species
1. Leaves plicate, 2–3 mm long. Not aquatic.   .   .   *uncinatus*
                                     (+*Brachythecium paradoxum*)
    Leaves not plicate, usually more than 3 mm long.
        Aquatic or semi-aquatic   2

2. Leaves entire, even at the apex . . . . . . *aduncus*
   Leaves denticulate, at least at the apex . . . . *fluitans*

D. uncinatus (Hedw.) Warnst.

The resemblance here at first glance is to *Hypnum* [133] and *Sematophyllum* [131] because of the small, very falcate leaves, closely imbricate, all along the stem, but the leaf anatomy is quite different.

The shoots are golden green, roughly pinnately branched, and 2 or more cm long, forming loose mats on damp earth or fallen logs in wet forest; very densely foliate indeed, with leaves 2–3 mm long × 0·4–0·5 mm wide at the base and tapering steadily from there to the long fine point. Even when moist, the leaves are *deeply plicate*, with 1–2 folds on either side of the single nerve and so strongly falcate as to be almost bent in a complete circle at times (circinate). The nerve extends up the leaf to ¾ or beyond and the margins are plane, slightly denticulate or entire. In mid-leaf, the cells are long and narrow, *c* 60 × 5 μm, shorter and with rather thick walls in the leaf base; the alar cells enlarged, square, hyaline and conspicuous, but not projecting to form auricles.

The seta is 1–5 cm long, crimson and *smooth*, with a curved rather narrowly cylindrical capsule of *c* 1·5–2·0 mm, widened to the mouth and with a conical or almost hemispherical operculum with an abrupt very short point on top. At the base of the seta, the perichaetial bracts are very long, sheathing, *plicate*, finely tapering and sometimes curled at the very tip.

DISTRIBUTION: TAS, VIC, NSW; widely distributed in Europe, Asia, Africa, America and New Zealand.

ILLUSTRATIONS: Plate 81; Handbook, Plate 69.

It is not common in Australia. The habitat alone is sufficient to distinguish this species from the rest of the genus, but the strongly plicate leaves too, will separate it from all but *Brachythecium paradoxum* [122], which is virtually inseparable morphologically when sterile. When fruiting, however, that species has a papillose seta and much shorter, not plicate, perichaetial bracts; there tends also to be a less abrupt change in size between the alar cells and those above.

D. fluitans (Hedw.) Warnst.

The stems here are a dull dark brown below with yellowish leaves at the growing tips, *c* 10 cm long and irregularly pinnately branched with branches in the region of 0·5–1·0 cm long. The leaves are 3–4 mm long × 0·3–0·5 mm wide near the base, very narrowly triangular or narrowly lanceolate, tapering to a long fine point and strongly falcate, the terminal leaves on stem and branches matted together in a thick hooked point. There is a single nerve, strong and distinct, extending well up the leaf. The margins are plane, entire or slightly crenulate–denticulate below, slightly but clearly denticulate at the apex. The cells are very long and narrow, 90–120 × 5–6 μm in mid-leaf, shorter and wider towards the base, with a few alar cells inflated and empty with strongly coloured walls forming a distinct group, especially on one side of the leaf, but often missed on dissection.

We have not seen sporophytes but, from descriptions, they are rather similar to *uncinatus* but with a straw-coloured seta and no annulus.

DISTRIBUTION: TAS, SA, VIC, NSW, ACT, ?? The world distribution is wide, rather similar to that of *uncinatus*. It is an aquatic or semi-aquatic plant of lakes, ditches and water trickles in open country.

ILLUSTRATIONS: Plate 82; Watson (1955), Fig. 122; Brotherus (1924–5), Fig. 677.

D. aduncus (Hedw.) Warnst. is a very similar species which differs in the completely entire leaf (the Handbook illustration is misleading here) and, reputedly, in the presence of an annulus in the capsule. The feature, mentioned in the Handbook, of the stem cells tearing away when the leaves are stripped off, seems to work better here than with New Zealand material: tails of cells are attached to the leaves at both nerve and margins when they are torn off. The distribution, once again, is very wide, especially in the northern hemisphere.

DISTRIBUTION: TAS, SA, VIC, NSW, ACT.

ILLUSTRATIONS: Handbook, Plate 70; Grout (1928–40), III, Plate 24.

D. fontinaloides (Hampe) Broth. (NSW) is recorded from the Blue Mountains.

D. strictifolius Broth. & Watts (NSW) is unknown to us but does not sound, from the type description, like a *Drepanocladus*.

PLATE 81. *Drepanocladus uncinatus* VIC—Fruiting shoot ×4 (below) with detail (above) ×15, cells ×1000

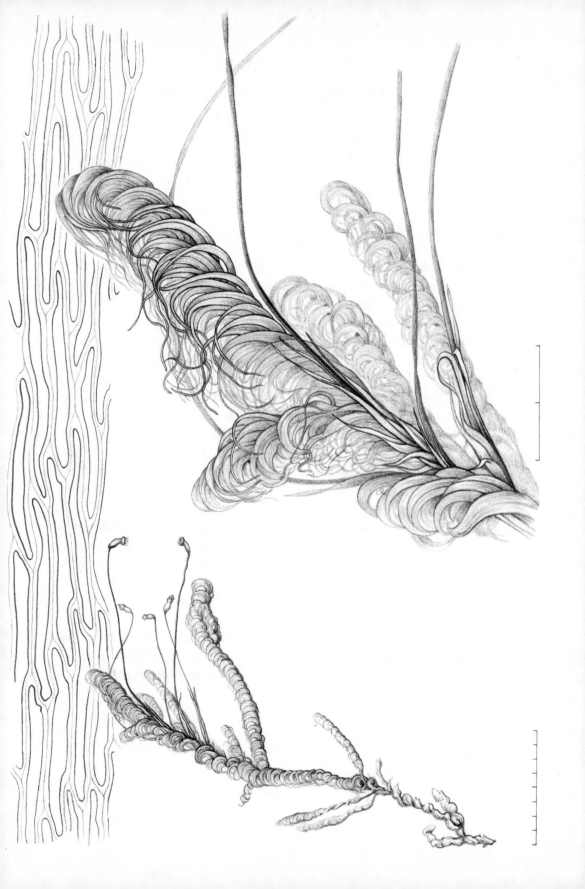

Leptodictyum (Schimp.) Warnst.

This genus is not greatly different from *Amblystegium* [116] in which it is sometimes included (e.g. Nyholm, 1965) but we retain it as a separate genus provisionally, for convenience.

L. riparium (Hedw.) Warnst.

There is a group of aquatic and semi-aquatic species, of which this is one, which can be very troublesome to separate, especially when not fruiting. It is not a common species, although probably overlooked. The shoots vary from olive green to golden- or brownish green, *c* 3–6 cm long or more, loosely branched in a roughly pinnate arrangement. The leaves are somewhat channelled below but flat above, lanceolate, *c* 2·3–3·1 mm × 0·5–0·7 mm, widest at no more than ⅕ way up from the base, tapering quite quickly below that to the insertion, and gradually tapering above it to a long, flexuose, *twisted, ribbon-like* point. The leaves crumple when dry and are very slow to soak out again. They spread widely, to give a shoot *c* 4–5 mm wide, and tend to be complanately arranged in the same plane as the branches to give a flattened, rather feathery frond. The nerve is single, faint and not very distinct, and the cells are long and narrow, mostly rather shortly pointed, *c* 75–120 × 10–15 μm in mid-leaf, rather shorter and more rectangular in the leaf base, especially towards the outer angles, but without clearly defined alar cells. The margins are *entire* but sometimes lightly crenulate with projecting cells.

We have not seen sporophytes in Australia. The capsule, held horizontally on a rather short seta, is short and fat, with a conical, not beaked, operculum.
DISTRIBUTION: SA, VIC, NSW, ACT; also in New Zealand and almost throughout the northern hemisphere. The Index does not record it for Australia.
ILLUSTRATIONS: Watson (1955), Fig. 119; Dixon (1924a), Plate 57 (as *Hypnum*).

This plant is very similar indeed to some forms of *Drepanocladus fluitans* [120] which, however, usually have the apical leaves slightly falcate. *Campylium polygamum* [118, q.v.] is even closer to *Leptodictyum*.

PLATE 82. *Drepanocladus fluitans* VIC—Shoot × 4 (above) with detail (below) × 15, cells × 1000

BRACHYTHECIACEAE
Includes: *Homalothecium australiense* Dix. (?wa)

Brachythecium B.S.G.

Apart from the falcate-leaved species, *paradoxum*, this genus is characterized by rather big straggly plants with straight ovate or triangular leaves, single nerved to well beyond mid-leaf, and with narrow cells except for the alar group which is more distinct or less, according to species. The capsule is rather short and the seta is frequently papillose—a crucial specific criterion which, unfortunately, is often not available. The conical operculum without a beak distinguishes the genus from the related *Eurhynchium, Rhynchostegium* etc.; *B. salebrosum* and/or *rutabulum* are almost inseparable vegetatively and are by far the commonest species.

Key to species

1. Leaves strongly falcate . . . . *paradoxum*
   Leaves straight 2

2. Leaves ending in a long, fine, hair-like point . . *albicans*
   Leaves pointed, but not hair-like at the tip . . 3

3. Seta smooth. Leaves clearly plicate. . . . . *salebrosum*
   Seta papillose. Leaves scarcely plicate. . . . . *rutabulum*

B. rutabulum (Hedw.) B.S.G.
It could reasonably be claimed that this is the biggest and coarsest of our pleurocarpous mosses. The shoots are loosely branched in prostrate or arching segments which are commonly 5 cm long or more, with fairly erect side branches, forming big rather loose yellowish green to deep green wefts among damp lush vegetation on the ground in forest or grassland. The leaves are very shining, slightly crumpled when dry, and lightly striate but scarcely plicate, c 2·3–3·0 × 1·6 mm wide for main stem leaves, 2·0–2·3 × 0·8–1·0 mm on branches, triangular–ovate, slightly narrowed to the base, very concave and narrowed to a rather abrupt short, spike-like

PLATE 83. *Brachythecium rutabulum* vic—Fruiting shoot extracted from mat × 4, leaf × 50, cells from leaf-margin × 1000

point. There is a single nerve to ½ way or beyond, and the margins are plane and usually denticulate almost throughout by projecting cell tips. The cells are typically 90–110×6 μm in mid-leaf, long and narrow, pointed at both ends but shorter and wider below, especially at the angles where they are rather thick-walled and porose although not conspicuously distinct there; beyond the angles there are very narrow slightly decurrent strips down the stem, only a few cells wide.

The seta is 2–3 cm long, *heavily papillose throughout*, and topped by a fat horizontal capsule, oval in profile, *c* 2 mm long × 1 mm wide, with a conical operculum, *c* 0·8 mm long, abruptly contracted to a very short sharp point. The peristome is double, with the usual double peristome mechanism for spore dispersal.

DISTRIBUTION: TAS, SA, VIC, NSW, ACT; widely distributed in both hemispheres.

ILLUSTRATIONS: Plate 83; Watson (1955), Fig. 133; Allison and Child (1971), Plate 31.

This is normally a big, very coarse-textured moss with rather spreading leaves. It is common among damp tall grass and on the bases of trees etc. Without sporophytes, it is hard to separate it from the next species.

B. salebrosum (Web. & Mohr) B.S.G.

Despite the claims of the Handbook, that this tends to be a smaller species, we have not found it so. In habitat, colouration, leaf shape and branching pattern it is very similar indeed to the preceding species but differs, from our rather restricted observations, in having the leaves not at all decurrent and usually appreciably plicate. The crucial feature is the *smooth* seta, when that is present.

DISTRIBUTION: TAS, VIC, NSW, ACT; almost world-wide.

ILLUSTRATIONS: Dixon (1924a), Plate 52; Nyholm (1965), Fig. 363.

This is at least as common in Australia as the preceding species, and has a not dissimilar world distribution except that it is missing from S. America.

B. albicans (Hedw.) B.S.G.

It seems certain that this species is introduced, originally from the northern hemisphere (Willis, 1955c, p. 11), but it is now not uncommon in VIC, and perhaps elsewhere. The silky shoots are slender and string-like, of a pallid olive green colour, 2–4 cm long × only 1 mm wide, and little branched. The leaves, narrowly triangular, *plicate*, densely imbricate and erect, measure *c* 2·7–3·0 × 0·8–1·0 mm, including a rather long fine point. There is quite a strong nerve reaching ¾ way up the leaf. The cells in mid-leaf are narrow, pointed at both ends, becoming quadrate in a big alar group. The margin at the base of the leaf is broadly reflexed and decurrent in quite a long and conspicuous tail. This decurrent strip does not seem to be mentioned in the literature.

DISTRIBUTION: TAS, VIC, ?? Also in Europe, N. Africa, and N. America, and introduced to New Zealand.

ILLUSTRATIONS: Watson (1955), Fig. 132; Grout (1928–40), III, Plate 7.

It is not uncommon in sandy lawns and pastures, and other dry earthy and sandy habitats. It is not unlike *Ischyrodon lepturus* [127, q.v.].

B. paradoxum (Hook.f. & Wils.) Jaeg. is distinguished by its plicate, strongly falcate leaves and small size. It appears to be a not uncommon alpine plant in VIC and is very like a small *Drepanocladus uncinatus* [120].

DISTRIBUTION: TAS, VIC, NSW, ACT; also in New Zealand, S. America and S. Africa.

ILLUSTRATIONS: Brotherus (1924–5), Fig. 688; Wilson and Hooker (1847), Plate 155.

B. rivulare B.S.G. (NSW) is predominantly a northern hemisphere species distinguished by the big inflated auricles but otherwise like *rutabulum*.

B. kayseri Geheeb and *B. novae-valesiae* Geheeb are both recorded only from NSW. We know nothing of either.

Eurhynchium B.S.G.

We have taken what many bryologists will consider the reactionary step of including in this genus species which the Index allocates to 4 different genera—*Eurhynchium, Oxyrrhynchium, Platyhypnidium,* and *Rhynchostegiella*—and have resisted a strong temptation to include *Rhynchostegium* as well. Effectively, the only difference between *Eurhynchium* and *Rhynchostegiella* is that, in the former, the tip of the nerve occasionally projects through the leaf abaxially to form a short spicule and this is said never to be the case in *Rhynchostegiella*, but this does not seem adequate grounds for a genus, even if the distinction were consistent. We have, in any case, plants which appear to be *Rhynchostegiella*, in which such a spicule is occasionally present. The unsatisfactory nature of the genus is immediately apparent when one tries to construct a key in which the two genera have to be separated. The separation of *Oxyrrhynchium* is little better, apparently depending mainly on slight differences in leaf shape; as for *Platyhypnidium*, Sainsbury (Handbook) has argued cogently that *P. austrinum* is quite out of place in that genus and we have followed him in retaining it in *Eurhynchium*.

Thus defined, all Australian species of *Eurhynchium* have leaves with single nerves usually to mid-leaf or beyond, with the nerve sometimes ending in a spicule projecting abaxially, the seta rough with papillae, a normal double peristome and an operculum with a long fine beak. None of them is common, but all are possibly much under-collected and may be a good deal commoner than the records suggest.

Key to species

1. Leaves (at least main stem leaves) wide, ovate. .   2
   Leaves narrow, lanceolate   3

2. Slender plants. Leaves dimorphous; branch leaves
   lanceolate, stem leaves widely triangular . .   *praelongum*
   Robust plants. Leaves all alike . . . . . .   *austrinum*

3. Leaves tapering to a long fine point . . . .   *muriculatum*
   Leaves shortly pointed or obtuse . . . . .   *cucullatum*

E. praelongum (Hedw.) B.S.G. (=*Oxyrrhynchium* of the Index)

The straggling dull dark green or dull yellow-green shoots of this species are very slender and rather densely pinnately branched in one plane giving a fine feathery frond extending 10 cm or more, with tufts of rhizoids at intervals, like a very slender *Thuidium*. The leaves are dimorphous; the stem leaves almost triangular, tapering to a long point, cordate at the base and with long narrow decurrent strips down the stem from the corners of the leaf. The branch leaves are narrow, acute, shortly pointed and lanceolate, not decurrent, and all leaves are nerved to almost mid-leaf, closely and sharply denticulate at the margins and moderately plicate. The cells are rather long and narrow, those at the angles wider and rectangular. We have seen no capsules.

DISTRIBUTION: TAS, SA, VIC; introduced? Throughout the northern hemisphere, S. America and New Zealand.

ILLUSTRATIONS: Watson (1955) Fig. 139; Dixon (1924a), Plate 54.

This is a rare or overlooked plant of damp grassland (including lawns) and may be introduced in Australia, at least in part of its range. The dimorphous leaves and very fine feathery fronds will separate it from all except *Thuidium laeviusculum* [114] which has quite different cells.

Oxyrrhynchium howeanum Broth. & Watts has been described from Lord Howe.

E. austrinum (Hook.f. & Wils.) Jaeg. (=*Platyhypnidium* of the Index.)
The coarse shoots of this semi-aquatic species have somewhat concave, pointed, ovate leaves, cordate at the base. It is a much larger plant than *praelongum* and the leaves are all alike. It is not unlike some forms of *Cratoneuropsis relaxa* [119] but differs in the lack of paraphyllia, and more weakly nerved leaves which are not at all curved backwards. We have seen very little Australian material.

DISTRIBUTION: TAS, VIC, NSW, ACT; probably elsewhere. Also in New Zealand.

ILLUSTRATIONS: Wilson (1854), Plate 89 (as *Hypnum*).

E. asperipes (Mitt.) Dix. is rather similar but has more sharply serrate leaves, shorter and broader, abruptly contracted to fine points which are incurved to give a slightly chain-like effect. When moist, the leaves are rather rigidly squarrose so that the plant somewhat resembles a small version of *Ptychomnion* [91]. Recently found in S.W. Victoria on logs in wet forest, it is also recorded from TAS, NSW, QLD and New Zealand.

E. muriculatum (Hook.f. & Wils.) Jaeg. (=*Rhynchostegiella* of the Handbook)
The dull green, irregularly pinnate shoots of this species are very small and narrow, with widely spreading often slightly complanate asymmetrical leaves, ovate–lanceolate, c 0·8–1·0 × 0·4 mm, slightly rolled in at the margins to give a narrow outline, and rather distantly set, giving a sparse appearance

to the foliage. The leaves taper to a fine narrow point, which is shrivelled when dry. The cells in mid-leaf taper to a point at both ends, and measure *c* 45–75 × 7–8 μm, those above rather shorter and those below, at the angles, shortly rectangular (15–30 × 9–12 μm) but not forming a distinct group.

The perichaetial bracts are wide and short (*c* 1 mm long), sheathing and tapering to fine points. The seta is sharply tuberculate all over, *c* 1 cm long or slightly more, and the capsule ovoid and fat, 1·6 × 0·8 mm, more or less horizontal. The spores are *c* 15–20 μm.

DISTRIBUTION: TAS, VIC, NSW, Lord Howe; also in New Zealand.

ILLUSTRATIONS: Wilson (1854), Plate 89 (as *Hypnum*).

This species is said, by Clifford and Willis (1951–2, p. 156), to be a common plant in shady habitats in TAS, but has so far seldom been recorded for VIC. It is one of the smallest of the pleurocarpous mosses, and is closest to *Rhynchostegium laxatum* [125] whose smooth seta is the only safe guide to distinguishing them, and to *Amblystegium serpens* [116] which has a much longer seta and much shorter cells. We have a specimen from ACT, apparently agreeing in all respects with *muriculatum* but with rather larger leaves than usual, 1·6 × 0·4 mm, and with the nerve sometimes ending in an abaxially projecting spicule. If not *muriculatum*, it is clearly very close indeed to it.

E. cucullatum (Mitt.) Stone & Scott, 1973

This is a very distinctive species which is easily recognized in the field once it is known. The typical habitat is muddy bases of shrub or tree trunks beside creeks liable to flooding, and the typical colouration is yellow-green and glossy with a bronze tinge; the stems are richly pinnately and fasciculately branched, forming an intricate mat. The leaves are even more distinctive, those on the small branches strongly inrolled, especially when dry, to make the leaf into a narrow tube (hence the superfluous, but more appropriate epithet of *convolutifolium* Hampe). When dry, they tend to be either slightly complanate or secund. They are concave, narrowly ovate, *c* 0·8–1·0 × 0·3–0·4 mm, rather broadly acute at the apex or sometimes obtuse, single nerved to about mid-leaf. The cells are pointed at both ends, *c* 35–50 × 7 μm in mid-leaf, wider towards both apex and base of the leaf, and with 3–4 rows of alar cells, in the leaf base, quadrate or rectangular in a patch extending obliquely up the margins. Most of the marginal cells have projecting tips giving a minutely denticulate edge to the leaf.

There is a very short tuberculate seta, *c* 6 mm long, with an almost erect capsule, *c* 1·2 × 0·6 mm, slightly curved and asymmetric. When mature, the capsule is mid-brown, with the neck, mouth and seta blackish. There is a double peristome, the outer teeth folding in half and penetrating behind the inner peristome, hooking it open on drying out. The operculum is shorter than in most other *Eurhynchium* species.

DISTRIBUTION: TAS, VIC, NSW, QLD; endemic. It is quite likely to turn up in other regions since the habitat is not a common collecting ground and the colouration is well camouflaged.

ILLUSTRATIONS: Plate 79.

Rhynchostegiella subconvolutifolia Broth. & Watts is probably the same thing.

We know nothing of *E. laevisetum* Geh. (NSW) and *E. devexum* (Bosw.) Par. (NSW).

Pseudoscleropodium (Limpr.) Fleisch. in Broth.

P. purum (Hedw.) Fleisch.

This introduced species grows in shaded grassy habitats, under trees in pine plantations, etc. It is not yet common.

The stems are of two sorts—robust main stems 2·0–2·5 mm wide, with narrower pinnate branches, often quite close, coming off dorso-laterally as in *Camptochaete* leaving the succession of leaves on the underside of the stem unbroken. The pale olive green or yellowish shoots extend to 7 cm long and the side branches to *c* 1·5 cm; all are very blunt. The main stems are also infrequently branched to produce further main branches which are more dorsal than lateral, so that the subsequent frond (also pinnately branched) comes to lie more or less on top of and parallel to the parent frond. This produces a multi-layered mat-like growth form which gives the plant its characteristic appearance.

The leaves are single-nerved to mid-leaf and diagnostic in shape: oval-oblong, with an *abrupt* short wide point which is usually slightly twisted and curled to one side, distinctly concave, glossy, slightly plicate and crumpled when dry, with the leaf tips curled back slightly, away from the stem. The cells are long and narrow, sigmoid at the ends and c 50–80 × 5 µm, rather thin-walled with slight thickenings at the tips. Towards the leaf-base the cells are rather wider and thicker-walled. We have not seen capsules.

DISTRIBUTION: VIC and probably other States. Widely distributed in the northern hemisphere and introduced to New Zealand and Australia.

ILLUSTRATIONS: Watson (1955), Plate 10, Fig. 143; Dixon (1924a), Plate 53 (as *Brachythecium*).

   This is not a difficult plant to recognize. The tiny projecting leaf tips are highly characteristic and so is the growth form. Both features are easily seen in the field.

Rhynchostegium B.S.G.

While it is rather more satisfactory than *Rhynchostegiella*, this segregate from *Eurhynchium* is by no means a strong genus and differs mainly in having a smooth, not papillose seta. The species may be common enough in Australia but we have rarely collected them and have had to rely mainly on our New Zealand collections and on material in the State herbaria.

R. tenuifolium (Hedw.) Reichdt.

The shoots of this small species are irregularly pinnately branched, pale yellowish green, with wide-spreading glossy leaves, commonly slightly complanate, giving a total shoot width of c 2 (–3) mm. The leaves are up to 1·5–2·5 mm long × up to c 1·5 mm wide, ovate, but with a narrow, fine and slightly twisted point which is sometimes long and hair-like, weakly nerved to about ⅔ way up the leaf; the margins plane and entire or serrulate, not at all decurrent. The cells are very long and narrow, pointed at both ends, c 60–100 × 6 µm (the Handbook gives 80–140 µm); shorter and wider at the angles but not forming a conspicuous alar group.

   It is autoicous. The perichaetial bracts have recurved, spreading tips. The capsule is short, plump and curved, c 1 mm long, on a *smooth seta* of c 2 cm or so; the operculum long and curved.

CHROMOSOME NUMBER: n = 22 (NSW).

DISTRIBUTION: TAS, WA, SA, VIC, NSW, ACT, QLD; also in S. America and New Zealand.

ILLUSTRATIONS: Handbook, Plate 72; Wilson (1859), Plate 176 (as *Hypnum collatum*).

This has something of the size, and often the colour, of small *Hypnum cupressiforme* [133] but the leaves are not at all falcate and the alar cells and capsule shape are quite different. In the absence of sporophytes it could easily be overlooked as small *Brachythecium rutabulum* [122].

R. laxatum (Mitt.) Par. differs from the previous species in the smaller cells (50–80 μm according to the Handbook) and the leaves more consistently ending in a fine hair-point, but there seem to be intermediates between the two species and clearly a thorough revision is needed.

CHROMOSOME NUMBER: n=22 (20+2 m) (NSW).

DISTRIBUTION: TAS, VIC, NSW, ACT, elsewhere? Also in New Zealand.

ILLUSTRATIONS: Wilson (1859), Plate 176 (as *Hypnum aristatum*).

The remaining species in the Index are:
R. dentiferum (Hampe) Jaeg. (VIC, NSW).
R. inaequale Dix. (QLD).
R. nano-pennatum (Broth.) Kindb. (QLD).
R. patulum Jaeg. (VIC, NSW, QLD).
R. pseudo-murale (Hampe) Jaeg. (VIC).
R. stramineoides (Sauerb.) Wijk & Marg. (NSW), all of which are endemic, and
R. rhaphidorrhynchum (C.Muell.) Jaeg. which is also known from S. Africa.
We have seen none of them.

FABRONIACEAE

Includes: *Anacamptodon wattsii* Broth. (NSW)
Helicodontiadelphus australiensis Dix.

Fabronia Raddi

There are not many genera in which the leaf margin is ciliate with long hairs, but it is a common feature of this genus. However, it is an astonishingly variable feature, even in a single species, and there seems little doubt that many of the published species are merely synonyms whose authors have been deceived by this polymorphism. The plants are delicate and tiny, often glistening with cilia and forming soft mats on bark or soil. The tiny capsule is short and erect, with a single peristome on a proportionately long seta arising from a short branch on the main stem. The number of species in Australia, as elsewhere, is uncertain and there is a great need for critical revision. *F. hampeana* in particular is a species which seems, from our material, to have been misunderstood.

F. australis Hook.

These are extremely delicate and pretty little plants which are not uncommon in a variety of well drained, periodically moist, stable habitats which are usually shaded from direct sun. They are peculiarly common on the trunks of cycads in the NSW coastal forests, but are also found on dry soil on or at the foot of rock ledges etc.

The stems are creeping, anchored by bunches of smooth rhizoids, which arise from the cells in the base of the leaf next to the nerve, and bearing numerous short erect branches, mostly *c* 0·5 cm tall. The shoots are soft and silky, very slender indeed (less than 1 mm wide including leaves) but forming quite thick and dense turfs of a slightly olive green, tinged with grey from hair-points and cilia.

The leaves are imbricate, narrowly ovate, *c* 0·6–0·8 × 0·2–0·3 mm with a hair-point extending to 0·2–0·4 mm, usually secund, all pointing towards the same (upper) side of the stems. There is a single weak nerve reaching about mid-leaf (sometimes apparently absent) and plane margins which may be completely entire or ciliate with long hairs, or any stage in between. The variation in this feature, which is presumably strongly affected by such environmental factors as humidity, has been responsible for much taxonomic confusion. The marginal cilia, when present, have a characteristic origin from a long narrow marginal cell which bends in the middle or below so that the upper part of the cell sticks out abruptly from the margin like a rag-nail, sometimes with support from an adjacent cell at the base. The margins are thus shredded. It is, in fact, an extreme example of the kind of denticulation

PLATE 84. *Fabronia australis* NSW—Fruiting shoot × 30, leaf (centre) × 50, cells × 1000; *F. australis* VIC leaf (right) × 50; *F. hampeana* WA leaf (left) × 50

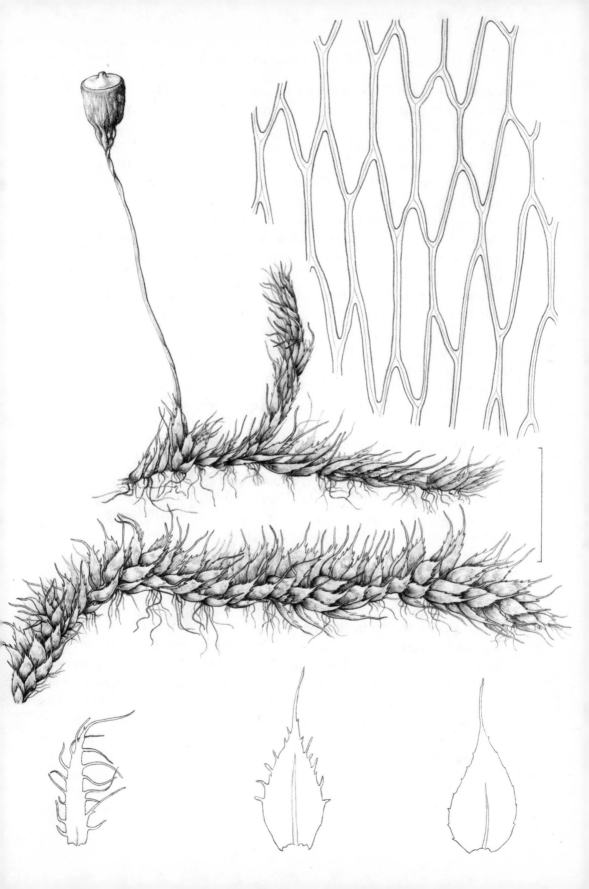

found on the margins of many pleurocarpous mosses. The hair-point at the leaf apex ends also in a long single cell, up to 170 μm long. Both cilia and hair-point tend to be hyaline or nearly so. Other cells in mid-leaf are pointed at both ends and vary greatly in size from plant to plant, from 30–190 × 8–12 μm; those in about 4 rows at the angles are quadrate forming two patches of short cells which cut diagonally across the corners of the leaf. The cell walls are all firm, and even (in places) thick.

Sporophytes are borne on specialized branches at the base of the current year's growth and consist of a rather slender seta, *c* 5 mm long, by 80 μm diameter, pale yellowish, bearing a capsule which varies from hemispherical to widely conical, *c* 0·6–0·8 mm long and wide. There is a flat peristome, inserted under the brown rim, with 16 teeth united in pairs, short and wide, heavily striated by lines of papillae on the outer face. There is no inner peristome. The spores are green, *c* 12 μm, coarsely warted–papillose. The operculum is *flat*, when dry, with a slight pimple in the middle. The membranous calyptra falls off early.

CHROMOSOME NUMBER: n = 20 (NSW)

DISTRIBUTION: TAS (apparently a new record), WA, SA, VIC, NSW, ACT, QLD; also in New Zealand.

ILLUSTRATIONS: Plate 84; Handbook, Plate 73.

The extreme variation from entire to ciliate leaf margins does not prevent this species from being easily recognizable by a combination of its size, growth form and rather large cells. The main difficulty is in separating it from *F. Lampeana*. It seems very probable that *F. australis* is merely a facies of the widespread *F. ciliaris* (Brid.) Brid. (= *F. octoblepharis*).

F. tayloriana Hampe is held to be *australis* by Clifford and Willis (1951–2, p. 153).

F. hampeana Sonder in Hampe

We believe that this species has so far only been recorded from WA and that the records from VIC and NSW are the ciliate form of *australis*. From the latter, *hampeana* differs in being much more ciliate, the hairs so dense that the narrow leaves are scarcely discernible through them. The capsule, too, differs in the rounded–conical operculum and perhaps in the rather stout seta which, in our material, is *c* 2 mm long × 100–115 μm in diameter. The leaf cells measure about 76–90 × 8–9 μm and the terminal cells of the leaf extend to

190 μm or more, but these are not distinctive. The single nerve may be very faint or apparently absent.

DISTRIBUTION: WA; ?endemic.

ILLUSTRATION: Plate 84; Hampe (1844), Plate 13.

This species, too, frequents cycad trunks. It is not improbable that intermediates between this species and *australis* exist in the intervening country, but the two are worth keeping separate until proved conspecific.

The other Australian species given in the Index are:

F. brachyphylla C.Muell. in Broth. (NSW, ACT, QLD) and *F. scottiae* C. Muell. (NSW, QLD), both supposedly endemic to Australia. Both species are entire-leaved (or nearly so); the former has broad obtuse leaves, at least on most shoots, and has recently been re-discovered by Prof. D. Catcheside; the latter has acuminate leaves with a long terminal cell.

F. leptura (Tayl.) Broth. we have treated as an *Ischyrodon*.

Ischyrodon C. Muell.

I. lepturus (Tayl.) Schelpe

The pallid green to fawn stringy shoots of this moss are quite common in dry stony and sandy soils, *c* 0·5–2·0 cm long or less, and very slender, *c* 0·2 mm wide, varying from slightly to quite densely branched. The leaves are narrowly ovate, concave, especially below, 0·6–0·8 × 0·3 mm with a single strong nerve nearly to the apex; the margins are plane, entire below and slightly crenulate–denticulate at the apex. The cells are long and narrow, pointed at both ends, *c* 90 × 9 μm in mid-leaf, narrower at the margins and shorter below; quite conspicuous patches of short squarish cells forming the alar groups extend obliquely across the corners of the leaf and pass right across the adaxial face of the very base of the nerve to join up with each other. Capsules are unknown in Australia.

DISTRIBUTION: WA, SA, VIC, NSW; also in New Zealand and S. Africa (Schelpe, 1970).

ILLUSTRATIONS: Brotherus (1924–5), Fig. 639; Sim (1926), p. 365 (both as *I. seriolus*).

Formerly the species was treated as *Fabronia leptura* (Tayl.) Broth. Its closest resemblance is to small *Brachythecium albicans* [122] from which it differs in the cells at the base of the leaf and in the smooth, not plicate, leaves.

ENTODONTACEAE

Includes: *Entodon mackayensis* C.Muell. (QLD)

E. pallidus Mitt. in Seem. (QLD)

E. pancherianus (Besch.) Jaeg. (Lord Howe)

E. tasmanicus Mitt. (TAS)

E. terrae-reginae Dix. (QLD)

Campylodontium flavescens (Hook.) Bosch & Lac.

Trachyphyllum inflexum (Harv.) Gepp in Hiern (QLD)

T. papuanum (Broth.) Broth. (QLD)

PLAGIOTHECIACEAE

Plagiothecium B.S.G.

This genus is distinguished from its close relatives by a combination of characters: entire, decurrent leaves, absence of pseudoparaphyllia, and having the outermost layer of stem cells large and thin-walled (Ireland, 1969).

P. denticulatum (Hedw.) B.S.G.

The colour of this world-wide species is its most striking feature in the field —a vivid, pale, very glossy green. The stems are c 2 cm long and quite wide, c 2·5–3·0 mm, sparsely branched with both branches and leaves flattened in one plane. The shoots tend to be rather sparse, forming an open, straggling weft. The leaves are densely imbricate and complanate, slightly concave but not folded, c 1·2–1·4 × 0·6–0·7 mm, widely ovate or oval, the leaves at the sides of the shoots very asymmetric. The nerve is rather faint and double, disappearing below mid-leaf; the margins plane and entire or very slightly denticulate near the apex which has a short, rather broad and usually slightly curved point. The cells in mid-leaf are *c* 60–90 × 9 µm, pointed at both ends, thin-walled; those at the base slightly wider, especially at the angles where they are shortly rectangular and often hyaline, *decurrent* in a long narrow strip. Near the leaf apex the cells are shorter. Characteristically the chloroplasts tend to aggregate at the ends of the cells to give a diagonally green-striped effect, rather as in *Goniobryum*.

The capsule is held horizontally, slightly curved and rather plump, *c* 2 mm long and 1 mm wide, rather feebly ribbed when dry; the operculum conical with a distinct beak. There is a normal double peristome.

DISTRIBUTION: TAS, VIC, ?NSW, ?QLD; almost world-wide.

ILLUSTRATIONS: Handbook, Plate 73; Ireland (1969), Plates 5, 6.

Although widely distributed in the cooler and moister regions of Australia, this is by no means an abundant species. It tends to grow on humus on the ground, or on tree bases or trunks in wet sclerophyll and rain-forests. The very strongly complanate leaf arrangement will separate it from *Eurhynchium* spp. [123]; the decurrent leaves and the nerve, even if short, from *Isopterygium* [130].

P. novae-valesiae Broth. is very similar in appearance but has pseudoparaphyllia, the leaves are scarcely decurrent and the cells are wider, 10–14 µm wide in one specimen we have seen. It is recorded only from NSW and seems unlikely to come within the range of variation of the genus, but further study is required to settle its status.

Catagonium C.Muell.

C. politum (Hook.f. & Wils.) Dus. ex Broth.

This moss is common and easily recognized by the pale yellowish green or sometimes dark green glossy prostrate *flattened shoots*, 1·5–2·0 mm wide, irregularly branched and usually quite densely overlain to produce a thickish mat. The leaves are virtually entire, triangular–oblong in outline when flattened out, contracted at the very base but normally tightly folded along the mid-line and spreading at 45° to the stem; *distichous and densely imbricated at the bases*, the leaves stacked into each other to give the appearance of a continuous ribbon-like stem. The leaves are nerveless but contracted at the apex into a short, recurved, spike-like tip; sometimes curled slightly to one side, to make the whole shoot a shallow inverted U-channel. The thin-walled cells are very long and very narrow, c (60–) 80–130 × 3–5 μm, the apical and basal cells shorter and wider, especially away from the margins, but with no alar cells.

The female branches are very short, c 1·5 mm long, at the base of the current year's growth. The male branches are still smaller, bud-like, under 1 mm. There is a short, deep maroon-coloured seta, c 1·5 cm long, with a capsule which is narrowly cylindrical, c 3 mm long, horizontal, curved down-wards, and expanded to a wide mouth and double peristome. The basal portion of the capsule (c ¼) is rather narrow and wrinkled, forming a definite apophysis. The membranous side-split calyptra covers little more than the operculum, which has a long inclined beak.

DISTRIBUTION: TAS, VIC, NSW, ACT, QLD, Lord Howe; also in S. Africa, S. America and New Zealand.

ILLUSTRATIONS: Plate 72; Handbook, Plate 73; Allison and Child (1971), Plate 32.

The systematic position of this moss is uncertain. Brotherus placed it in the Neckeraceae, presumably from its vegetative characters, while the Handbook preferred Plagiotheciaceae on the basis of sporophyte. It fits convincingly into neither. It is a very common moss in wet forest, less common in dry, on stones and soil and among grass, more rarely epiphytic. In very wet habitats a dark green form occurs. Once it is known and the tightly folded leaves recognized for that and not mistaken for half-width unfolded leaves, it can be confused with almost nothing else, even when the frond is not shiny, as sometimes happens.

Orthorrhynchium elegans (Hook.f. & Wils.) Reichdt. is like a glorified *Catagonium*, several times wider and with a spectacular, mirror-like metallic gloss, perhaps the shiniest of all mosses. We have seen it occasionally in NSW and frequently in QLD. Outside Australia it is known from S.E. Asia, Oceania, and New Zealand, where it is a common epiphyte in rain-forest.

Isopterygium Mitt.

One sometimes has the feeling that this is a genus defined by elimination, for there is virtually no nerve, usually no alar cells, and nothing distinctive about the capsule or peristome. It is held to be related to *Plagiothecium*, mainly because of the long narrow pointed cells and glossy complanate leaves, but the complanate arrangement is often not evident. The leaves, however, are never decurrent in *Isopterygium*, the outermost stem cells are small and thick-walled, and pseudoparaphyllia are often present. The Index lists 11 species from Australia, of which we only know one, and Willis (1955a, p. 161) has recorded another.

I. limatum (Hook.f. & Wils.) Broth.
The small very glossy shoots are mostly only *c* 1 cm long, not much branched, and only 0·5–1·0 mm broad, including the leaves. The leaves, pale glistening yellowish green, are often strongly curved—sometimes as much as in *Hypnum*—or sometimes nearly straight; typically *c* 1 mm × 0·3 mm, slightly concave, triangular–lanceolate, tapering to a long fine point. The margins are plane and entire, and there is virtually no nerve. The cells are very narrow and long, 70–90 × 5 μm, pointed at both ends, slightly sigmoid and often with porose walls, shorter and wider at the very base where there is about one row of almost quadrate cells along the insertion, but elsewhere throughout the leaf little altered, and not at all different at the angles. The rhizoids are finely papillose.

The capsule has a high rounded–conical operculum.
DISTRIBUTION: TAS, VIC; also in New Zealand.
ILLUSTRATIONS: Handbook, Plate 74.

This appears to be a rare species, growing on humus-rich substrates in rain-forest. It is very like a tiny *Hypnum cupressiforme* [133] without alar cells and with papillose rhizoids.

I. acuminatum Boswell (TAS) is similar but with straight complanate leaves with hair-points (Sainsbury, 1956b, p. 45). Material in Herb. MEL is merely a growth form of *limatum* (J. Lewinsky, pers. comm.).

I. albescens (Hook.) Jaeg. was discovered in VIC by Willis (1955a). It differs from *limatum* in the complanate, not falcate, leaves. It is also found in NSW,

QLD, Lord Howe, E. and S.E. Asia, New Zealand and Oceania. It is illustrated in Bartram (1933), Fig. 188.

I. amblyocarpum (Hampe) Broth., from Apollo Bay, VIC, and from NSW, we have not seen.

The remaining species all seem to be tropical:

I. amoenum Broth. (NSW).
I. arachnoideum Broth. (NSW).
I. howeanum Broth. & Watts (Lord Howe).
I. latifolium Broth. (NSW).
I. minutirameum (C.Muell.) Jaeg. (QLD).
I. neocaledonicum Thér.
I. norfolkianum (C.Muell.) Jaeg.
I. subarachnoideum Broth. (NSW).
I. neocaledonicum is found in New Caledonia, *minutirameum* in S.E. Asia and
     Oceania. The other species are endemic.

SEMATOPHYLLACEAE

Includes: *Acanthorrhynchium papillatum* (Harv.) Fleisch. (QLD)
            Acroporium erythropodium (Hampe) Broth. (QLD)
            A. scalarirete Dix. (QLD)
            Glossadelphus dimorphus Dix. (QLD)
            G. serrifolius (Broth. & Watts) Broth. (QLD)
            Macrohymenium mitratum (Doz. & Molk.) Broth. (QLD)
            Meiothecium secundifolium Dix.
            M. wattsii (Broth.) Broth. (NSW, QLD)
            Rhaphidostegium lucidloides Dix. (NSW)
            Taxithelium instratum (Brid.) Broth. in Renn. (QLD)
            T. kerianum (Broth.) Broth. (QLD)
            T. novae-valesiae (Broth.) Broth. (NSW, QLD)
            T. petrophilum Williams (QLD)
            T. wattsii Broth. (QLD)

and also *T. selenithecium* (C.Muell.) Par., omitted from the Index but
recorded from QLD on the authority of Bartram (1952, p. 245).

Trichosteleum elegantulum Broth. & Watts (QLD)

T. hamatum (Doz. & Molk.) Jaeg. (QLD)

T. muscicola Broth. (Lord Howe)

T. pallidum Dix. (QLD)

Warburgiella cupressinoides C.Muell. ex Broth. (QLD)

Sematophyllum Mitt.

The splitting off of *Rhaphidorrhynchium* (Schimp.) Fleisch. to include the
falcate-leaved species, is to be deplored unless grounds for the separation can
be found which will command more respect; for example *leucocytus* and
amoenum, which are very similar in all other respects, and not always very
different in leaf arrangement, thereby find themselves in different genera.
There are, in fact, much more discordant elements left within *Sematophyllum*,
such as *homomallum* or *contiguum*, and there are any number of ways of sub-
dividing the genus on the basis of one or two arbitrary characters.

In the broad sense, as used in this treatment, the genus is characterized by
smooth rhizoids in bunches on the prostrate stems, ±nerveless leaves,
conspicuous alar cells which in most cases include a few inflated balloon-like
cells at the angles, leaves usually secund, often falcate, in some cases pointing
downwards and in others upwards, small capsules with double peristomes
and a seta which is smooth or nearly so.

Key to species

1. Leaves, at least of ultimate branches, falcate and
    pointing downwards to the substratum . . . 2
    Leaves straight or curved upwards, away from the
    substratum 4
2. Leaves of main (primary) steps straight or only
    slightly falcate 3
    Leaves of main stems strongly and clearly falcate. *amoenum* and
    subcylindricum

3. Plant small with leaves to 0·5 mm long. On logs and trees . . . . *leucocytus*

Plants large with leaves over 2 mm long. On wet rock . . . . *uncinatum*

4. Leaves whitish or bleached-yellow, tapering to long hair-like points . . . . *contiguum*

Leaves bright or dark green or bronze; points short, not hair-like 5

5. Alar cells inflated, in 1 row. Plants green; on very wet ground . . . . *jolliffii*

Alar cells in several rows. Plants usually bronze; on soil, rock, or tree bases, but not consistently wet *homomallum*

S. amoenum (Hedw.) Mitt. (*Rhaphidorrhynchium amoenum* of the Index) In the field, there is no way of distinguishing this species with certainty from small *Hypnum cupressiforme*. The leaves are falcate and curled towards the underside of the stem, varying from nearly straight in some leaves to ¾ of a circle in others, *c* 1 mm × 0·2 mm, ovate–oblong below and with a long fine, slightly channelled point. The margins are plane and crenulate, toothed in the subula, and there is no nerve. The cells in mid-leaf are very narrow, pointed at both ends, *c* 45–75 × 4–6 µm, with quite firm walls thickened especially at the ends of the cells; those at the insertion are yellow with 2–3 big hyaline inflated alar cells.

It is monoicous and sporophytes are common. The capsule varies in size, 0·8–1·6 × 0·2–0·6 mm, and is almost cylindrical, generally curved horizontally or even hanging; the smooth seta, 0·5–1·0 cm tall, is reddish and twisted (left hand). The smooth green spores usually measure 8–15 µm in diameter. The operculum has a long fine point almost as long as the rest of the capsule and usually slanting or curved. There is a double peristome with single quite evident cilia between the processes and with the outer teeth beautifully ornamented along the edges by projecting papillose bars. The almost square outermost cells of the capsule wall have angular, collenchymatous thickenings at the corners.

CHROMOSOME NUMBER: n = 11 (10 + m) (NSW, QLD); 11 (NSW); 22 (NSW).

DISTRIBUTION: TAS, WA, SA, VIC, NSW, ACT, QLD; also in New Zealand.

ILLUSTRATIONS: Mueller (1864), Plate 14 (as *Hypnum callidioides*).

The inflated alar cells will distinguish this common species from *Hypnum cupressiforme* [133]; usually the capsule is smaller too, but this is not a safe character to rely on and it is better to trust the slender curved beak on the operculum if it is present or the bordered peristome teeth. It is a very common species in wet and dry forest, mainly on trees and logs, less commonly on the ground.

Rhaphidorrhynchium callidioides (C.Muell.) Broth. is conspecific with *S. amoenum* as Sainsbury (1956b) realized, and *R. calliferum* (Geh. & Hampe) Fleisch. probably is also (Studies, p. 308) although we have seen no material.

S. subcylindricum (Broth. ex Fleisch.) Sainsb. is very similar, vegetatively, to the preceding species; it is said to differ in the recurved leaf margins, but this is not a constant feature. Fruiting material, however, is more easily separated. The capsule is typically almost erect, cilia of the inner peristome usually absent and the spores thick-walled and papillose, very variable in size but commonly 18–25 µm, rising to 34 µm. The capsule wall thickenings too, tend to be slightly different from those of *amoenum* with extra thickenings on the longitudinal walls, and less on the transverse walls, approaching the condition in *leucocytus*. *S. subcylindricum* is not recorded from Australia in the Index, but we have found material in VIC and TAS which must be referable to it, and it is also recorded by Clifford and Willis (1951–2, p. 156).
DISTRIBUTION: TAS, VIC; also in New Zealand, S.E. Asia.
ILLUSTRATIONS: ?

S. leucocytus (C.Muell.) Sainsb.
There is a curious bleached, whitish tinge about the small dense mats of this species, which is not hard to recognize and which distinguishes it from other species like *amoenum*. The leaves are narrow, long and tapering; those on the prostrate main stems usually *almost straight* but those on the erect or side branches falcate as in *amoenum*, but tending to be plastered together at the branch tips to form thick points like a camel-hair brush, only falcate; they measure *c* 1·0–1·4 × 0·2–0·3 mm; the margins are slightly denticulate almost throughout the leaf but especially in the long *flat* subula. The cells are very narrow, *c* 50–75 × 4–5 µm; the alar cells inflated.

It is a dioicous species and the perichaetia are much sparser than in *amoenum*. The perichaetial leaves have long tapering points, and are sharply denticulate above. The seta, *c* 1·5–2·0 cm long, is twisted (left hand) and usually *slightly papillose*, in part, with low rounded papillae. The capsule has a distinct *angular ridge* round the very base, like a collapsed apophysis and the outermost cells of the capsule wall have the thickenings of the longitudinal walls pronounced, giving a *striped* appearance under the microscope. The spores are mainly 18–24 μm, green and finely papillose, often oblong or oval and very variable in shape as well as size.

DISTRIBUTION: TAS, VIC; also in New Zealand. (The Handbook records it also from S. America, but the Index does not.)

ILLUSTRATIONS: Mueller (1864), Plate 15 (as *Hypnum trachychaetum*).

This is not a particularly common species, in our limited experience; it grows on logs etc. in fern gullies and rain-forest and differs from *amoenum* in the whitish colour, papillose seta, the straightish leaves on the main stem, and in the different thickenings in the capsule wall as well as the distinctive capsule base (sometimes found in *amoenum*). The sporophyte characters are the most satisfactory to distinguish the species.

Rhaphidorrhynchium trachychaetum (F.Muell.) Broth. is clearly conspecific.

S. contiguum (Mitt.) Mitt. in Seeman

The prostrate pale yellow mats have much the colouration of *leucocytus*, but the leaves, 0·8–1·8 × 0·2–0·5 mm, are *straight* throughout, scarcely at all falcate, although the branches are commonly slightly curved and pointing upwards i.e. secund to the dorsal side of the stem. The leaves are slightly channelled below but flat above with a long fine point, slightly denticulate in the upper ½. The cells are similar to those of *amoenum*. The seta is *c* 1 cm long, twisted (left hand), smooth, and deep crimson, carrying a curved horizontal or pendulous capsule, *c* 1 × 0·4 mm, with a long operculum and a peristome as in *amoenum*; commonly there is a slight *struma* on the *upper* side of the neck. The green spores are 8–18 μm in our material (10–12 μm in the Handbook), sparsely and finely papillose.

DISTRIBUTION: TAS, SA, VIC, NSW, ACT, Lord Howe and probably QLD; also in New Zealand and Oceania, S. Africa and S. America(?).

ILLUSTRATIONS: Wilson (1859), Plate 177 (as *Hypnum*).

The straight, secund leaves and pallid colouring are sufficient to separate this species from others of the genus. It is a plant of rain-forest, fern gullies and river banks, on logs, stones etc.

S. aciculum (Dix.) Dix. has also been recorded from Australia (NSW, QLD, Lord Howe). It differs from the previous species, according to the Handbook, in having "subquadrate and scarcely inflated" alar cells, but Sainsbury doubted if it was distinct from *contiguum*. We have seen no material on which to form an opinion, but the alars of *contiguum* may be either inflated or quadrate, even on the same plant, so the distinction between the two species is not impressive on paper.

S. homomallum (Hampe) Broth.

Of all the Australian species of the genus, this is perhaps the easiest to recognize because of the characteristic dull bronze-green colour, and erect, secund leaves. The shoots are very glossy, with ovate–lanceolate leaves tapering to a fine but short, not hair-like, point, c $1·4–1·7 \times 0·3–0·5$ mm, slightly concave, with the margins plane or slightly recurved. The cells are c 60 μm × 7 μm (becoming shorter at the apex), tapering at both ends, with quite firm walls which are thicker and porose near the leaf base. The alar cells are in a large orange group of c 4 rows of 4 enlarged thick-walled cells increasing in size towards the leaf base, where the basal 3–4 cells are usually inflated (despite the Handbook's claim to the contrary).

Bud-like male branches are sometimes very abundant in the leaf axils on the upper side of the stem.

The seta is about 5–7 mm long, yellowish orange and twisted (left hand) with an almost erect cylindrical capsule of c $1·2 \times 0·5$ mm, very slightly curved. The capsule wall has a striped appearance as in *leucocytus*. The spores are green, 24–27 μm, finely granular-papillose.

DISTRIBUTION: TAS, WA, SA, VIC, NSW and probably QLD; also in New Zealand, Oceania, and S.E. Asia.

ILLUSTRATIONS: Brotherus (1924–5), Fig. 738; Hampe (1844), Plate 6 (as *Leskea*).

This is a very common plant in wet and dry sclerophyll forest, coastal scrub, swamp forest etc. It occurs on rock but seems to be at least equally common on tree bases and branches. It is perhaps the most discordant of the species we know in this genus.

S. jolliffii (Hook.f.) Dix. (=*Rhaphidorrhynchium jolliffii* of the Index)
We have seen this species only rarely in Australia. It is similar in size to the preceding species but occurs in swamps and on wet rock, and is pale dirty

golden-yellow in colour. It is distinguished by having, on most leaves, only *one row* of 3–6 inflated thin-walled and coloured alar cells.

DISTRIBUTION: TAS, VIC; also in New Zealand.

ILLUSTRATIONS: Wilson (1859), Plate 177 (as *Hypnum*).

S. uncinatum Stone & Scott, 1973 (=*tenuirostre* of the Handbook)

This, too, is a big semi-aquatic species of montane springs and bogs, and has a reddish-bronze tinge. The stems are long and quite broad for the genus, to *c* 5 cm × 2 mm, *falcate and hooked* at the tips, dull glistening yellow. The leaves, 2·0–2·7 × 0·7–0·9 mm, are ovate and *concave*, tapering to a sharp point which is usually short but may be quite long and fine, slightly but distinctly falcate and strongly secund. They tend to be crumpled and undulate when dry. The leaves are much as in *amoenum*, with similar cells but yellow leaf bases and inflated thin-walled alar cells.

The perichaetial leaves are wide, short, and entire, tapering to a short point which is not at all hair-like. There is a smooth, red, slightly twisted (left hand) seta, to *c* 2 cm, and a horizontal to pendulous capsule; the spores are 15–20 μm, smooth.

DISTRIBUTION: TAS, VIC, possibly NSW although Burges (1952) says the record is doubtful; also in New Zealand and S. America.

ILLUSTRATIONS: Hooker, W. J. (1818–20), II, Plate 111 (as *Hypnum*).

This is a rare species, perhaps closest to *Brachythecium paradoxum* [122] in habitat and appearance, but differs, of course, in the inflated alar cells, in the absence of nerve, and in the smooth seta.

The following, of which we have seen no material, are also on the Australian list; the first is illustrated in Mueller (1864):

Rhaphidorrhynchium congruens (Hampe) Broth. (VIC).

R. tuloferum (Hampe) Broth. (VIC).

Sematophyllum caespitosum (Hedw.) Mitt.

Wijkia Crum
(=*Acanthocladium* of the Handbook)

W. extenuata (Brid.) Crum (=*A. extenuatum* (Brid.) Mitt.)

Once known, this is recognized to be an extremely common moss in forests, especially wet ones, on ground and rotting logs, but it is not a striking species at first glance and is quite variable. The stems are *red*, quite closely pinnately branched at right angles, either on both sides to form flat fronds, or on one side with most of the branches erect, the fronds 4–12 cm long or even more with the main branches and stems tending to end in a spike-like point. The leaves are U-channelled in transverse section; those of the main stem and branches 0·8–1·2 × 0·3–0·4 mm, with an additional 0·7 mm or more of fine hair-point, only 2–3 cells wide, abruptly produced from the broad apex of the oblong–ovate leaf. The margins are slightly toothed at the apex and along the hair-point, those in the upper half rolled in to form a rather tubular upper half of the leaf. Despite the hair-point, the nerve is lacking or very short and double. The leaves of the branchlets are narrow, more tapering and with a shorter point. In mid-leaf, the cells are long and narrow, 40–70 × 6 μm, very thick-walled and porose, especially near the ends. The lower cells are shorter, yellow at the insertion of the leaf; those at the angles inflated and balloon-like, empty, hyaline and thick-walled in a large, conspicuous group. The cells are often papillose on both surfaces in mid-leaf and above, with the papillae often uniseriate. Sometimes, on the other hand, all cells are completely smooth.

Fruit is quite common, from short, lateral sexual branches on the main stem. The seta, 3–4 cm long, is bright red, with a slightly curved cylindrical capsule *c* 2 mm long with a wide mouth. There is a normal double peristome and a conical operculum.

CHROMOSOME NUMBER: $n = 11$ (NSW), 20 (NSW, TAS).

DISTRIBUTION: TAS, VIC, NSW, ACT, QLD, Lord Howe; also in New Zealand and Oceania.

ILLUSTRATIONS: Plate 85; Handbook, Plate 74.

The red stems, spiky pinnate fronds with the branches spreading at right angles, and the hair-pointed but nerveless leaves are sufficient to separate this species from almost all others, despite its great variability, and it is usually not a difficult species to recognize in the field. Once known, it is not likely to be confused with any other. Not uncommonly, forms are found with masses of short erect branches, bearing very reduced round leaves. These may be propagules but that is not yet known. The Handbook records a similar form in New Zealand where it is, apparently, less common.

W. crossii (Broth. & Geh. ex Broth.) Crum (NSW) and *W. rigidifolia* (Dix.) Crum (QLD), both endemic, are unknown to us.

The transfer from the illegitimate but more familiar *Acanthocladium* is made and discussed by Crum (1971).

HYPNACEAE

Includes: *Ctenidium pubescens* (Hook.f. & Wils.) Broth. (NSW)
           Ectropothecium condensatum Broth. & Watts. (NSW)
           E. howeanum Broth. & Watts. (Lord Howe)
           E. leucochloron (Hampe) Broth. (Lord Howe)
           E. moritzii Jaeg. (QLD)
           E. riparioides Bartr. (QLD)
           E. sandwichense (Hook. & Arnott) Mitt. in Seem. (QLD)
           E. sydneyense Dix. (QLD)
           Taxiphyllum taxirameum (Mitt.) Fleisch. (QLD)
           Trachythecium verrucosum (Jaeg.) Fleisch. (QLD)
           Vesicularia montagnei (Bel.) Broth. (Lord Howe)
           V. rivalis Broth. (NSW, QLD)
           V. slateri (Hampe) Broth. (NSW, QLD)

Hypnum Hedw.

Although it once contained the majority of pleurocarpous mosses, this genus has since been whittled away until relatively few species remain in it. Of recent years, this splitting has been extended to the species level with renewed attempts to fragment the polymorphous *H. cupressiforme* but, as far as we know, without objective biometric data in support. The recent paper by Ando (1972), however, allows Australia only three species and two more are recorded in the Index.

PLATE 85. *Wijkia extenuata* VIC—Whole fruiting shoot × 1·5 (below) with detail × 7, leaf × 50, upper cells and alar cells (below) × 1000

H. cupressiforme Hedw.

There may possibly be more polymorphous mosses than this species, but, if so, they are not nearly so well known. "*Hyp. cup.*" has been the despair of many experienced bryologists and all beginners, for the range of variation is staggering. We have the impression that it is not quite so variable in Australia as in Europe but that may only mean that it has been less collected. The larger forms are very easily recognizable with fair certainty in the field, but the smaller forms are virtually indistinguishable macroscopically from *Sematophyllum* spp. The rhizoids are smooth in both. In almost all cases, the shoots have a very characteristic appearance like a plaited pigtail when viewed from above, because of the way the falcate leaves are bent downwards to alternate sides, but this feature is scarcely detectable in the var. *filiforme*. Perhaps the commonest, middle-sized, form of the plant has pinnately branched stems, to 2 cm, with pale green or olive green leaves c 1·2 mm long, tapering to a fine hooked point; the nerve absent or, more usually, short and double; margins plane, crenulate–denticulate or entire. The cells are long and narrow, pointed at both ends, c 60–90 × 6 μm, with the alar cells, which are diagnostic, numerous, rather square, *dark*, usually with granular contents, rather thick-walled and often coloured, sometimes with a few enlarged and hyaline cells at the extreme angle.

The variety *filiforme*, by contrast, has very fine, much less falcate, narrow leaves and very slender stems, sometimes growing out from bigger falcate-leaved shoots. It often fruits and is not uncommon as an epiphyte in forests; it is correctly treated by the Index as a mere form. The variety *tectorum*, at the other extreme, rarely fruits, is very robust and plump, olive green to deep golden brown, with leaves 2·5–3·0 mm, ovate and very concave with a rather short and broad denticulate point.

There is a variable seta, 1–2 cm long, slender, red and twisted when dry (left hand), conspicuously sheathed at the base by long perichaetial bracts. The capsule, c 1–2 mm long, is cylindrical and horizontal with a rather fine-beaked operculum. Usually the beak is straight and short. The outer peristome teeth are densely transversely striated but lack the papillose marginal ornamentation of *Sematophyllum*.

PLATE 86. *Hypnum cupressiforme* VIC—Shoot × 4; (upper) near var. *tectorum* VIC—Shoot × 4; (lower right) near var. *filiforme* VIC—Fruiting shoot × 4, cells × 1000

CHROMOSOME NUMBER: n = 10 (TAS).

DISTRIBUTION: TAS, WA, SA, VIC, NSW, ACT, QLD, Lord Howe; cosmopolitan.

ILLUSTRATIONS: Plate 86; Allison and Child (1971), Plate 32; Watson (1955), Plate 12.

This is a very common and widespread species in all sorts of habitats from rain-forest to dry sclerophyll forest and open grassland, at all altitudes and on all substrates—rock, soil, wood, humus—or epiphytic. There seems to be no limit to its variability. Even with a great deal of experience, it is impossible to be certain of separating it in the field from *Sematophyllum amoenum* [131], although the shorter, straighter operculum is fairly reliable; in the laboratory the dark alar cells, instead of balloon-like alar cells, will separate it easily.

H. chrysogaster C.Muell. has orange- or yellow-coloured alar cells. Ando (1972) accepts it as a good species from TAS, but it is treated in the Handbook as a variety of *cupressiforme*. We do not know of a reliable distinguishing feature for it.

H. subchrysogaster (Broth.) Par. is from QLD.

H. walterianum (Hampe) Jaeg. (TAS, NSW) is merged into *cupressiforme* both by Ando and by Sainsbury (1956b, p. 47).

ADDITIONS AND CORRECTIONS

Sphagnaceae

Sphagnum cymbifolioides C.Muell. (=*subsecundum* of the Index)
Investigations at the British Museum have shown that the Australian "*S. subsecundum*" does not match northern hemisphere specimens and is best called by its earliest name, *S. cymbifolioides*, pending further investigation. (A. Eddy, pers. comm.)

Fissidentaceae

Fissidens bifrons Schimp. ex C.Muell.
This most distinctive species has a short, wide leaf consisting almost entirely of vaginant lamina, the dorsal lamina reduced to 2–4 cells wide and the upper lamina almost absent. The leaf apex is abruptly recurved at the very tip and the leaves are widely spaced on straggly stems up to 1 cm long, giving a peculiar appearance. The fertile stems are shorter, with larger leaves partly bordered, the border sometimes intramarginal.
DISTRIBUTION: WA, VIC; also in S. Africa.
ILLUSTRATIONS: Brotherus (1924–5), Fig. 120.

This species has recently been found in Australia and is probably widespread. It has been considered conspecific with *splachnifolius* Hornsch. by Sim (1926) but that species has the capsules erect (Sim, p. 188) whereas our material has them slightly curved.

Ditrichaceae

Cheilothela chilensis (Mont.) Broth.
Anatomically there is no more distinctive moss than this in the Australian flora. The opaque, fleshy leaves, which are incurved when dry, have something of the solid appearance of *Polytrichum* leaves but are bistratose, almost

throughout, with high vesicle-like mamillae on both surfaces. The shoots are *c* 0·5–2·0 cm tall, often very yellowish, especially on the old growth. It has recently been found in subalpine and alpine peaty grassland in Victoria and has also been found in the Snowy Mountains by McVean (1969).
DISTRIBUTION: VIC, NSW; also in New Zealand and S. America.
ILLUSTRATIONS: Handbook, Plate 10.

Dicranaceae

Trematodon amoenum (C.Muell.) Stone & Scott comb. nov. Basionym: *Bruchia amoena* C.Muell. *Flora* **71**, 11. 1888.

Pottiaceae

Aloina bifrons (De Not.) Delgadillo
Delgadillo (1973) considers *A. sullivaniana* to be synonymous with this species, which has priority. However, he describes *bifrons* as having no costa (nerve) whereas Australian specimens of *sullivaniana* have a broad nerve, 4–6 cells thick, of heterogeneous cells although stereids are rare. The synonymy cannot therefore be accepted without further investigation.

Barbula crinita Schultz (=*pseudopilifera* of the Index)
Weber (1972) has examined material of *B. pilifera* (Hook.) Brid. from S. Africa and S. America and proclaimed it conspecific with Australasian *B. pseudopilifera*. There is a controversy about whether *pilifera* should be in *Barbula* or *Tortula* since it is somewhat intermediate in structure but we have, at least temporarily, preferred to retain it in *Barbula*. In that genus, however, the combination *B. pilifera* was already in use for another plant and the earliest alternative name, given by the Index, is *B. crinita*.

Hookeriaceae

Achrophyllum Vitt & Crosby
The proposal for conservation of *Pterygophyllum* [109] has now been rejected (*Taxon* 24: 249. 1975).

Bibliography

The principal references on taxonomy of Australian mosses are marked with an asterisk.

ALLISON, K. W. (1960). Contributions to the knowledge of the New Zealand bryophyte flora. *Trans. R. Soc. N.Z.* **88**(1), 9–12. (*Fissidens epiphytus* sp. nov.)

ALLISON, K. W. (1963). New and rare mosses in New Zealand. *Trans. R. Soc. N.Z. (Bot.)* **2**(11), 133–41. (*Fissidens variolimbata* sp. nov.)

ALLISON, K. W. and CHILD, J. (1971). "The Mosses of New Zealand." 155pp. University of Otago Press, Dunedin. (Details and illustrations of most common spp.)

ANDO, H. (1972). Distribution and speciation in the genus *Hypnum* in the circum-Pacific region. *J. Hattori bot. Lab.* **35**, 68–98. (*H. chrysogaster, subchrysogaster, cupressiforme* from Aust.)

ANDREWS, A. L. (1945). Taxonomic notes V. The genus *Tetrapterum. Bryologist* **48**, 190–93. (Includes *T. cylindricum*, Aust.)

ANDREWS, A. L. (1947). Taxonomic notes VI. The Leucobryaceae. *Bryologist* **50**, 319–26.

ANDREWS, A. L. (1949). Taxonomic notes VIII. The genus *Acrocladium. Bryologist* **52**, 72–77. (*A. chlamydophyllum* (Australasia) is distinct from *A. auriculatum* (S. America))

ANDREWS, A. L. (1951). Taxonomic notes X. The family Leptostomaceae. *Bryologist* **54**(4), 217–23. (Family reduced to part of Bryaceae)

ASHTON, D. H. and McRAE, R. F. (1970). Distribution of epiphytes on beech trees at Mt. Donna Buang, Victoria. *Victorian Nat.* **87**, 253–61 (after p. 324)

BAILEY, F. M. (1883). "A Synopsis of the Queensland Flora; Containing Both the Phaenogamous and Cryptogamous Plants." Govt. Printer, Brisbane. (Mosses pp. 724–36; index pp. 864–6)

BAILEY, F. M. (1886). First supplement to Bailey, 1883. Govt. Printer, Brisbane. (Mosses pp. 65–69)

BAILEY, F. M. (1888). Second supplement to Bailey, 1883. Govt. Printer, Brisbane. (Mosses pp. 67–73)

BAILEY, F. M. (1890a). Third supplement to Bailey, 1883. Govt. Printer, Brisbane. (Mosses pp. 95–100)

BAILEY, F. M. (1890b). "Catalogue of the Indigenous and Naturalised Plants of Queensland." Govt. Printer, Brisbane. 116 pp. including addenda to 3rd supplement of Synopsis. (Mosses pp. 60–63, 114–15)

BAILEY, F. M. (1891a). Contributions to the Queensland Flora. *Botany Bull. Dep. Agric. Qd* No. 2 (March 1891). (Descriptions of: *Archidium, Ceratodon purpureus, Phascum cylindricum, Tortula wildii, Bryum pusillum, Plagiobryum wildii, Wildia solmsiellacea, Lepidopilum australe, Meteorium flexicaule, Porotrichum deflexum, Fissidens calodictyon*, and other records without descriptions)

BAILEY, F. M. (1891b). (As Bailey, 1891a), No. 4 (Dec. 1891). (Mosses p. 21.)

BAILEY, F. M. (1892). (As Bailey, 1891a), No. 5 (May 1892). (Mosses p. 29. Descriptions of *Amblystegium* and *Polytrichum* and list of spp. with localities)

BAILEY, F. M. (1893). (As Bailey, 1891a), No. 8 (Dec. 1893). (Mosses pp. 87–90. Descriptions of: *Archidium brisbanicum, Leucoloma clavinerve, Fissidens splachnoides, Bryum tryonii, B. immarginatum, Hookeria karsteniana, Pterobryum recurvulum, Thuidium nano-delicatulum*, some of which are nomina nuda)

BAILEY, F. M. (1896). (As Bailey, 1891a), No. 13 (April 1896). (Mosses pp. 17–19. Descriptions of: *Gymnostomum calcareum, Tortula nervosa, Funaria calvescens, Rhynchostegium tenuifolium*, with list of other species and localities)

BAILEY, F. M. (1898). Contributions to the Flora of Queensland. *Qd agric. J.* **3**(3), 203. (Descriptions of *Leucobryum strictifolium, Calymperes panduraefolium, Leptohymenium papuanum*)

★BAILEY, F. M. (1913). "Comprehensive Catalogue of Queensland Plants." Govt. Printer, Brisbane. (Mosses pp. 656–72)

BARTRAM, E. B. (1933). Manual of Hawaiian mosses. *Bull. Bernice P. Bishop Mus.* **101**, 1–275.

BARTRAM, E. B. (1944). Henry Neville Dixon. *Bryologist* **47**(4), 137–44. (Includes a useful 6p. bibliography)

BARTRAM, E. B. (1951). West Australian mosses. *Trans. Br. bryol. Soc.* **1**(5), 465–70. (*Campylopus angustilimbatus, Dicranum contortifolium, Pottia scabrifolia, Pohlia cuspidata, Anomobryum filescens* spp. nov.)

BARTRAM, E. B. (1952). North Queensland mosses collected by L. J. Brass. *Farlowia* **4**(1), 235–47. (List of spp. and collecting data, including: *Fissidens terrae-reginae, F. subkurandae, Campylopus brassii, Dicranoloma spiniforme, Leucoloma circinatulum, Ectropothecium riparioides, Eriopus brassii* spp. nov.)

BASTOW, R. A. (1886a). Tasmanian mosses, their identification, etc. *Pap. Proc. R. Soc. Tasm.* for 1885, pp. 318–20. (Part 1)

BASTOW, R. A. (1886b). Moss flowers, split-moss, bog-moss, and earth-moss. *Pap. Proc. R. Soc. Tasm.* for 1885, pp. 337–41. (Part 2, referring to Andreaeaceae, Sphagnaceae, Bryaceae Tribe Phascae)

BASTOW, R. A. (1886c). Mosses of Tasmania (continued), Tribe 2, Weissiae. *Pap. Proc. R. Soc. Tasm.* for 1885, pp. 395–99. (Part 3)

BASTOW, R. A. (1887). Tasmanian mosses. *Pap. Proc. R. Soc. Tasm.* for 1886, pp. 38–102 including large illustrated fold-out key. (Final part, 4)

BASTOW, R. A. (1892). Description, collection, and preservation of mosses. *Victorian Nat.* **9**, 123–27.

BASTOW, R. A. (1905). Cryptogamic botany at Braybrook. *Victorian Nat.* **22**, 144. (5 mosses listed)

BASTOW, R. A. (Aug. 1907). Mosses. *The School Paper (Classes V & VI)*, pp. 103–06. (Article for school children of Victoria; no scientific importance)

BEAUGLEHOLE, A. C. (1947). Checklist of the indigenous flora of the Lower Glenelg. *Victorian Nat.* **64**, 78–86. (16 Mosses, p. 78)

BEAUGLEHOLE, A. C. and LEARMONTH, N. F. (1957). The Byaduk caves. *Victorian Nat.* **73**, 204–10. (List of 8 mosses including *Tortella dakinii* and also *Anoectangium bellii* new to Australia)

BESCHERELLE, E. (1876). Note sur un *Phascum* pleurocarpe de la Tasmanie. *Revue bryol.* **3**, 29. (*Pleurophascum grandiglobum*)

BIBBY, P. (1954). Cryptogams of the 1948 Archbold Cape York (Queensland) Expedition. *J. Arnold Arbor.* **35**, 260–65. (*Camptochaete brisbanica, Eriopus* sp.)

BOPP, M. and BÖHRS, H-L. (1965). Versuche zur Analyse der Protonema-entwicklung der Laubmoose. III. Die Regeneration der Caulonemen von *Funaria hygrometrica. Planta*, **67**, 357–74.

BOSWELL, H. (1892). New exotic mosses. *J. Bot., Lond.* **30**, 97–99. (*Macromitrium prolixum* NSW, *Isopterygium acuminatum* TAS, *Acrocladium trichocladium* AUST, *Hypnum devexum* NSW, spp. nov.)

BRIDEL-BRIDERI, S. E. von (1826–7). "*Bryologia Universa* seu Systematica ad Novam Methodum Dispositio, Historia et Descriptio Omnium Muscorum Frondosorum Hucusque Cognitorum cum Synonymia ex Auctoribus Probatissimis." 2 vols. Barth, Leipzig.

★BROTHERUS, V. F. (1890). Some new species of Australian mosses described. I. *Öfvers. finska Vetensk Soc. Förh.* **33**, 1–22 (23 spp. from Austr.)

★BROTHERUS, V. F. (1893). (As Brotherus, 1890), II. *Öfvers. finska Vetensk Soc. Förh.* **35**, 34–56 (spp. nos. 24–52; 24 of them from Austr.)

★BROTHERUS, V. F. (1895). (As Brotherus, 1890), III. *Öfvers. finska Vetensk Soc. Förh.* **37**, 149–72 (spp. nos. 53–89; 22 of them from Austr.)

★BROTHERUS, V. F. (1898). (As Brotherus, 1890). IV. *Öfvers. finska Vetensk Soc. Förh.* **40**, 159–93 (spp. nos. 90–141; 8 of them from Austr.)

★BROTHERUS, V. F. (1900). (As Brotherus, 1890). V. *Öfvers. finska Vetensk Soc. Förh.* **42**, 91–129. 1 Plate. (spp. nos. 142–189; 28 of them from Austr.)

★BROTHERUS, V. F. (1924–5). Unterklasse Bryales II. Spezieller Teil. *In* A. Engler & K. Prantl "Die natürlichen Pflanzenfamilien" Teil I, Bd. 10, pp. 143–478; Bd. 11, pp. 1–542. Edn. 2, Engelmann, Leipzig.

★BROTHERUS, V. F. (1916). Descriptions of some new species of Australian, Tasmanian, and New Zealand mosses. VI. *Proc. Linn. Soc. N.S.W.* Ser. 2. **41**(3), 575–96. (spp. nos. 190–226; 34 of them from Austr.)

★BROTHERUS, V. F. and Watts, W. W. (1912). The mosses of the Yarrangobilly Caves District, N.S.W. *Proc. Linn. Soc. N.S.W.* **37**(2), 363–82. (*Dichodontium wattsii, Tortula (Syntrichia) brunnea, T. (S.) subbrunnea, Philonotis (Euphilonotis) austrofalcata, P. (E.) fontanoides, Amblystegium (Euamblystegium) Novae-Valesiae,*

Sciaromium (Aloma) elimbatum, S. (A.) Forsythii, Drepanocladus (Warnstorfia) strictifolius, Ectropothecium (Cupressina) condensatum, Rhynchostegiella subconvolutifolia spp. nov.)

⋆BROTHERUS, V. F. and WATTS, W. W. (1915). The mosses of Lord Howe Island. *Proc. Linn. Soc. N.S.W.* **40**(2), 363–85. (22 spp. nov.)

⋆BROTHERUS, V. F. and WATTS, W. W. (1918). The mosses of North Queensland. *Proc. Linn. Soc. N.S.W.* **43**, 544–67. (*Pterobryidium* gen. nov., 14 spp. nov.)

BROWN, R. (1811). Some observations on the parts of fructifications in mosses; with characters and descriptions of two new genera of that order. *Trans. Linn. Soc. Lond.* **10**, 312–24 (*Dawsonia polytrichoides, Leptostomum inclinans*)

⋆BURGES, A. (1932). Notes on the mosses of New South Wales. I. Additional records and description of a new species of *Buxbaumia*. *Proc. Linn. Soc. N.S.W.* **57**, 239–44. (Completion of Watts & Whitelegge's Catalogue for NSW; *B. colyerae* sp. nov.)

⋆BURGES, A. (1935). Notes on the mosses of New South Wales. II. Additional records. *Proc. Linn. Soc. N.S.W.* **60**, 83–93.

BURGES, A. (1949). The genus *Dawsonia*. *Proc. Linn. Soc. N.S.W.* **74**, 83–96.

BURGES, A. (1952). Census of the N.S.W. Mosses; Unpub. 84 pp. Botany School, University of Sydney. (Including an earlier one by Watts)

BURBIDGE, N. T. and GRAY, M. (1965). The plants of the Australian Capital Territory. *Tech. Pap. Div. Pl. Ind. C.S.I.R.O. Aust.* 21. (Includes key to moss genera. Bryophyta—Musci, pp. 16–30)

BURR, I. L. (1939). The development of the antheridium, archegonium and sporogonium of *Cyathophorum bulbosum* (Hedw.) C.M. *Trans. R. Soc. N.Z.* **68**(4), 437–56.

CARDOT, J. (1908). Notes bryologiques IV. Le *Dicranum Novae-Hollandiae* Hsch. *Bull. Herb. Boissier.* Ser 2. **8**, 173–74.

CATCHESIDE, D. G. (1958). Some bryophytes collected in Arnhem Land. *In* "Records of the American–Australian Scientific Expedition to Arnhem Land" (R. L. Specht and C. P. Mountford, eds) Vol. 3. Botany and Plant Ecology, Section 8, p. 169. Melbourne University Press, Melbourne. (*Pseudephemerum axillare*)

CATCHESIDE, D. G. (1967). *Tortula pagorum* in Australia. *Muelleria* **1**(3), 227–30.

CLARKE, G. C. S. (1973). Type specimens in Manchester Museum Herbarium: Musci. *Mus. Publs Manchr Mus.* New Series, no. NS. 2.73. pp. 1–20.

CLIFFORD, H. T. (1952). Victorian Musci Part I: Introduction and Andreaeaceae. *Proc. R. Soc. Vict. NS.* **64**(1), 4–9.

CLIFFORD, H. T. (1955). On the distribution of *Rhacomitrium crispulum* (H.f. & W.) H.f. & W. *Bryologist* **58**, 330–34. (Relationship of this sp. to *R. heterostichum*, and list of 39 closely related or synonymous spp.)

⋆CLIFFORD, H. T. and WILLIS, J. H. (1951–2). The genera of Victorian mosses, and new records of species for the State. *Victorian Nat.* **68**(8–9), 135–8, 151–8. (Records of 35 spp. not previously known from Victoria)

Bibliography

COGHILL, G. (1904). The Buffalo Mountains Camp-Out. Thursday, 24th December, 1903, to Monday, 4th January, 1904. *Victorian Nat.* **20**, 144–59. (List of mosses p. 159)

COSTIN, A. B. (1954). "A Study of the Ecosystems of the Monaro Region of New South Wales with Special Reference to Soil Erosion." Govt. Printer, Sydney. 860 pp. (List of mosses in Table 14 pp. 72–4. Also mosses in bog communities in Chapters 17–19)

CROCKER, R. L. and EARDLEY, C. M. (1939). A South Australian *Sphagnum* bog. *Trans. R. Soc. S. Aust.* **63**(2), 210–14. Pl. 8.

CROSBY, M. R. (1968). *Micromitrium* Aust., an earlier name for *Nanomitrium* Lindb. *Bryologist* **71**, 114–7. (Includes comb. nov.)

CROSBY, M. R. (1972). *Pterygophyllum* Brid.—nomen rejiciendum et illegitimum. *Taxon* **21**, 205–09.

CRUM, H. (1971). Nomenclatural changes in the Musci. *Bryologist* **74**, 165–74 (*Wijkia* nom. nov.)

CRUM, H. (1973). A taxonomic account of the Erpodiaceae. *Nova Hedwigia* **23**, 201–24 (description of *Wildia solmsiellacea* QLD)

CRUNDWELL, A. C. (1970). Notes on the nomenclature of British Mosses. I. *Trans. Br. bryol. Soc.* **6**(1), 133–38. (Several points of relevance to Australian mosses e.g. spelling of *Rhacomitrium*)

CRUNDWELL, A. C. (1971). (As Crundwell, 1970), II. *Trans. Br. bryol. Soc.* **6**(2), 323–26.

★CRUNDWELL, A. C. and NYHOLM, E. (1964). The European species of the *Bryum erythrocarpum* complex. *Trans. Br. bryol. Soc.* **4**(4), 597–637. (Many of these spp. are now known to occur in Australia)

DELGADILLO, C. (1973). A new species, nomenclatural changes, and generic limits in *Aloina*, *Aloinella*, and *Crossidium* (Musci). *Bryologist* **76**, 271–77.

DIXON, H. N. (1912). On some mosses of New Zealand. *J. Linn. Soc. (Bot.)* **40**, 433–59. Plates 20, 21. (*Leucobryum teysmannanum* NSW, *Thuidium hastatum*, *Campylopus insititius*)

★DIXON, H. N. (1913–29). "*Studies* in the Bryology of New Zealand, with Special Reference to the Herbarium of Robert Brown, of Christchurch, New Zealand." 6 parts. Govt. Printer, Wellington.

DIXON, H. N. (1915). New and rare Australasian mosses, mostly from Mitten's herbarium. *Bull. Torrey bot. Club* **42**, 93–110. Pl. 9. (*Dicranoloma angustiflorum* TAS)

DIXON, H. N. (1916). Miscellanea bryologica. V. *J. Bot., Lond.* **54**, 352–59. (*Dicranoloma dichotomum* = *D. billardieri*; Australasian *Cryphidium* spp.)

DIXON, H. N. (1924a). "The Student's Handbook of British Mosses." 3rd Ed. Sumfield and Day, Eastbourne.

DIXON, H. N. (1924b). Miscellanea bryologica. IX. *J. Bot., Lond.* **62**, 228–36. (*Fissidens humilis* stands against *F. coarctatus* nom. nud. in synon.; *Rhynchostegiella cucullata* comb. nov.)

DIXON, H. N. (1929). Critical mosses. *Revue bryol.* N.S. **2**, 21–29. (*Rhaphidostegium luciduloides* NSW, *R. aciculum* NSW, QLD, Lord Howe, spp. nov.; *Whiteleggea australis* Broth. = *Hampeella pallens*; *Rhaphidostegium ovale* Broth. = *R. caespitosum*)

DIXON, H. N. (1936). Decas generum novorum muscorum. *J. Bot., Lond.* **74**, 1–10. Plate 610. (*Helicodontiadelphus australiensis* sp. nov., Kangaroo Valley, NSW)

DIXON, H. N. (1937). Notulae bryologicae. I. *J. Bot., Lond.* **75**, 121–9. (*Goniobryum* spp. reduced to give a monotypic genus; *Tortula luteola* = *T. pseudopilifera* comb. nov.)

DIXON, H. N. (1938). Mosses in North Queensland. *N. Qd Nat.* **6**, 2, 4. (List of plants belonging to different world floristic regions)

*DIXON, H. N. (1942). Additions to the mosses of North Queensland. *Proc. R. Soc. Qd* for 1941, **53**(2), 23–40. (20 spp. nov.)

*DIXON, H. N. (1950). Notes on the moss collections of the Royal Botanic Garden, Edinburgh. Part II. *Notes R. bot. Gdn Edinb.* (1948) **20**, 93–102. (*Fissidens pachyneuron* NSW, VIC, *F. homomallulus* NSW, *Schlotheimia funiformis* ?WA, *Pterobryum australiense* NSW, *Fabronia leptura* WA, *Homalothecium australiense* ?WA, *Meiothecium secundifolium, Ectropothecium sydneyense* NSW spp. nov.)

DIXON, H. N. and RODWAY, L. (1923). On *Phascum tasmanicum*. *Pap. Proc. R. Soc. Tasm.* for 1922, pp. 25–26 (including fig.)

DOIGNON, P. (1953). Les *Stereodon* exotiques. *Revue bryol. lichen.* **22**, 34–51. (*Hypnum cupressiforme* sensu lato)

DUBY, J. E. (1870). Choix de Cryptogames exotiques nouvelles ou mal connues. *Mém. Soc. Phys. Hist. nat. Genève* **20**, 351–64. 4 pls. (*Dicranum dichotomum, D. menziesii, Campylopus australiensis* sp. nov., *C. erythropoma* sp. nov.)

ERDTMAN, G. (1943). "An Introduction to Pollen Analysis." 239 pp. Chronica Botanica, Waltham, Mass.

ERLANSON, C. O. (1930). The attraction of carrion flies to *Tetraplodon* by an odoriferous secretion of the hypophysis. *Bryologist* **33**, 13–14.

FARRELL, T. P. and ASHTON, D. H. (1973). Ecological studies on the Bennison High Plains. *Victorian Nat.* **90**, 286–98. (Account of *Sphagnum* regeneration complex)

FLEISCHER, M. (1906). Neue Familien, Gattungen und Arten der Laubmoose. I. Teil. *Hedwigia* **45**, 53–87. (*Müllerobryum whiteleggei* (Broth.) comb. nov. QLD)

FLEISCHER, M. (1913). Seltene sowie einige neue indische Archipelmoose nebst *Calymperopsis* gen. nov. *Biblthca bot.* **80**, 1–11. (*C. wattsii* sp. nov.)

FLEISCHER, M. (1914). Kritische Revision von Carl Müllerschen Laubmoosgattungen I. *Hedwigia* **55**, 280–85. (*Cryphaea nova-valesiae* E. Aust. = *Papillaria filipendula*; *C. brevidens* and *C. tenella* = *C. tenella* E. Aust.; *C. tasmanica* Broth. TAS = *Dendrocryphaea; Dendropogon viridissimus* QLD, and *D. Muelleri* E. Aust. = *Cyptodon muelleri*)

FLEISCHER, M. (1918). (As Fleischer, 1914), II. *Hedwigia* **59**, 212–19. (*Leptodon australis* NSW = *L. smithii f. australis; Dusenia subproducta* QLD and *D. australis*

NSW = *Forsstroemia* sp.; *Cladomnium tasmanicum* TAS = *Glyphothecium* sp.; *Lepyrodon lagurus* TAS = *Lepyrodon* sp.)

FLEISCHER, M. (1920). (As Fleischer, 1914). III. *Hedwigia* **61**, 402–08. (*Braunia campbellii* VIC = *Hedwigidium campbellii*; *B. weymouthii* TAS = *Hedwigidium wey-mouthii*; *Hedwigia novae-valesiae* NSW, *H. occidentalis* SA, *H. juratzkae* NSW, *H. microcyathea* VIC all = *Hedwigidium albicans*)

FLEISCHER, M. (1922). Kritische Revision der Carl Müllerschen Laubmoosgattungen IV. *Hedwigia* **63**, 209–16. (*Patellidium cartilagineum* NSW = *Bryum c.*; *Campylodontium pallidissimum* QLD = *Entodon p.*; *Cyathophorum pennatum* TAS = *C. bulbosum*; *Hypopterygium pallens* TAS = *Lopidium p.*; *H. pinnatum* = *Lopidium p.*; *H. hyalino-limbatum* NSW = *Lopidium h.*; *H. nematosum* NSW = *Lopidium n.*; *Mniadelphus assimilis* TAS, *M.subminutifolius* S.E. Aust., *M.subrotundus* Aust., *M.whiteleggeanus* NSW, *M.subsinuosus* TAS are all transfered to *Distichophyllum*)

FORSYTH, W. (1900). Contribution to a knowledge of the mosses of New South Wales. *Proc. Linn. Soc. N.S.W.* for 1899, **24**, 674–86. (Localities for 61 spp., of which 43 are new to NSW)

FRASER, L. and VICKERY, J. W. (1938). The ecology of the upper Williams River and Barrington Tops districts. II. The rain-forest formations. *Proc. Linn. Soc. N.S.W.* **63**, 139–184. (Mosses mentioned in text)

GANGULEE, H. C. (1969). "Mosses of Eastern India and Adjacent Regions." Fasc. 1. Privately Published, Calcutta.

GARNET, J. R. (1971). "The Wildflowers of Wilson's Promontory National Park." Lothian Publishers, Melbourne. pp. 83–85. (85 moss species listed)

GARNET, J. R. and WILLIS, J. H. (1949). Additions to the recorded flora of Lake Mountain (January, 1949). II. Cryptogams (excluding algae and fungi). *Victorian Nat.* **66**, 158–59.

GEHEEB, A. (1876). Sur une petite collection de mousses d'Australie récoltées par un amateur. *Revue bryol.* **3**, 2–4. (Sydney region)

GEHEEB, A. (1877). Sur quelques nouvelles espèces de mousses d'Australie et d'Afrique. *Revue bryol.* **4**, 43. (List from Toowoomba, QLD)

GEHEEB, A. (1896). Essai d'une monographie du genre *Dawsonia*, par le Dr. C. Schliephacke et A. Geheeb. *Revue bryol.* **23**(4), 73–9.

★GEHEEB, A. (1897). Nouvelles additions aux flores bryologiques de l'Australie et de la Tasmanie. *Revue bryol.* **24**(5), 65–79. (List of 100 species, none new, with commentaries on taxonomy and localities)

GEHEEB, A. and HAMPE, E. (1881). *See* Hampe and Geheeb (1881).

GOEBEL, K. (1906). Archegoniatenstudien 10. Beiträge zur Kenntnis australischer und neuseeländischer Bryophyten. *Flora, Jena* **96**, 1–202.

GROUT, A. J. (1928–40). "Moss Flora of North America North of Mexico". 3 Vols. Privately published, Newfane, Vermont. (Reprinted, Hafner, New York, 1972)

HAMPE, E. (1844). "*Icones Muscorum* Novorum vel Minus Cognitorum." Henry and Cohen, Bonn. (Descriptions and detailed illustrations including: *Leskea homomalla, Fabronia hampeana, Bryum preissianum, B. australe, Trichostomum cirrhatum, T. calcicola* all from WA)

HAMPE, E. (1846–7). Musci Dill. *In* "*Plantae Pressianae* sive Enumeratio Plantarum Quas in Australasia Occidentali et Meridionali-occidentali Annis 1838–41 Collegit Ludovicus Preiss." (J. G. C. Lehmann, ed), Vol. II, pp. 113–20. Meissner, Hamburg. (Mostly from WA)

★HAMPE, E. (1856). Plantae Muellerianae. Musci frondosi in Australasia felici lecti (continuatio). *Linnaea* **28**, 203–15. (Approx. 17 spp. nov.; cf. Müller and Hampe, 1853)

★HAMPE, E. (1860). Muscorum frondosorum Florae Australasiae auctore Dr F. Müller mox edendae species novas. *Linnaea* **30**, 623–46. (46 spp. nov.)

★HAMPE, E. (1870). Species muscorum novas ex Herbario Melbourneano Australiae. *Linnaea* **36**(5), 513–26.

HAMPE, E. (1871). See Müller, C. (1871).

★HAMPE, E. (1872). Musci novi Australiae ex Herbario Melbournio, a Doctore F. von Müller missi. *Linnaea* **37**, 513–19. (*Encalypta Novae—Valisiae* [sic], *E. aristata, Ditrichum brachycarpum, Dicranum trichophyllum, Guembelia cyathocarpa, Bryum (Webera) erythrocaule, Catharinea (Psilopilum) pyriformis, Neckera (Pilotrichella) dimorpha* C.M., *Drepano-Hypnum fontinaloides, Hypnum pseudostramineum*, spp. nov.)

★HAMPE, E. (1874). Species muscorum novas ex Herbario Melbourneo [sic] Australiae. *Linnaea* **38**, 661–72. (15 spp. nov.)

★HAMPE, E. (1876). Musci novi Musei Melbournei. Continuatio. *Linnaea* **40**, 301–26 (37 spp. nov.)

★HAMPE, E. (1880). Musci Frondosi Australiae continentalis, praesertim e Baronis de Mueller collectionibus. *In* Mueller, F. "Fragmenta Phytographiae Australiae." Vol. 11. Supplement 3, pp. 45–52.

HAMPE, E. and GEHEEB, A. (1881). Musci frondosi in Tasmania et Nova-Seelandia a Dr. O. Beccari, anno 1878, lecti. *Revue bryol.* **8**, 25–8. (*Dicranum kroneanum* TAS, *Fissidens tortuosus* TAS, *Pterygophyllum levieri* TAS, *Rhaphidostegium calliferum* and *Mniadelphus beccarii* are the spp. nov. in a list of 39 spp.)

HARDY, A. D. (1905). Excursion to the Otway Forest. *Victorian Nat.* **21**, 149–62. (Includes list of a few mosses p. 162)

HEDWIG, J. (1801). "*Species Muscorum* Frondosorum Descriptae et Tabulis Aenis LXXVII Coloratis Illustratae." Barth, Leipzig. (Posthumously published and edited by Schwaegrichen who also published seven volumes of supplements to it. Hedwig's work is the starting point of bryological nomenclature)

HOFFMAN, G. R. (1964). The effect of certain sugars on spore germination in *Funaria hygrometrica* Hedw. *Bryologist* **67**(3), 321–29.

†HOOKER, J. D. (1844–45). See Wilson, W. and Hooker, J. D. (1845).

†HOOKER, J. D. (1845–47). See Wilson, W. and Hooker, J. D. (1847).

†HOOKER, J. D. (1852–55). See Wilson, W. (1854).

†HOOKER, J. D. (1855–59). See Wilson, W. (1859).

† Dates according to Stafleu, 1967

Bibliography

*HOOKER, J. D. and WILSON, W. (1844). Musci Antarctici; being characters with brief descriptions of the new species of mosses discovered during the voyage of H.M. Discovery ships Erebus and Terror in the southern circumpolar regions, together with those of Tasmania and New Zealand. *Lond. J. Bot.* **3**, 533–56.

*HOOKER, W. J. (1818–20). "*Musci Exotici*; Containing Figures and Descriptions of New or Little Known Foreign Mosses and Other Cryptogamic Subjects". Vols 1 and 2. Longmans, London.

HOOKER, W. J. (1840, 1845). "*Icones Plantarum*; or Figures, with Brief Descriptive Characters and Remarks, of New or Rare Plants, Selected from the Author's Herbarium". (Series I. 1840, Vol. 3, Plate 248 *Tridontium tasmanicum* TAS; 1845, Vol. 8, Plate 737 *Phascum cristatum, P. exiguum* WA; Plate 738 *Schistidium arcuatum, S. pulchellum* WA; Plate 739 *Weissia pallens, Orthodontium sulcatum,* and *Fabronia tomentosa,* all WA)

HOOKER, W. J. and GREVILLE, R. K. (1824a). Sketch of the characters of the species of mosses, belonging to the genera *Orthotrichum,* (including *Schlotheimia, Micromitrion* and *Ulota,*) *Glyphomitrion,* and *Zygodon. Edinb. J. Sci.* **1**, 110–33, Plates 4–6.

HOOKER, W. J. and GREVILLE, R. K. (1824b). On the genus *Tortula,* of the order Musci. *Edinb. J. Sci.* **1**, 287–302, Plate 12 (*T. calycina, T. australasiae*)

IRELAND, R. R. (1969). A taxonomic revision of the genus *Plagiothecium* for North America, north of Mexico. *Publs Bot. Natn. Mus. Nat. Sci. Can.* **1**, 1–118.

IWATSUKI, Z. (1972). Geographical isolation and speciation of bryophytes in some islands of Eastern Asia. *J. Hattori bot. Lab.* **35**, 126–41. (*Rhizogonium spiniforme* from N. Australia)

JAEGER, A. and SAUERBECK, F. (1876–9). "Genera et Species Muscorum Systematice Disposita seu *Adumbratio Florae Muscorum* Totius Orbis Terrarum." Vols 1, 2. Zollikofer, St. Gallen (Switz.). (Reprinted from *Ber. Thätigk. St. Gallischen Naturwiss. Ges.,* 1870–75)

JOHNSON, K. L., WRIGHT, G. M. and ASHTON, D. H. (1968). Ecological studies of Tunnel Cave, Mt. Eccles. *Victorian Nat.* **85**, 350–56.

KARCZMARZ, K. (1966). Taxonomic studies on the genus *Acrocladium* Mitt. *Nova Hedwigia* **11**(1–4), 499–505. (Comparisons of *A. auriculatum* and *A. chlamydophyllum*)

KAVINA, K. (1915). Ein Beitrag zur Torfmoosflora Australiens. *Sber. K. böhm. Ges. Wiss. Math. Naturw. Kl.* 1915(9), 1–8. (*Sphagnum dominii* sp. nov. QLD)

KAY, C. B. (1959). Control of the growth of aquatic bryophytes in water-conducting channels in south-eastern Australia. M.Sc. Thesis, University of Melbourne.

KINDBERG, N. C. (1901). Grundzüge einer Monographie über die Laubmoos-Familie Hypopterygiaceae. *Hedwigia* **40**, 275–303.

KINDBERG, N. C. (1902). Grundzüge einer Monographie der Laubmoos-Gattung *Thamnium. Hedwigia* **41**, 203–68. (*pumilum* TAS, NSW, QLD, *pandum* TAS, QLD, *novae-walesiae* NSW, *homalioides* TAS, *arbuscula* TAS, *vagum* NSW, QLD, *gracile* TAS, *deflexum* TAS, NSW, QLD, *leichhardtii* NSW, QLD, *ramulosum* VIC, TAS, *excavatum* QLD)

KOPONEN, T. (1968). Generic revision of Mniaceae Mitt. (Bryophyta). *Ann. bot. Fenn.* **5**, 117–51.

LABILLARDIÈRE, J.-J. (1806–7). "Novae Hollandiae Plantarum Specimen." Huzard, Paris, Vol. 2. Reprinted Cramer, 1966. (Illustrations of: *Leskea pennata* and *Hypnum comosum*; pp. 106–8, plate 253)

LAWTON, E. (1972). The genus *Rhacomitrium* in America and Japan. *J. Hattori bot. Lab.* **35**, 252–62. (*R. crispulum* from Australia and N.Z. discussed.)

LEE, R. D. (1952). Re-discovering the moss *Bryobartramia robbinsii*. *Victorian Nat.* **69**, 9–11. (Includes photographs)

LESLIE, J. R. (1924). The cryptogams of the Hurstbridge excursion. *Victorian Nat.* **41**, 152. (List of 8 mosses)

LESLIE, J. R. (1925). Mosses of Wilson's Promontory. *Victorian Nat.* **42**, 116–17. (*c* 37 spp. of mosses recorded)

LINDBERG, S. O. (1870a). Nya mossor. *Öfvers. K. Vetensk Akad. Förh.* **12**, 70–84. (*Mesochaete undulata* gen. et sp. nov. p. 70)

LINDBERG, S. O. (1870b). Contributions to British bryology. *J. Linn. Soc. Bot.* **11**, 460–68. (Includes diagnosis of *Mesochaete* and *M. undulata*, gen. et sp. nov. NSW, p. 463)

LINDBERG, S. O. (1875). On a new moss from Tasmania. *Lond. J. Bot.* **13**, 167–68 (*Pleurophascum grandiglobum* gen. et sp. nov.)

LOTHIAN, N. (1955). Mosses in South Australia. *S. Aust. Nat.* **30**, 25. (List of 17 mosses from Sliding Rock and Encounter Bay)

McVEAN, D. N. (1969). Alpine vegetation of the central Snowy Mountains of New South Wales. *J. Ecol.* **57**, 67–86.

MALTA, N. (1923a). Studien über die Laubmoosgattung *Zygodon* Hook. et Tayl. (1–4). *Latv. Augstsk. Rak.* **5**, 187–92.

MALTA, N. (1923b). (As Malta, 1923a), (5–9). *Latv. Augstsk. Rak.* **6**, 273–95.

MALTA, N. (1924a). (As Malta, 1923a), (10). *Latv. Augstsk. Rak.* **9**, 111–53.

MALTA, N. (1924b). (As Malta, 1923a), (11–12). *Latv. Augstsk. Rak.* **10**, 303–35.

★MALTA, N. (1926). Die Gattung *Zygodon* Hook. et Tayl. Eine monographische Studie. *Acta Horti bot. Univ. latv.* (*Darbi*). **I**, 1–185.

★MALTA, N. (1933). A survey of the Australasian species of *Ulota*. *Acta Horti bot. Univ. latv.* (*Raksti*). **7**, 1–24. (*U. lutea* Mitt. TAS, VIC, *cochleata* Vent. TAS, *viridis* Vent. TAS; also *laticiliata, membranata, dixonii*, spp. nov. TAS)

MARGADANT, W. D. (1968). "Early Bryological Literature. A Descriptive Bibliography of Selected Publications Treating Musci During the First Decades of the Nineteenth Century and Especially of the Years 1825, 1826 and 1827." 277 pp. Hunt Botanical Lab., Penn., U.S.A.

MARGADANT, W. D., MILLER, H. A. and MATTERI, C. M. (1972). Proposal for the conservation of the generic name *Pterygophyllum* Brid. *Taxon* **21**, 536.

MARGINSON, M. A. and MURRAY-SMITH, S. (1969). Further investigations in the Kent Group (Bass Strait Islands). *Victorian Nat.* **86**(5), 254–68. (5 mosses listed)

MARTIN, W. (1946). Geographic range and internal distribution of the mosses indigenous to New Zealand. *Trans. R. Soc. N.Z.* **76**(2), 162–84. (Including Australian species)

MARTIN, W. (1949). Distribution of the mosses indigenous to New Zealand. Supplement 1. *Trans. R. Soc. N.Z.* **77**(3), 355–60. (List of mosses common to N.Z. and Australia)

MARTIN, W. (1951). Notes on the moss flora of New Zealand. *Trans. Br. bryol. Soc.* **1**(5), 471–74.

MARTIN, W. (1957). G. O. K. Sainsbury, F.L.S. *Bryologist* **60**, 363–67.

MATTERI, C. M. (1968). Las especies de *Philonotis* (Bartramiaceae) del sur de Argentina. *Revta Mus. argent. Cienc. nat. Bernardino Rivadavia.* **3**(4), 185–234. (*P. scabrifolia* illustrated)

MATTERI, C. M. (1972). Las Hookeriaceae (Musci) Andino-Patagonicas. II. *Revta Mus. argent. Cienc. nat. Bernardino Rivadavia.* **4**(2), 243–80. (*Sauloma tenella* illustrated)

MATTERI, C. M. (1973). Revision de las Hypopterygiaceae (Musci) austrosud-americas. *Boln. Soc. argent. Bot.* **15**, 229–50.

*MEIJER, W. (1952). The genus *Orthodontium*. *Acta bot. neerl.* **1**, 3–80. (Also separately issued)

MILLER, H. A. (1971). *Pterygophyllum* Brid. is a good name. *Taxon* **20**, 382–83.

*MITTEN, W. (1856). A list of the musci and hepaticae collected in Victoria, Australia, by Dr. F. Mueller. *Hooker's J. Bot. & Kew Gdns Miscell.* **8**, 257–66.

*MITTEN, W. (1860). Description of some new species of musci from New Zealand and other parts of the southern hemisphere, together with an enumeration of the species collected in Tasmania by William Archer Esq.; arranged upon the plan proposed in the "Musci Indiae Orientalis". *J. Linn. Soc. Bot.* **4**, 64–100.

*MITTEN, W. (1882). Australian Mosses, enumerated by Wm. Mitten Esq. *Trans. Proc. R. Soc. Vict.* **19**, 49–96. (see Mueller, F. 1882)

Montagne, J. F. C. (1844). Plantes cellulaires. *In* C. H. Hombron and J. B. Jacquinot (1844–53) "Voyage au Pole Sud et dans l'Océanie sur les Corvettes l'Astrolabe et la Zélée, Exécuté par Ordre du Roi pendant les Années 1837–1840, sous le Commandement de M. J. Dumont D'Urville, Capitaine de Vaisseau." Botanique. Vol. 1, pp. 281–335. Gide & Cie, Paris. (*Hypnum aciculare*, *Cyathophorum pennatum*, *Hypopterygium rotulatum* from TAS)

MORRIS, P. F. (1929). Ecology of Marysville and Lake Mountain. *Victorian Nat.* **46**, 34–42. (4 mosses recorded)

MUELLER, F. VON. (1854). "Second General Report of the Government Botanist on the Vegetation of the Colony." Dated 5th October 1854. Govt. Printer, Melbourne. 20 pp. Appendix: Second Systematic Index of the plants of Victoria comprising those which were examined between September, 1853, and October, 1854. (70 mosses listed on pp. 17–18)

MUELLER, F. VON. (1858). "Fourth Systematic Index of the Plants of Victoria, Comprising Those Collected and Examined in 1857 and 1858; by Ferdinand Mueller P.H. & M.D., Government Botanist." *In* Annual Rept. of Govt. Botanist, Victoria. (Mosses pp. 12–13)

MUELLER, F. VON. (1864). "Analytical Drawings of Australian Mosses." Fasc. 1. (only fascicle printed) Govt. Printer, Melbourne. 20 plates.

MUELLER, F. VON. (1880). "Fragmenta Phytographiae Australiae." Govt. Printer, Melbourne. Vol. 11. 1. Supplement 3. Musci Frondosi Australiae Continentalis, Praesertim e Baronis de Mueller Collectionibus, by E. Hampe, pp. 45–52. 2. Additamenta pro supplementi voluminis undecimi. (pp. 107–15)

MUELLER, F. VON. (1882). Australian Mosses, enumerated by William Mitten Esq. *See* Mitten, W. (1882).

★Müller, C. (1848–1851). "*Synopsis Muscorum* Frondosorum Omnium Hucusque Cognitorum." Parts 1 & 2. Foerstner, Berlin.

★MÜLLER, C. (1851). Die, von Samuel Mossman im Jahre 1850, in Van Diemen's Land, Neuseeland und Neuholland gemachte Laubmoossammlung. *Bot. Ztg* **9**, 545–52, 561–67. (List of 51 collections including: *Funaria sphaerocarpa* NSW, *Dissodon calophyllus* TAS, *Mnium* (*Rhizogonium*) *Mossmanianum* TAS, *Catharinea* (*Polytrichadelphus*) *innovans* TAS, *Leptotrichum cylindricarpum* TAS, *Bartramia* (*Vaginella*) *Mossmaniana* TAS, *Grimmia* (*Dryptodon*) *emersa* TAS, *Neckera* (*Euneckera*, *Rhystophyllum*) *hymenodonta* TAS, *Pilotrichum* (*Cryphaea*, *Dichotomaria*) *micro-cyatheum* TAS, *Pilotrichum* (*Cryphaea*, *Eucryphaea*) *ovalifolium* NSW, *Hypnum* (*Omalia*, *Cupressina*) *Mossmaniana* TAS—all spp. nov.)

★MÜLLER, C. (1856). Symbolae ad Synopsin Muscorum. *Bot. Ztg* **14**, 415–21, 436–40, 455–59. (*Hypnum stuartii* sp. nov. TAS)

★MÜLLER, C. (1868). Beitrag zur östaustralischen Moosflor. *Linnaea* **35**, 613–26. (Includes *Ångströmia* (*Diobelon*) *tricruris* QLD, *A.* (*Dicranella*) *Dietrichiae* QLD, *Barbula subcalycina* QLD, *Macromitrium* (*Eumacromitrium*) *Scottiae* NSW, *Hypopterygium Scottiae* NSW, QLD, *Fabronia* (*Eufabronia*) *Scottiae* NSW, *Lasia australis* NSW, QLD, *Neckera* (*Papillaria*) *Scottiae* NSW, QLD, *Hypnum* (*Illecebrina*) *chlorocladum* QLD, *H.* (*Cupressina*) *umbilicatum* NSW, QLD, *H.* (*Taxicaulis*) *candidum* QLD, *Trichostomum* (*Eutrichostomum*) *Leptotheca* ?WA, *T.* (*E.*) *rubiginosum* VIC—all spp. nov.)

★MÜLLER, C. (1871). Musci Australici praesertim Brisbanici novi. *Linnaea* **37**, 143–62. +Appendix: Hypna duo Australiae adnumerat Ernst Hampe. (*Endotrichella* gen. nov., 24 spp. nov.)

MÜLLER, C. (1874a). Musci polynesiaci praesertim Vitiani et Somoani Graeffeani. *J. Mus. Godeffroy* **3**(6), 51–90. (*Syrrhopodon fimbriatus* C.M., hom illeg., changed to *S. fimbriatulus*; *Hypopterygium* (*Lopidium*) *nematosum* sp. nov. from NSW)

MÜLLER, C. (1874b). Novitates Bryothecae Müllerianae publicavit Carolus Müller Hal. 2. Musci Novo-Granatenses Wallisiani adjectis nonnullis aliis muscis novis andinis vel tropico-Americanis vel australasiacis. *Linnaea* **38**, 572–620. (*Bryum* (*Dicranobryum*) *semperlaxum* QLD, p. 582)

MÜLLER, C. (1883). Die auf der Expedition S.M.S. "Gazelle" von Dr. Naumann gesammelten Laubmoose. *Engler's Bot. Jb.* **5**, 76–88. (*Sphagnum naumannii* and *Macromitrium repandum* spp. nov. from QLD)

MÜLLER, C. (1887). Sphagnorum novorum descriptio. *Flora* **70**, 403–22. (*S. wilcoxii*, *S. whiteleggei*, *S. comosum*, NSW)

MÜLLER, C. (1888). Musci cleistocarpici novi. *Flora* **71**, 1–14. (Several spp. nov.)

MÜLLER, C. (1889). Laubmoose (Musci Frondosi). *In* "Die Forschungsreise S.M.S. Gazelle in den Jahren 1874 bis 1876 unter Kommando des Kapitän zur Zee Freiherrn von Schleinitz." (F. C. Naumann, ed., 1888–90) IV. Theil. Botanik, pp. 1–64. Admiralty Hydrographic Dept., Berlin. (*Sphagnum naumanni* "sp. nov." and *Dicranum introflexum*, *Macromitrium repandum* QLD, on pp. 59–60)

MÜLLER, C. (1897a). Synopsis generis *Harrisonia*. *Öst. bot. Z.* **47**(11), 387–98; (12), 417–20. (p. 392 *H. webbianus* sp. nov. WA, p. 397 *H. australis* TAS, VIC, NSW)

MÜLLER, C. (1897b). *Triquetrella* genus Muscorum novum conditum et descriptum. *Öst. bot. Z.* **47**(12), 420–24. (*T. scabra* VIC, *T. filiformis* SA, *T. richardsiae* SA, NSW, *T. fragilis* VIC, all spp. nov. and *T. preissianae* Hampe from WA)

★MÜLLER, C. (1897c). Symbolae ad Bryologiam Australiae I. *Hedwigia* **36**, 331–65. (79 spp. nov. of which 60 are Australian)

★MÜLLER, C. (1898). (As Müller, 1897c), II. *Hedwigia* **37**, 76–171. (Spp. Nos. 80–300 of which 140 are Australian)

★MÜLLER, C. (1901). "Genera Muscorum Frondosorum." Kummer, Leipzig.

★MÜLLER, C. (1902). Symbolae ad Bryologiam Australiae III. *Hedwigia* **41**, 119–34. (Spp. Nos. 301–339 of which 18 are Australian)

MÜLLER, C. and BROTHERUS, V. F. (1900). Ergebnisse einer Reise nach dem Pacific (H. Schauinsland 1896/7). Musci Schauinslandiani. Ein Beitrag zur Kenntnis der Moosflora der Pacifischen Inseln. *Abh. naturw. Ver. Bremen* **16**(3), 493–512. (List of collections, some from Aust. Many new spp. mostly from N.Z. and *Leucobryum turgidulum* sp. nov. NSW)

★MÜLLER, C. and HAMPE, E. (1853). Musci frondosi Australasiae ab Dre. Ferd. Müller lecti. *Linnaea* **26**, 489–505. (84 spp. of which there are 27 Australian spp. nov.)

MURDOCH, J. R. (1910). Mosses *in* Excursion to Toolangi. *Victorian Nat.* **26**, 150 (list of *c* 18 spp.)

NYHOLM, E. (1965). "Illustrated Moss Flora of Fennoscandia." Gleerup, Lund. II. Fasc. 5.

NYHOLM, E. (1971). Studies in the genus *Atrichum* P. Beauv. A short survey of the genus and the species. *Lindbergia* **1**, 1–33.

OCHI, H. (1967). Notes on moss flora V. *Hikobia* **5**(1–2), 14–38. (*Bryum laevigatum*, *B. coronatum*)

OCHI, H. (1968a). On the status of *Bryum handelii* Broth. (Musci). *J. Jap. Bot.* **43**, 480–85. (Illustrations of *B. blandum* from TAS)

OCHI, H. (1968b). A revised list of mosses of the family Bryaceae in Japan and the adjacent regions. *J. Fac. Educ. Tottori Univ. nat. Sci.* **19**(1), 24–40. (*Bryum plumosum* QLD, NSW, new to Australia)

OCHI, H. (1969). Notes on moss flora VI. *Hikobia* **5**(3–4), 153–71. (*Bryum capillare*)

★OCHI, H. (1970). A revision of the subfamily Bryoideae in Australia, Tasmania, New Zealand and the adjacent islands. *J. Fac. Educ. Tottori Univ. nat. Sci.* **21**(1), 7–67.

OCHI, H. (1971). What is true *Bryum truncorum*? *Bryologist* **74**, 503–06.

OCHI, H. (1972). Some problems of distributional patterns and speciation in the subfamily Bryoideae in the regions including Eurasia, Africa and Oceania. *J. Hattori bot. Lab.* **35**, 50–67.

★OCHI, H. (1973). Supplement to the family Bryoideae (Musci) in Australia and New Zealand. *Hikobia* **6**(3–4), 217–23.

★Paris, E. G. (1903–1906). "*Index Bryologicus* sive Enumeratio Muscorum ad Diem Ultimam Anni 1900 Cognitorum Adjunctis Synonymia Distributioneque Geographica Locupletissimis." 2nd Ed. 5 vols. Hermann, Paris.

PROSKAUER, J. (1958). On the peristome of *Funaria hygrometrica*. *Am. J. Bot.* **45**, 560–63.

RAMSAY, H. P. (1964). The chromosomes of *Dawsonia*. *Bryologist* **67**(2), 153–62.

RAMSAY, H. P. (1966a). Cytological Studies of Australian Mosses. Ph.D. Thesis, University of Sydney. Held in Library there.

RAMSAY, H. P. (1966b). Sex chromosomes in *Macromitrium*. *Bryologist* **69**(3), 293–311.

RAMSAY, H. P. (1967a). Intraspecific Polyploidy in *Hypopterygium rotulatum* (Hedw.) Brid. *Proc. Linn. Soc. N.S.W.* **91**(3), 220–30.

RAMSAY, H. P. (1967b). *In* Löve, A. IOPB Chromosome Number reports. XIV. *Taxon* **16**, 552–61.

RAMSAY, H. P. (1973). Unusual sporocytes in *Dicnemoloma pallidum* (Hook.) Wijk & Marg. *Bryologist* **76**, 178–82.

★RAMSAY, H. P. (1974). Cytological studies of Australian mosses. *Aust. J. Bot.* **22**, 293–348.

READER, F. M. (1898a). Contributions to the flora of Victoria. 5. Description of new species of moss. *Victorian Nat.* **15**, 31. (*Dawsonia victoriae*; description translated from Müller, *Hedwigia* **36**, 335)

READER, F. M. (1898b). Contributions to the flora of Victoria. 6. Descriptions of new mosses. *Victorian Nat.* **15**, 59. (*Polytrichum longipilum, P. nodicoma*; translated from Müller, *Hedwigia* **36**, 344, 346)

★REED, C. F. and ROBINSON, H. E. (1972). Index to Die natürlichen Pflanzenfamilien (Musci-Hepaticae). Editions 1 and 2. *Contr. Reed Herb. Baltimore* No. 21.

REICHARDT, H. W. (1870). Fungi, hepaticae et musci frondosi. *In* "Reise der Österreichischen Fregatte Novara um die Erde in den Jahren 1857–1859 unter den Befehlen des Commodore B. von Wüllerstorf-Urbair." Botanischer Theil. Bd. 1. Sporenpflanzen. (E. Fenzl, ed.) Vol. 1, pp. 133–96; Pl. 20–36. (Mosses pp. 166–96; Pls. 28–36)

RENAULD, F. (1901). Nouvelle classification des *Leucoloma. Revue bryol.* **28**(4–5), 66–70, 85–87. (*Dicranoloma menziesii* TAS, *D. serratum* QLD)

RICHARDS, P. W. (1963). *Campylopus introflexus* (Hedw.) Brid. and *C. polytrichoides* De Not. in the British Isles; a preliminary account. *Trans. Br. bryol. Soc.* **4**(3), 404–17.

ROBINSON, H. (1970). A revision of the moss genus, *Trichostomopsis. Phytologia* **20**, 186–91.

*RODWAY, L. (1913–14). Tasmanian Bryophyta. Vol. 1. Mosses. *Pap. Proc. R. Soc. Tasm.* for 1912, pp. 3–24, 87–138; for 1913, pp. 177–263. (Issued later in 1914 as a separate publication, 163 pp. Royal Soc., Hobart)

*RODWAY, L. (1915, 1916). Additions to the Tasmanian flora. *Pap. Proc. R. Soc. Tasm.* for 1915, pp. 104–07; for 1916, pp. 44–47. (Several mosses described)

*ROTH, G. (1911). "Die aussereuropäischen Laubmoose." Band I. Heinrich, Dresden. 272 pp., 24 plates.

*ROTH, G. (1913). Nachtrag I zu Band I der aussereuropäischen Laubmoose von 1910/11. *Hedwigia* **53**, 81–98. Plates I, II. (*Trachycarpidium novae-valesiae* NSW, *Acaulon robustum* NSW, *Trematodon adaequans* NSW, spp. nov.)

*ROTH, G. (1914). Nachtrag II zu Band I der aussereuropäischen Laubmoose von 1910/11. *Hedwigia* **54**, 267–74. Plate 10. (*Archidium rothii* QLD, *Pleuridium austro-subulatum, Astomum wattsii, A. novae-valesiae, Acaulon austro-muticum*, all NSW, *Physcomitrella austro-patens* VIC, *Bruchia minuta* TAS, spp. nov.)

SAINSBURY, G. O. K. (1932). The study of Australian Mosses. *Victorian Nat.* **48**, 255–60. **49**, 46–9, 77–80, 108–11. (Notes on how to begin and descriptions of 22 spp.)

SAINSBURY, G. O. K. (1935). Vegetative reproduction in New Zealand mosses. *J. Proc. R. Soc. N.S.W.* **69**, 86–104.

SAINSBURY, G. O. K. (1938). Vegetative reproduction in New Zealand mosses. II. *Bryologist* **41**, 11–18.

SAINSBURY, G. O. K. (1945). New and critical species of New Zealand mosses. *Trans. R. Soc. N.Z.* **75**(2), 169–86. (*Tortula flavinervis* and *Triquetrella curvifolia* VIC)

SAINSBURY, G. O. K. (1947). Additions to the mosses of Victoria. *Victorian Nat.* **63**(10), 222–23. (*Sainsburia novae-zealandiae, Triquetrella curvifolia, Tortula flavinervis*)

*SAINSBURY, G. O. K. (1948). Bryobartramiaceae, a new moss family. *Bryologist* **51**, 9–13.

Sainsbury, G. O. K. (1952). Vegetative reproduction in New Zealand mosses. III. *Bryologist* **55**, 71–79.

SAINSBURY, G. O. K. (1953a). Two new species of Tasmanian mosses. *Victorian Nat.* **70**, 30–31. (*Dicranoloma perichaetiale, Blindia tasmanica*)

*Sainsbury, G. O. K. (1953b). Notes on Tasmanian mosses from Rodway's Herbarium. *Pap. Proc. R. Soc. Tasm.* **87**, 83–91.

*SAINSBURY, G. O. K. (1955a). A handbook of the New Zealand Mosses. *Bull. R. Soc. N.Z.* No. 5. 490 pp.

*SAINSBURY, G. O. K. (1955b). Notes on Tasmanian mosses from Rodway's Herbarium. 2. *Pap. Proc. R. Soc. Tasm.* **89**, 1–11; *(1955c). 3. **89**, 13–20; *(1955d). 4. **89**, 21–35; *(1955e). 5. **89**, 37–43; *(1955f). 6. **89**, 45–53; *(1956a). 7. **90**, 35–39; *(1956b). 8. **90**, 41–47.

SAINSBURY, G. O. K. (1956c). A new species of *Pottia* from Central Australia. *Revue bryol. lichen.* **25**, 237–38.

SALMON, E. S. (1900). Bryological Notes (8). *Eccremidium exiguum* (Hook.f. & Wils.). *Revue bryol.* **27**(6), 85–6.

SARAFIS, V. (1971). A biological account of *Polytrichum commune*. *N.Z. Jl Bot.* **9**(4), 711–24.

SCHELPE, E. A. C. L. E. (1970). A provisional check-list of the bryophyta of the Cape Peninsula. *Contr. Bolus Herb.* **2**, 49–70. (*Ischyrodon lepturus* (Tayl.) Schelpe comb. nov.)

SCHODDE, R. and KRAEHENBUEHL, D. (1957). Hindmarsh Falls Excursion. *S. Aust. Nat.* **31**, 39–43. (9 mosses tentatively identified)

SCHOFIELD, W. B. (1974). Bipolar disjunctive mosses in the Southern Hemisphere, with particular reference to New Zealand. *J. Hattori bot. Lab.* **38**, 13–32.

*SCHULTZE-MOTEL, W. (1970a). Monographie der Laubmoosgattung *Andreaea*. I. Die costaten Arten. *Willdenowia* **6**, 25–110. (Excellent review and key, with photographs; includes *A. australis* NSW, VIC, *A. subulata* NSW, VIC, TAS, *A. nitida* VIC, TAS)

SCHULTZE-MOTEL, W. (1970b). Ephemeropsidaceae—ein neuer Name für eine Familie der Laubmoose. *Taxon* **19**(2), 251–2. (Nemataceae considered invalid)

SCHWAEGRICHEN, C. F. (1811–42). "Species Muscorum . . . Supplementum." 7 vols. (4 supplements in 11 parts.) Barth, Leipzig. See Hedwig (1801).

SCOTT, G. A. M. (1971). A bibliography of New Zealand bryology. *N.Z. Jl Bot.* **9**, 750–71.

SHIRLEY, J. (1888). Field naturalists' excursion to Caboolture. September 22nd and 23rd, 1888. *Proc. R. Soc. Qd.* **5**, 137–42. (List of 34 mosses on p. 141, determined C. J. Wild)

SIM, T. R. (1926). The Bryophyta of South Africa. *Trans. R. Soc. S. Afr.* **15**, 1–475.

SIMMONDS, J. H. (1888). Field Naturalists' excursion to Woolston. *Proc. R. Soc. Qd* **5**, 173–78. (36 spp. collected. pp. 177–8, determined C. J. Wild)

SIMMONDS, J. H. (1889). Excursion of Field Naturalists' section to Brookfield. December 10th, 1888. *Proc. R. Soc. Qd* **6**, 65–70. (List of 8 mosses p. 69)

SMITH, G. G. (1962). The flora of granite rocks of the Porongurup Range, South Western Australia. *Jl R. Soc. West. Aust.* **45**, 18–23.

SMITH, G. G. (1969). *Sphagnum subsecundum* in Western Australia. *West. Aust. Nat.* **11**(3), 56–9.

SMITH, G. L. (1971). Conspectus of the genera of Polytrichaceae. *Mem. N.Y. bot. Gdn* **21**(3), 1–83.

SMITH, G. L. (1972). Continental drift and the distribution of Polytrichaceae. *J. Hattori bot. Lab.* **35**, 41–9.

SMITH, J. E. (1808). Characters of *Hookeria*, a new genus of mosses, with descriptions of ten species. *Trans. Linn. Soc. Lond.* **9**, 272–82. (*H. pennata* New Holland, *H. rotulatum, H. arbuscula*)

★STAFLEU, F. (1967). "Taxonomic Literature. A Selective Guide to Botanical Publications with Dates, Commentaries and Types." Regnum Vegetabile Vol. 52. Int. Bureau Pl. Taxon., Utrecht. 556 pp. (Invaluable for details of much early bryological literature)

STIRLING, J. (1886). The Cryptogamia of the Australian Alps. Part 1. *Proc. R. Soc. Vict.* **22**, 49–56.

STONE, I. G. (1961a). The gametophore and sporophyte of *Mittenia plumula* (Mitt.) Lindb. *Aust. J. Bot.* **9**(2), 124–51.

STONE, I. G. (1961b). The highly refractive protonema of *Mittenia plumula* (Mitt.) Lindb. (Mitteniaceae). *Proc. R. Soc. Vict.* **74**, 119–24. 3 Plates

STONE, I. G. (1971). The sporophyte of *Tortula pagorum* (Milde) De Not. *Trans. Br. bryol. Soc.* **6**(2), 270–77. (Detailed illustrated anatomical account, with comparisons of similar species)

★STONE, I. G. (1973a). Two new species of *Archidium* from Victoria, Australia. *Muelleria* **2**(4), 191–213.

STONE, I. G. (1973b). A new species of *Brachydontium* from Australia. *J. Bryol.* **7**(3), 343–51.

STONE, I. G. (1975). *Trachycarpidium* in Queensland, Australia. *Muelleria* **3**(2), 122–29. (*T. brisbanicum comb. nov.*, formerly *Astomum*. Illustrated)

STONE, I. G. and SCHELPE, E. A. C. L. E. (1973). Two new generic records of mosses for southern Africa. *Jl S. Afr. Bot.* **39**(2), 131–32. (*Bryobartramia, Eccremidium*)

STONE, I. G. and Scott, G. A. M. (1973). Name changes in Australian mosses. *J. Bryol.* **7**, 603–05.

SULLIVAN, D. (1887). Mosses of Victoria, with brief notes. *Victorian Nat.* **4**, 106–10. (Notes on how and where to collect. Descriptions of some Nematodonteae. Intended to be first of a series, but never continued.) (Sullivan was head teacher, State School, Moyston)

★SYED, H. (1973). A taxonomic study of *Bryum capillare* Hedw. and related species. *J. Bryol.* **7**(3), 265–326.

SYMON, D. E. (1971). Pearson Island Expedition 1969. 3. Contributions to the Land Flora. *Trans. R. Soc. S. Aust.* **95**(3), 131–32. (9 mosses listed)

TADGELL, A. J. (1924). Mount Bogong and its Flora. *Victorian Nat.* **41**, 56–80. (List of 16 mosses p. 70)

TATE, R. (1880). A list of the charas, mosses, liverworts, lichens, fungs [sic], and algals of extra-tropical South Australia (extracted from "Supplementum Fragmentorum Phytographiae Australiae"). *Trans. R. Soc. S. Aust.* **4**, 5–24. (List of 36 mosses on pp. 7–8)

★TAYLOR, T. (1846). The distinctive characters of some new species of Musci, collected by Professor William Jameson, in the vicinity of Quito, and by Mr. James Drummond at Swan River. *Lond. J. Bot.* **5**, 41–67.

THÉRIOT, I. (1922). Le problème du *Leucobryum candidum*. *Bull. Soc. bot. Genève*. Ser. 2. **13**, 217–25. (*L. confusum* TAS sp. nov.)

THÉRIOT, I. (1936). Reliquiae Boissierianae. *Bull. Soc. bot. Genève* **26**, 76–91. (*Campylopus lenormandi* sp. nov. VIC)

THOMSON, G. K. (1974). Natural history of the Hogan Group (Bass Strait Islands). 4: Bryophyte Flora. *Trans. Proc. R. Soc. Tasm.* **107**, 99–104. (18 mosses listed)

THROWER, S. L. (1964). An investigation of translocation in *Dawsonia intermedia* C.M. *Trans. Br. bryol. Soc.* **4**(4), 664–67.

★TOUW, A. (1971). A taxonomic revision of the Hypnodendraceae (Musci). *Blumea* **19**(2), 211–354. (Excellent, fully illustrated and thorough account)

TROUGHTON, J. H. and SAMPSON, F. B. (1973). "Plants. A Scanning Electron Microscope Survey." Wiley, Sydney.

VENTURI, G. DE (1896). Notice sur quelques espèces d'*Orthotrichum* de l'Australie. *Revue bryol.* **23**(4), 65–67. (*O. pseudopumilum, O. praeperistomatum, O. rupestriforme* spp. nov. TAS)

VITT, D. H. (1973). A revisionary study of the genus *Macrocoma*. *Revue bryol. lichen.* **39**(2), 205–20.

VITT, D. H. and CROSBY, M. R. (1972). *Achrophyllum*—a new name for a genus of mosses. *Bryologist* **75**(2), 174–75. (*Pterygophyllum* replaced)

WARNSTORF, C. (1890). Beiträge zur Kenntniss exotischer Sphagna. *Hedwigia* **29**, 179–258. Figs 4–14.

WARNSTORF, C. (1891). (As Warnstorf, 1890), **30**(1), 12–46. Figs 1–5. (3), 127–180. Figs 14–24.

WARNSTORF, C. (1893). (As Warnstorf, 1890), **32**(1), 1–17. Pls 1–4. (*S. serrulatum, S. pseudo-rufescens, S. macrocephalum* spp. nov. TAS)

WARNSTORF, C. (1897). (As Warnstorf, 1890), **36**, 145–76.

WARNSTORF, C. (1898). (As Warnstorf, 1890), *Bot. Zbl.* **76**(13), 417–23. (*S. wattsii* sp. nov. NSW)

WARNSTORF, C. (1900a). Weitere Beiträge zur Kenntniss der Torfmoose. *Bot. Zbl.* **82**, 7–14, 39–45, 65–76. (*S. grandifolium, serratifolium, brotherusii, drepanocladum* spp. nov. NSW)

WARNSTORF, C. (1900b). Neue Beiträge zur Kenntniss europäischer und exotischer Sphagnumformen. *Hedwigia* **39**, 100–10. (*S. trichophyllum* sp. nov. TAS)

WARNSTORF, C. (1907). Neue europäische und aussereuropäische Torfmoose. *Hedwigia* **47**, 76–124. (*S. decipiens, wardellense* spp. nov. NSW)

★WARNSTORF, C. (1911). Sphagnales—Sphagnaceae (Sphagnologia universalis). *In* "Das Pflanzenreich" (A. Engler, ed.) Heft **51**, 546 pp. Engelmann, Leipzig.

★WARNSTORF, C. (1916a). *Pottia*-Studien als Vorarbeiten zu einer Monographie des Genus "*Pottia*" Ehrh. sensu stricto. *Hedwigia* **58**, 35–152.

WARNSTORF, C. (1916b). Bryophyta nova europaea et exotica. *Hedwigia* **57**, 62–131. (*Sphagnum weymouthii* and *S. rodwayi* TAS, spp. nov., but already described in 1914)

Bibliography

WATSON, E. V. (1955). "British Mosses and Liverworts." Cambridge University Press.

WATTS, W. W. (1899). Notes on some recently described species of N.S. Wales mosses. *Proc. Linn. Soc. N.S.W.* **24**, 374–76.

WATTS, W. W. (1900). Notes on some new mosses from New South Wales. *Proc. Linn. Soc. N.S.W.* for 1899. **24**, 632–33. (Localities for 13 spp. new to the state)

WATTS, W. W. (1905). Some Melbourne mosses. *Victorian Nat.* **21**, 140–42.

WATTS, W. W. (1906). Australian mosses. Some locality pictures. *Bryologist* **9**, 34–36, 41.

WATTS, W. W. (1912). The Sphagna of Australia and Tasmania. *Proc. Linn. Soc. N.S.W.* **37**(2), 383–89. (No spp. nov. described)

WATTS, W. W. (1916). Some cryptogamic notes from the Botanic Gardens, Sydney. *Proc. Linn. Soc. N.S.W.* **41**, 377–86. (Mosses pp. 384–86, Plate 20. *Fissidens humilis* sp. nov. NSW, *Leptostomum inclinans*, new for NSW, *Hampeella pallens* NSW, QLD)

WATTS, W. W. Unpublished MS census of NSW mosses. (Incorporated into Burges, A. 1952)

★WATTS, W. W. and WHITELEGGE, T. (1902, 1906). Census Muscorum Australiensium. *Proc. Linn. Soc. N.S.W.* **27**, supplement pp. 1–90. **30**, supplement pp. 91–163. (Deals only with acrocarps. See Burges, A. 1932, 1935 for continuation.)

WEBER, W. A. (1968). Preliminary list of mosses of the Australian Capital Territory (duplicated list of 128 spp.)

WEBER, W. A. (1972). The identity of *Barbula pseudopilifera* (Musci: Pottiaceae). *Lindbergia* **1**, 214–16.

WELCH, W. H. (1970). Hookeriaceae species and distribution in Africa, Europe, Asia, Australia and Oceania. *Proc. Indiana Acad. Sci.* **79**, 377–87. (Lists of species by geographical regions)

WEYMOUTH, W. A. (1894, 1896). Some additions to the moss flora of Tasmania. *Pap. Proc. R. Soc. Tasm.* for 1893, pp. 200–10, Plates 4, 5; for 1895, pp. 106–20.

WEYMOUTH, W. A. (1903). Some additions to the bryological flora of Tasmania. Part III. *Pap. Proc. R. Soc. Tasm.* for 1902, pp. 115–32.

★WEYMOUTH, W. A. and RODWAY, L. (1922). Bryophyte Notes. *Pap. Proc. R. Soc. Tasm.* for 1921, pp. 173–75. (New to TAS, determined H. N. Dixon: *Trematodon mackayii, Pottia heimii, P. melbourniana, Ditrichum punctulatum, Dicranum trichopodum, Mnium rostratum, Macromitrium rodwayi* Dixon sp. nov., *Leucobryum brachyphyllum*)

WHITEHOUSE, H. L. K. (1966). The occurrence of tubers in European mosses. *Trans. Br. bryol. Soc.* **5**(1), 103–16.

WHITELEGGE, T. and BROTHERUS, V. F. (1892). (Notes and Extracts: list of collection of 20 spp. from Lord Howe Isd.) *Proc. Linn. Soc. N.S.W.* **7**, 277.

WIJK, R. VAN DER (1957). Precursory studies in Malaysian Mosses I. Revision of the genus *Dawsonia* R. Brown. *Revue bryol. lichen.* **26**(1–2), 8–19.

WIJK, R. VAN DER and MARGADANT, W. D. (1958–62). New Combinations in Mosses 1–8. *Taxon* **7**, 287–90; **8**, 70–75, 106; **9** 50–52, 189–91; **10** 24–26, **11**, 221–23.

★WIJK, R. VAN DER, MARGADANT, W. D. and FLORSCHÜTZ, P. A. (1959–1969). "Index Muscorum." 5 vols. Int. Bureau Pl. Taxon., Utrecht. (*Regnum Vegetabile* Vols. 17, 26, 33, 48, 65)

WILD, C. J. (1888a). Notes on Some Queensland mosses. *Proc. R. Soc. Qd* **5**, 116–19. (Notes and amendments to lists in Bailey, F. M. 1883–1890)

WILD, C. J. (1888b). Bryological notes. *Proc. R. Soc. Qd* **5**, 148–50. (Descriptions of *Symblepharis perichaetialis, Fissidens tenellus, F. oblongifolius,* new to QLD)

WILD, C. J. (1889a). Bryological notes. *Proc. R. Soc. Qd* **6**, 76–79. (Descriptions of *Splachnobryum baileyi, Meteorium baileyi, Isopterygium robustum, Macromitrium pusillum, Rhizogonium parramattense*)

WILD, C. J. (1889b). Bryological notes. *Proc. R. Soc. Qd* **6**, 104–05. (Descriptions of *Octoblepharum albidum* and *Syrrhopodon fimbriatus*)

WILLIS, J. H. (1949). Botanical pioneers in Victoria. 1–3. *Victorian Nat.* **66**, 83–9, 103–9, 123–8. (Includes Sullivan, Reader, Watts, Bastow)

★WILLIS, J. H. (1950). The chequered story of two Tasmanian mosses. *Victorian Nat.* **67**, 30–5. (*Tayloria gunnii, T. tasmanica.* Illustrated)

WILLIS, J. H. (1951). A new species of Victorian moss. *Victorian Nat.* **68**, 83–4. (*Fissidens hunteri* sp. nov.)

★WILLIS, J. H. (1952). Systematic notes on Victorian mosses—1. *Victorian Nat.* **69**, 15–18. (*Sphagnum*)

WILLIS, J. H. (1953a). Nemataceae, a moss family new to Australia. *Nature, Lond.* **172**, 127–28.

WILLIS, J. H. (1953b). Mitchell Gorge mosses. *Victorian Nat.* **69**, 131. (5 spp. listed)

WILLIS, J. H. (1953c). The myth of *Macromitrium* in Western Australia (a moss note). *Victorian Nat.* **69**, 159–60. (*M. incurvifolium* and *involutifolium* deleted from WA list) ·

★WILLIS, J. H. (1953d). Systematic notes on Victorian mosses—2. *Victorian Nat.* **70**, 55–57. (*Sphagnum*)

WILLIS, J. H. (1953e). The archipelago of the Recherche. Part 3a. Land Flora. Austr. geog. Soc., Melbourne. (19 mosses listed)

★WILLIS, J. H. (1954a). Systematic notes on Victorian mosses—3. *Victorian Nat.* **70**, 169–72. (*Acaulon, Pottia*)

WILLIS, J. H. (1954b). Mosses new to Western Australia. *Victorian Nat.* **71**, 8–12. (*Pterygoneurum kemsleyi* sp. nov.)

★WILLIS, J. H. (1955a). Systematic notes on Victorian mosses—4. *Victorian Nat.* **71**, 157–63. (Amendments to lists of genera and species)

★WILLIS, J. H. (1955b). Some further notes on *Sphagnum. Victorian Nat.* **71**, 189–90.

★WILLIS, J. H. (1955c). Systematic notes on Victorian mosses—5. *Victorian Nat.* **72**, 5–11. (*Trematodon alpinus, Tortella dakinii* spp. nov., *Ditrichum rufo-aureum* comb. nov., *Bryum subcurvicollum, Brachythecium albicans, Cratoneuropsis*)

★WILLIS, J. H. (1955d). New and interesting moss records for Australia. *Victorian Nat.* **72**, 73–8. (WA, NT, QLD, NSW, TAS)

★WILLIS, J. H. (1955e). The present position of muscology in Victoria. (A Centennial Review) *Muelleria* **1**(1), 55–9.

★WILLIS, J. H. (1957a). Systematic notes on Victorian mosses—6. *Victorian Nat.* **74**, 23–5. (Distribution records and synonymy)

★WILLIS, J. H. (1957b). New records of mosses for Australian States (Victoria, Tasmania and the Northern Territory). *Victorian Nat.* **74**, 101–5.

WILLIS, J. H. (1958). Additional notes on Northern Territory mosses. *Victorian Nat.* **74**, 189. (Adds *Funaria gracilis* and *Barbula torquata* to the list)

WILLIS, J. H. (1959). Plants of the Recherche Archipelago, W.A. *Muelleria* **1**(2), 97–101. (*Tortula papillosa* only moss listed)

WILLIS, J. H. (1970). Mosses *In* "The Vegetation of Hattah Lakes National Park". National Parks Authority, Victoria, cyclostyled booklet. (18 mosses listed on p. 30)

WILLIS, J. H. (1972a). *In* Plant list, for King Island, Bass Strait. *Victorian Nat.* **89**(6), 287–99. (*c* 49 mosses listed)

WILLIS, J. H. (1972b). Checklist of mosses in far northwestern Victoria. Sunraysia Naturalists' Research Trust Annual Rept. (9th report). pp. 15, 16. (List of 31 spp.)

★WILSON, W. (1846a). Remarks on some rare mosses of the Southern Hemisphere. *Lond. J. Bot.* **5**, 142–44. Plates 3, 4. (Swan River plants; good discussion of *Goniomitrium enerve, G. acuminatum, Anictangium repens*; no new spp.)

★WILSON, W. (1846b). Remarks on the new species of Musci from Quito and Swan River, indicated by Dr. Taylor in the London Journal of Botany, Vol. V. p. 41. *Lond. J. Bot.* **5**, 447–55, Plates 15, 16. (A critical revision and reduction of Taylor's new mosses)

★WILSON, W. (1854). Musci. *In* "The Botany of the Antarctic Voyage . . ." (J. D. Hooker, 1852–5) Part II. Flora Novae-Zealandiae. Vol. 2, pp. 57–125. Reeve, London.

★WILSON, W. (1859). Musci. *In* "The Botany of the Antarctic Voyage . . ." (J. D. Hooker, 1855–9) Part III. Flora Tasmaniae. Vol. 2, pp. 160–221.

★WILSON, W. and HOOKER, J. D. (1845). Musci. *In* "The Botany of the Antarctic voyage of H.M. Discovery Ships *Erebus* and *Terror* in the Years 1839–43" (J. D. Hooker, 1844–45) Part 1. Flora Antarctica. Vol. 1, pp. 117–43. Reeve, London.

★WILSON, W. and HOOKER, J. D. (1847). Musci. *In* "The Botany of the Antarctic Voyage . . ." (J. D. Hooker, 1845–47) Part I. Flora Antarctica. Vol. 2, pp. 395–423. Reeve, London.

★ZANTEN, B. O. VAN (1973). A taxonomic revision of the genus *Dawsonia* R. Brown. *Lindbergia* **2**, 1–48.

ZETTERSTEDT, J. E. (1867). Några mossor från Nya Holland. *Öfvers. K. Vetensk. Akad. Förh.* **24**(8), 571–76.

Glossary

abaxial: the surface of a leaf facing away from the stem (*outer, dorsal,* or *lower* side)

acrocarpous: with the sporophyte terminal on a main stem or branch, e.g. Plate 2 (cf. *pleurocarpous*)

acuminate: tapering gradually to a fine point usually with the margins slightly concave in outline, e.g. Plate 38

acute: tapering to a point usually with the margins straight in outline, e.g. Plate 12

adaxial: the surface of a leaf facing the stem. Usually equivalent to the *inner, ventral* or *upper* side

alar cells: specialized cells in the corners of a leaf where it adjoins the stem, e.g. Plate 85

annulus: the ring, generally of enlarged cells, between the bottom of the operculum and the rim of the capsule mouth; it usually peels off during dehiscence

antheridium: the sac-like male reproductive organ, containing spermatozoids within a wall one cell thick; usually ellipsoidal in shape, e.g. Plate 26

apiculate: ending in a short abrupt point (apiculus), e.g. Plate 48

apophysis: the region between the theca of the capsule and the seta, often provided with stomata; much swollen in some species, e.g. *Tayloria* (Plate 50), *Funaria apophysata* (cf. *neck*)

appendiculate: (of cilia) with short transverse bars

appressed: (of leaves) erect and closely held against the stem, e.g. Plate 18

archegonium: the female reproductive organ, consisting of a basal *venter* enclosing the *egg,* and a long hollow *neck*

areolae: the subdivisions of the surface of a spore or other organ, produced by sculpturing

areolation: the pattern of leaf cell walls determined by the size and shape of the cells and the configuration of their walls

attenuate: tapered to a drawn-out point

auricle: a small lobe at the basal margin of a leaf (elsewhere sometimes more loosely used to refer to the group of alar cells)

autoicous: having the male and female sex organs on the same plant, but on separate stems or branches

basal membrane: the undivided part of a single peristome or of an inner peristome

bistratose: consisting of two layers of cells, i.e. two cells thick

bordered: having a margin of cells distinct in size or shape from the rest of the leaf, e.g. Plate 36

bracts: modified leaves surrounding the reproductive organs (cf. *perichaetium*, *perigonial*), e.g. Plates 2, 5

bulbil: tiny bulb-like bud, usually in a leaf axil and falling off as a means of vegetative reproduction

calcicolous: preferring to grow on a substratum containing lime

calyptra: the hood over the capsule top, formed from the wall of the upper portion of the archegonium, e.g. Plate 4

campanulate: bell-shaped, e.g. *Bruchia* calyptra, Plate 14

capsule: the uppermost, spore-containing part of the diploid generation, consisting of theca, operculum and apophysis

channelled: (of leaves) with the margins raised, or the nerve sunk, to form a longitudinal channel, e.g. Plate 22

chlorophyllose: chlorophyll-containing; applied especially to the narrow green cells of a *Sphagnum* leaf

cilia: hair-like structures found between the processes of the inner peristome

circinate: bent or curved to form a complete or incomplete circle, e.g. Plate 81

clavate: club-shaped, e.g. capsule, Plate 46

cleistocarpous: with the capsule breaking open irregularly, not by a predetermined line of dehiscence

collenchymatous: (of cells) having the walls more thickened at the corners than at the sides, e.g. Plate 1a

columella: the sterile central axis of the capsule surrounded by the spores

coma: the terminal cluster of branches at the top of a *Sphagnum* stem (adj. *comal*) or a tuft of enlarged leaves at the stem apex in other mosses (adj. *comose*), e.g. Plate 51

complanate: with the shoots flattened or the leaves arranged more or less in one plane, e.g. Plate 72

cordate: heart-shaped with the broadest part at the attachment end

crenulate: with small rounded serrations on the leaf margin

cucullate: (of a leaf) hood-shaped, with the apex curved in to meet the sides, e.g. Plate 36

cuspidate: with a stiff, acute point

cygneous: (of seta) curved downwards like the neck of a swan, e.g. Plate 20

deciduous: falling off, not persistent

decumbent: (of stems) prostrate towards the base but with the tips erect, e.g. Plate 22

decurrent: with the leaf base extending down the stem beyond the main part of the insertion, forming ridges or wings on each side

dehiscence: the process of splitting open

dendroid: with an erect branching stem, resembling a small tree, e.g. Frontispiece, Plate 54

denticulate: finely toothed, e.g. Plate 83

dioicous: with male and female sex organs on different plants

distal: terminal; that part of an organ farthest from the point of attachment

distichous: (of leaves) in two opposite rows on the stem, e.g. Plate 10

dwarf males: tiny male gametophytes borne on the female plant, e.g. Plate 26

elliptic: elongated–oval in outline, narrowed to the rounded ends and widest at or near the middle, e.g. *Pottia* capsule, Plate 49

emarginate: with a notch at the end

emergent: half-uncovered, having the capsule extending slightly above the perichaetium

endemic: confined to a given geographic region; in this text, usually confined to Australia

endostome: the inner peristome

entire: not toothed

epiphragm: the membrane which closes the mouth of the capsule in *Polytrichum* and its allies

erecto-patent: (of leaves) making an angle of 30°–45° with the stem; half-way between spreading and appressed, e.g. moist shoot, Plate 29

excurrent: with the nerve extending beyond the leaf apex

exothecial cells: the epidermal cells of the capsule wall

exserted: having the capsule extending beyond the perichaetial bracts

failing: (of nerve) ceasing before reaching the leaf apex, e.g. Plate 43

falcate: curved like a sickle, e.g. leaves, Plate 2

falcate–secund: with each leaf falcate and all the leaves turned to the same side of the stem, e.g. Plates 2, 81

fasciculate: in bundles, e.g. of branches as in *Sphagnum*

fibrils: small, fibre-like strands in *Sphagnum*

filiform: filamentous, thread-like, long and slender

flexuose, flexuous: gently wavy; the counterpart of undulate but usually applied to filiform instead of broad structures, e.g. leaves, Plate 15

funiform: rope-like, e.g. shoots, Plate 44

gametophyte: the haploid generation, consisting of protonema and the stems it gives rise to with their accompanying leaves, rhizoids, sex organs etc.

gemmae: deciduous, morphologically distinct, particles of vegetative reproduction, unicellular, multicellular, bud-like, leaf-like, or branch-like; used in the broad sense to include brood-bodies and other specialized terms, e.g. Plates 31, 32

glabrous: smooth, without hairs

glaucous: with a bluish or whitish bloom

granular: (of nerves) roughened adaxially by projecting cells or cell-clusters which are attached, not gemmae

gymnostomous: (of capsule) without a peristome, e.g. Plate 47

hair-point: a filiform, usually hyaline leaf tip, usually formed by the excurrent nerve, e.g. Plate 11

hoary: grey or whitish from massed hyaline leaf tips

hyaline: clear and colourless, like glass; applied both to leaf tips and to cells, e.g. the empty, large cells of a *Sphagnum* leaf or stem

imbricate: closely overlapping

immersed: (of capsule) more or less covered by leaves, e.g. *Acaulon*, Plate 48

imperfect: (of a peristome) lacking some of the full complement of all possible parts in the inner peristome

incrassate: having thick cell walls, e.g. Plate 11

innovation: a branch or fresh shoot from a stem; especially from below the perichaetium (subfloral innovation), e.g. Plate 1

involute: having the margins rolled inwards, e.g. Plate 20

isodiametric: of equal dimensions in all directions

julaceous: (of leafy shoots) smoothly cylindrical, worm-like or catkin-like, e.g. Plate 18

lamellae: plates or walls of cells usually perpendicular to the surface of a leaf, e.g. Plate 3

lamina: the expanded part of the leaf, as distinct from the nerve

lanceolate: longer than broad, tapering to the apex from below the middle, usually widest at about one third the distance above the base, e.g. leaf, Plate 17

lax: (of cells) relatively thin-walled, tending to collapse when dry, e.g. Plate 50

lid: operculum (q.v.)

ligulate: strap-shaped, longer and narrower than lingulate (q.v.), e.g. leaf, Plate 10

limb: the upper portion of a leaf, above the sheathing base

linear: long and narrow with the edges parallel or nearly so, e.g. cells, Plate 82

lingulate: tongue-shaped, e.g. leaf, Plate 35

lumen: the space enclosed by the cell wall

mamilla: a single large swelling or curvature of the unthickened cell wall covering the cell and including an extension or bulge of the lumen. Also used of a low, rounded projection as on an operculum, e.g. Plate 11

mamillate: with a nipple-like projection on top of a mamilla

mamillose: with mamillae

micrometre: $\frac{1}{1000}$ of a millimetre, represented by the symbol μm; formerly micron [μ]

monoicous: having the male and female organs on the same plant

μm: see *micrometre*

neck: the narrow end of the apophysis of a capsule, next the seta; there is no hard and fast line between neck and apophysis and we have tended to use them almost interchangeably

nodulose: having small knobs

obovate: shaped like the longitudinal section of an egg, attached by the narrow end, broadest above the middle, e.g. leaf, Plate 28

obtuse: blunt, the sides of the apex making an angle of more than 90°, e.g. leaf, Plate 27

operculum: the lid of the capsule, covering the peristome, falling off when ripe

orthostichy: a rank of leaves, one directly above the other, up a stem (e.g. distichous with 2 ranks; tristichous with 3 ranks)

ovate: shaped like the longitudinal section of an egg, with the broad end basal

over-square: wider than long, e.g. cells, Plate 43

papillae: small, rounded, acute, cone-shaped or forked projections or thickenings on the outer wall of a cell, e.g. Plates 29, 34, 40

papillose: with papillae

paraphyllia: minute branched or unbranched filaments or leaf-like structures borne on the stem among the leaves, e.g. Plates 76, 77

paraphyses: hair-like or club-shaped structures borne among the sex organs, e.g. Plate 6

percurrent: with the nerve reaching the apex but not extending beyond it, e.g. Plate 37

perfect peristome: a double peristome with all possible parts, including an inner peristome with a basal membrane, processes, and appendiculate cilia

perichaetium: a cluster of perichaetial bracts, the specialized leaves surrounding the archegonia and later the base of the seta.

perigonial bracts: the specialized leaves surrounding the antheridia, e.g. Plate 62

peristome: the fringe of teeth usually present round the mouth of a capsule

pinnate: (of stem) branched like a feather with the branches in two rows on opposite sides of the stem, equal in length or near so, and often in the same plane, e.g. Plate 76

plane: flat

pleurocarpous: having the archegonia and later the sporophytes on specialized short side branches, not at the apices of main stems or branches (cf. acrocarpous), e.g. Plate 79

plicate: folded along the long axis, forming pleats or furrows, e.g. leaf, Plate 62

polysety: the production of more than one sporophyte at the apex of a single branch

porose: with pores, either thin spots in otherwise thick cell walls, e.g. Plates 24, 27, or, in *Sphagnum*, actual holes to the outside

processes: the name given to the teeth of the inner peristome

protonema: branched algal-like filaments, or plate-like growths arising from the spores and on which the leafy parts of the moss develop, e.g. Plate 49

pseudopodium: a stem of the gametophyte, often leafless, bearing gemmae or a capsule

pseudoparaphyllia: similar to filamentous or foliose paraphyllia but restricted to the base of branches or branch primordia

quadrate: (of cells) appearing square or nearly so in surface view, e.g. Plate 34

recurved: curved backward or downward, e.g. leaves, Plate 80

reflexed: abruptly bent backwards or downwards, e.g. peristome teeth, Plate 45

retort cells: specialized hyaline cells in some *Sphagnum* stems and branches which have a pore standing out on a mamilla near the upper end of the cell

revolute: rolled back from the margin or the apex, e.g. leaves, Plate 7

rhizoids: filaments, usually branched and mainly with oblique cross walls, serving for anchorage and/or absorption; most commonly brown but can be colourless, red, violet etc.

rhomboidal: (of cells) quadrilateral in surface view, or nearly so, with the lateral angles obtuse; "diamond"-shaped, e.g. Plate 75

rostrate: narrowed to form a long tip or point; beaked, e.g. operculum, Plate 13

saxicolous: growing on rock

secund: turned to one side

serrulate: minutely toothed, with teeth pointing forwards, e.g. Plate 50

sessile: without a stalk

seta: the stalk or part of the sporophyte which carries the capsule

setaceous: bristle-like

sheath: a leaf-base which more or less surrounds and encloses the stem or seta, e.g. Plate 16

sigmoid: (of cell outlines) with a slight *s*-twist; curved in opposite directions at the two ends, e.g. Plate 81

sinuose: (of cell walls) with a wavy margin or outline, e.g. Plate 68

spathulate: with a narrow base, gradually becoming wider to the broad, rounded apex, e.g. paraphysis, Plate 6

sporophyte: the diploid generation consisting of capsule, seta and a foot embedded in the gametophyte; the part of the plant which produces the spores

squarrose: having the leaves bent stiffly at right angles to the stem, e.g. Plate 80

stegocarpous: (of capsule) having a distinct, dehiscent operculum

stereid cells: small thick-walled cells evident in the cross section of the nerve in some species

striolate: minutely striate

struma: a swelling on one side at the base of a capsule, e.g. Plate 26

subula: a fine sharp point

subulate: with a fine sharp point, tapering from base to apex, e.g. leaf, Plate 16

sympodial: a system of branching with stems of limited growth, which are replaced by side branches, e.g. Plate 18

synoicous: with archegonia and antheridia mingled in one cluster

terete: smoothly cylindrical, circular in TS

theca: the main body of a capsule containing the spores, between the neck and the peristome

tomentose: covered with hairs or rhizoids, e.g. Plate 63

triquetrous: with three angles (used of tristichous stems with keeled leaves)

truncate: ending abruptly as if cut off at the end

TS: transverse section

tubers: rhizoid gemmae

undulate: wavy (up and down), with the surface alternately concave and convex, e.g. Plate 72

unistratose: of one layer of cells; one cell thick

urn: theca (q.v.)

vaginant lamina: the sheathing, double lamina of the leaf of *Fissidens*

vaginula: the cup found round the base of the seta, formed by the lower half of the archegonium; the complement of the *calyptra* (q.v.)

Index

All references are to page numbers; principal references are in italics, illustrations in bold type. Further lists of additional species names are catalogued in the text at the end of the relevant genus.

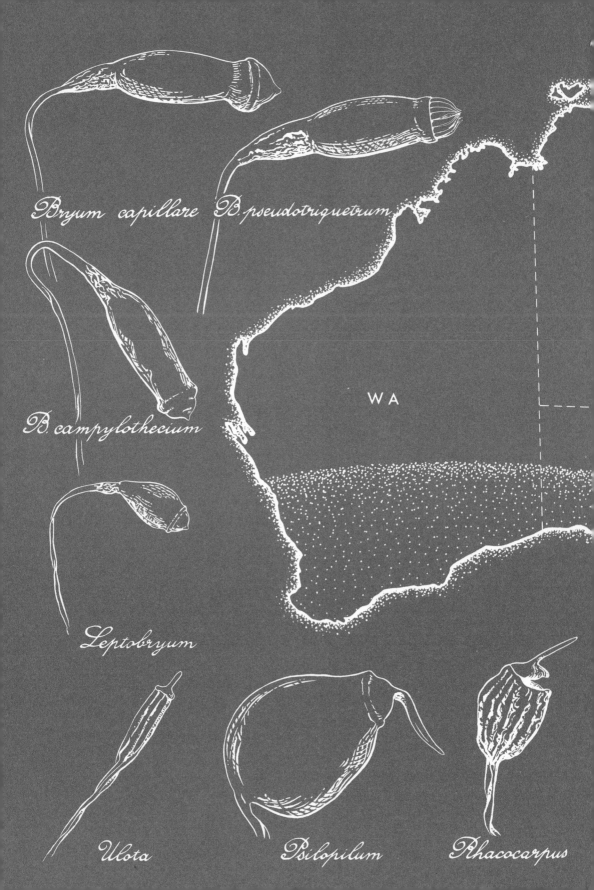

Bryum capillare   *B. pseudotriquetrum*

B. campylothecium

W A

Leptobryum

Ulota   *Psilopilum*   *Rhacocarpus*